SPECIAL FUNCTIONS
AN INTRODUCTION TO THE CLASSICAL FUNCTIONS OF MATHEMATICAL PHYSICS

SPECIAL FUNCTIONS
AN INTRODUCTION TO THE CLASSICAL FUNCTIONS OF MATHEMATICAL PHYSICS

SPECIAL FUNCTIONS
AN INTRODUCTION TO THE CLASSICAL FUNCTIONS OF MATHEMATICAL PHYSICS

NICO M. TEMME

Centrum voor Wiskunde en Informatica
Center for Mathematics and Computer Science
Amsterdam, The Netherlands

A Wiley-Interscience Publication
JOHN WILEY & SONS, Inc.
New York • Chichester • Brisbane • Toronto • Singapore

Library of Congress Cataloging in Publication Data:

Temme, N. M.

Special functions : an introduction to the classical functions of mathematical physics / Nico M. Temme

p. cm.

Includes bibliographical references and index.

ISBN 0-471-11313-1 (cloth : alk. paper)

1. Functions, Special. 2. Boundary value problems. 3. Mathematical physics. I. Title.

QC20.7.F87T46 1996

515.5—dc20 95-42939

CIP

CONTENTS

PREFACE

This book gives an introduction to the classical well-known special functions which play a role in mathematical physics, especially in boundary value problems. Usually we call a function "special" when the function, just as the logarithm, the exponential and trigonometric functions (the elementary transcendental functions), belongs to the toolbox of the applied mathematician, the physicist or engineer. Usually there is a particular notation, and a number of properties of the function are known. This branch of mathematics has a respectable history with great names such as Gauss, Euler, Fourier, Legendre, Bessel and Riemann. They all have spent much time on this subject. A great part of their work was inspired by physics and the resulting differential equations. About 70 years ago these activities culminated in the standard work *A Course of Modern Analysis* by Whittaker and Watson, which has had great influence and is still important.

This book has been written with students of mathematics, physics and engineering in mind, and also researchers in these areas who meet special functions in their work, and for whom the results are too scattered in the general literature. Calculus and complex function theory are the basis for all this: integrals, series, residue calculus, contour integration in the complex plane, and so on.

The selection of topics is based on my own preferences, and of course, on what in general is needed for working with special functions in applied mathematics, physics and engineering. This book gives more than a selection of formulas. In the many exercises hints for solutions are often given. In order to keep the book to a modest size, no attention is paid to special functions which are solutions of periodic differential equations such as Mathieu and Lamé functions; these functions are only mentioned when separating the wave equation. The current interest in q-hypergeometric functions would justify an extensive treatment of this topic. It falls outside the scope of the present work, but a short introduction is given nevertheless.

Today students and researchers have computers with formula processors at their disposal. For instance, MATLAB and MATHEMATICA are powerful packages, with possibilities of computing and manipulating special functions. It is very useful to exploit this software, but often extra analysis and knowledge of special functions are needed to obtain optimal results.

At several occasions in the book I have paid attention to the asymptotic and numerical aspects of special functions. When this becomes too specialistic in nature the references to recent literature are given. A separate chapter discusses the stability aspects of recurrence relations for several special functions are discussed. It is explained that a given recursion cannot always be used for computations. Much of this information is available in the literature, but it is difficult for beginners to locate.

Part of the material for this book is collected from well-known books, such as from HOCHSTADT, LEBEDEV, OLVER, RAINVILLE, SZEGÖ and WHITTAKER & WATSON. In addition to these I have used Dutch university lecture notes, in particular those by Prof. H.A. Lauwerier (University of Amsterdam) and Prof. J. Boersma (Technical University Eindhoven).

The enriching and supporting comments of Dick Askey, Johannes Boersma, Tom Koornwinder, Adri Olde Daalhuis, Frank Olver, and Richard Paris on earlier versions of the manuscript are much appreciated. When there are still errors in the many formulas I have myself to blame. But I hope that the extreme standpoint of Dick Askey, who once advised me: *never trust a formula from a book or table; it only gives you an idea how the exact result looks like,* is not applicable to the set of formulas in this book. However, this is a useful warning.

NICO M. TEMME

Amsterdam, The Netherlands

SPECIAL FUNCTIONS
AN INTRODUCTION TO THE CLASSICAL FUNCTIONS OF MATHEMATICAL PHYSICS

1

Bernoulli, Euler and Stirling Numbers

A well-known result from calculus is the alternating series

$$\frac{1}{2}\pi = 2\sum_{n=1}^{\infty} \frac{(-1)^{n-1}}{2n-1},$$

which can be used for the computation of the number π, although the series converges very slowly. However, summing the first 50 000 terms gives the remarkable result

$$\begin{aligned} 2\sum_{n=1}^{50\,000} \frac{(-1)^{n-1}}{2n-1} \\ = 1.5707\ \underline{8}6326\ 79489\ \underline{7}6192\ 31321\ \underline{1}9163\ 975\underline{20\ 5}2098\ 58\underline{331\ 46876}. \end{aligned}$$

Using a criterion for convergence of this type of series, we may conclude that this answer is correct to only six significant digits. When you compare the answer on the right-hand side with a 50-d approximation of $\frac{1}{2}\pi$, you will reach the surprising conclusion that nearly all digits in the above approximation are correct, except for those underlined. In this chapter this intriguing aspect will be explained with the help of simple properties of Euler numbers and Boole's summation method. Another example is the series

$$\ln 2 = \sum_{n=1}^{\infty} \frac{(-1)^{n+1}}{n}.$$

You may try to sum the first 50 000 terms with high precision, and compare the answer with an accurate approximation of $\ln 2$.

In this chapter we discuss basic properties of the Bernoulli, Euler and Stirling numbers, with applications to the summation methods of Euler and Boole. These methods are based on the polynomials of Euler and Bernoulli.

Such topics are extensively discussed in classical books on the calculus of differences, the subject that played a prominent part in numerical analysis. A short introduction to difference equations is given §1.1.2.

Just as many other special numbers, polynomials and functions, the special numbers and polynomials of this chapter can be introduced by generating functions. Usually these are power series of the form

$$F(x,t) = \sum_{n=0}^{\infty} f_n(x) t^n,$$

where each f_n is independent of t. The radius of convergence with respect to (complex) values of t may be finite or infinite. We say that $F(x,t)$ is the function which the sequence $\{f_n\}$ generates, and F is called the *generating function*. Often, F and the coefficients f_n are analytic functions in a certain domain.

1.1. Bernoulli Numbers and Polynomials

The Bernoulli numbers are named after Jakob Bernoulli, who mentioned the numbers in his posthumous *Ars conjectandi* of 1713; see BERNOULLI (1713). He discussed *summae potestatum*, sums of equal powers of the first n integers. For instance, we know from elementary calculus that

$$\begin{aligned}
\sum_{i=1}^{n-1} i &= \frac{1}{2}n(n-1) = \frac{1}{2}n^2 - \frac{1}{2}n,\\
\sum_{i=1}^{n-1} i^2 &= \frac{1}{3}n^3 - \frac{1}{2}n^2 + \frac{1}{6}n,\\
\sum_{i=1}^{n-1} i^3 &= \frac{1}{4}n^4 - \frac{1}{2}n^3 + \frac{1}{4}n^2,\\
\sum_{i=1}^{n-1} i^4 &= \frac{1}{5}n^5 - \frac{1}{2}n^4 + \frac{1}{3}n^3 - \frac{1}{30}n,
\end{aligned}$$

and so on. Bernoulli was, in particular, interested in the numbers multiplying the linear terms n at the right-hand sides: $-\frac{1}{2}, \frac{1}{6}, 0, -\frac{1}{30}, \ldots$. EULER (1755) called them Bernoulli numbers $B_1, B_2, B_3, B_4, \ldots$. As we know from the general result

$$\sum_{i=0}^{n-1} i^p = \frac{1}{p+1}\sum_{k=0}^{p}\binom{p+1}{k} B_k n^{p+1-k},$$

they show up in other terms also; see Exercise 1.3.

The Bernoulli numbers occur in practically every field of mathematics, in particular, in combinatorial theory, finite difference calculus, numerical analysis, analytical number theory, and probability theory. We discuss their role in the summation formula of Euler.

1.1.1. Definitions and Properties

Instead of introducing the *Bernoulli numbers* B_n as above, we use a generating function for their definition:

$$\boxed{\frac{z}{e^z - 1} = \sum_{n=0}^{\infty} \frac{B_n}{n!} z^n, \qquad |z| < 2\pi.} \tag{1.1}$$

Because the function

$$f(z) = \frac{z}{e^z - 1} - 1 + \frac{1}{2} z$$

is even (prove it!), all Bernoulli numbers with odd index ≥ 3 vanish:

$$B_{2n+1} = 0, \quad n = 1, 2, 3, \ldots .$$

The first nonvanishing numbers are

$$B_0 = 1, \quad B_1 = -\frac{1}{2}, \quad B_2 = \frac{1}{6}, \quad B_4 = -\frac{1}{30}, \quad B_6 = \frac{1}{42},$$

$$B_8 = -\frac{1}{30}, \quad B_{10} = \frac{5}{66}, \quad B_{12} = -\frac{691}{2730}, \quad B_{14} = \frac{7}{6}, \quad B_{16} = -\frac{3617}{510}$$

The *Bernoulli polynomials* are defined by the generating function

$$\boxed{\frac{ze^{xz}}{e^z - 1} = \sum_{n=0}^{\infty} \frac{B_n(x)}{n!} z^n, \qquad |z| < 2\pi.} \tag{1.2}$$

The first few polynomials are

$$\begin{aligned}
B_0(x) &= 1, \\
B_1(x) &= x - \frac{1}{2}, \\
B_2(x) &= x^2 - x + \frac{1}{6}, \\
B_3(x) &= x^3 - \frac{3}{2}x^2 + \frac{1}{2}x, \\
B_4(x) &= x^4 - 2x^3 + x^2 - \frac{1}{30}.
\end{aligned}$$

A further step yields the *generalized Bernoulli polynomials*:

$$e^{xz}\left(\frac{z}{e^z-1}\right)^\sigma = \sum_{n=0}^{\infty} \frac{B_n^{(\sigma)}(x)}{n!} z^n, \quad |z| < 2\pi, \tag{1.3}$$

where σ is any complex number. By taking $x = 0$ we obtain the *generalized Bernoulli numbers* $B_n^{(\sigma)} = B_n^{(\sigma)}(0)$, which are polynomials of degree n of the complex variable σ.

We now give some relations which easily follow from the definitions through the generating functions.

$$\int_0^1 B_n(x)\,dx = 0, \quad n = 1, 2, 3, \ldots . \tag{1.4}$$

$$B_n(x) = \sum_{k=0}^{n} \binom{n}{k} B_k x^{n-k}, \quad B_n(x+y) = \sum_{k=0}^{n} \binom{n}{k} B_k(x) y^{n-k}. \tag{1.5}$$

$$B_n(0) = B_n, \quad B_n(1) = (-1)^n B_n. \tag{1.6}$$

$$B_n(1-x) = (-1)^n B_n(x), \quad B_n(-x) = (-1)^n \left[B_n(x) + nx^{n-1}\right]. \tag{1.7}$$

$$B_n(\tfrac{1}{2}) = -\left(1 - 2^{1-n}\right) B_n. \tag{1.8}$$

$$\frac{d}{dx} B_n(x) = nB_{n-1}(x), \quad B_n(x+1) - B_n(x) = nx^{n-1}. \tag{1.9}$$

$$\sum_{k=0}^{n} \binom{n+1}{k} B_k(x) = (n+1)x^n. \tag{1.10}$$

The proof of (1.4) follows for example by integrating the left-hand side of (1.2) with respect to x. The properties (1.5)–(1.10) all hold for $n = 0, 1, 2, \ldots$. Property (1.10) gives for $x = 0$ the identity for the Bernoulli numbers:

$$\sum_{k=0}^{n} \binom{n+1}{k} B_k = 0, \tag{1.11}$$

with which the numbers can be generated by means of a simple recursion. Symbolic manipulation on the computer may be very useful here. Numerical computations with finite precision will yield very inaccurate results, due to instability of (1.11).

In Exercise 1.1c you can prove that

$$\tan z = \sum_{n=0}^{\infty} (-1)^{n+1} \frac{T_{2n+1} z^{2n+1}}{(2n+1)!}, \tag{1.12}$$

where the relation between the *tangent numbers* T_n and the Bernoulli numbers B_n is defined by

$$B_n = \frac{-nT_{n-1}}{2^n(2^n-1)}, \qquad n = 1, 2, \ldots .$$

T_n is an integer with $T_{2n} = 0,\ n > 0$. This follows from differentiating $\tan z$: all even derivatives at $z = 0$ vanish and the odd derivatives are integers. The same holds for the coefficients of the MacLaurin expansion. We have

$$T_0 = 1, \quad T_1 = -1, \quad T_3 = 2, \quad T_5 = -16, \quad T_7 = 272, \quad T_9 = -7936.$$

Finally, we mention

$$\int_a^x B_n(t)\,dt = \frac{1}{n+1}\left[B_{n+1}(x) - B_{n+1}(a)\right].$$

This property can be used in the proof of the memorable formulas

$$\boxed{\begin{aligned} B_{2n-1}(x) &= 2(-1)^{n+1}(2n-1)! \sum_{m=1}^{\infty} \frac{\sin(2\pi m x)}{(2\pi m)^{2n-1}}, \\ B_{2n}(x) &= 2(-1)^{n+1}(2n)! \sum_{m=1}^{\infty} \frac{\cos(2\pi m x)}{(2\pi m)^{2n}}, \end{aligned}} \tag{1.13}$$

where $n = 1, 2, 3, \ldots$ and $0 \le x \le 1$. For a proof we may begin with the first line with $n = 1$. This gives a well-known result from the theory of Fourier series for the function $B_1(x) = x - \frac{1}{2}$. Then induction and the above integral relation should be used. The special case $x = 0$ gives in (1.5) an interesting result for the even Bernoulli numbers:

$$\boxed{B_{2n} = 2(-1)^{n+1}(2n)! \sum_{m=1}^{\infty} (2\pi m)^{-2n}, \qquad n = 1, 2, 3, \ldots .} \tag{1.14}$$

It is of interest, since with this result the series $\sum_{m=1}^{\infty} m^{-s}$ (the Riemann zeta function, which will be discussed in the following chapter) can be expressed in terms of Bernoulli numbers when s is an even positive integer. When $s = 2, 4, 6$ we thus have

$$\begin{aligned} \sum_{m=1}^{\infty} \frac{1}{m^2} &= \frac{\pi^2}{6}, \\ \sum_{m=1}^{\infty} \frac{1}{m^4} &= \frac{\pi^4}{90}, \\ \sum_{m=1}^{\infty} \frac{1}{m^6} &= \frac{\pi^6}{945}. \end{aligned} \tag{1.15}$$

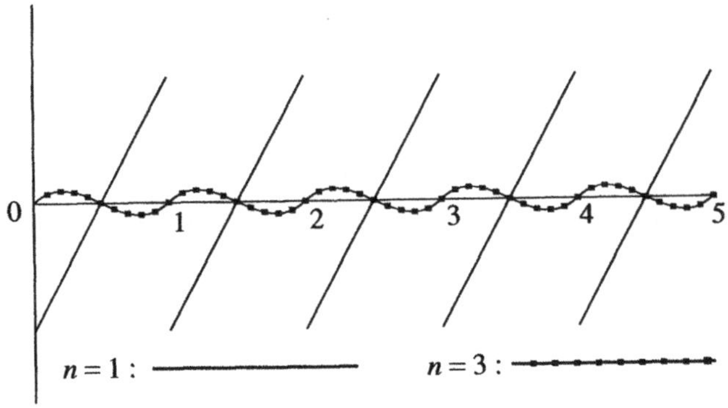

Figure 1.1. The functions $\widetilde{B}_n(x)$, $n = 1$ and $n = 3$.

For odd $s-$values a similar relation is never found.

The Fourier series for Bernoulli polynomials in (1.13) can be defined for all real values of x. Outside the interval $[0,1]$ the series do not represent polynomials, of course, but periodic functions of x. These periodic functions are very important, and we introduce a special notation $\widetilde{B}_n(x)$ by defining for $n = 0, 1, 2, \ldots$

$$\widetilde{B}_n(x) = B_n(x), \quad 0 \le x < 1,$$

and

$$\widetilde{B}_n(x+1) = \widetilde{B}_n(x), \quad x \in \mathbb{R}. \tag{1.16}$$

The functions $\widetilde{B}_n(x)$ have continuous derivatives up to order $n-1$. This easily follows from the earlier properties, for instance from (1.13). They become smoother as n increases. As will follow from §1.1.3, the periodic functions $\widetilde{B}_n(x)$ play an important part in *Euler's summation formula*.

In Figures 1.1 and 1.2 we show the first four functions $\widetilde{B}_n(x), n = 1, 2, 3, 4$.

1.1.2. A Simple Difference Equation

One of the results of the previous subsection (see (1.9)) reads

$$f(x+1) - f(x) = nx^{n-1},$$

a *difference equation* with solution $f(x) = B_n(x)$. It follows that the Bernoulli polynomials can be used to construct a solution of the difference equation

$$f(x+1) - f(x) = P_n(x),$$

where $P_n(x)$ is a polynomial. When $P_n(x) = \sum_{k=0}^{n} a_k x^k$, we can write the general solution in the form

$$f(x) = \sum_{k=0}^{n} \frac{a_k}{k+1} B_{k+1}(x) + \pi(x),$$

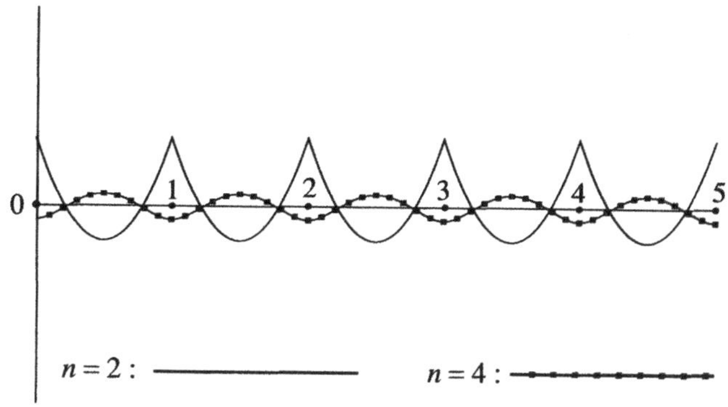

Figure 1.2. The functions $\widetilde{B}_n(x)$, $n = 2$ and $n = 4$.

where $\pi(x)$ is an arbitrary periodic function of x of period 1.

The function $f(x) = \omega^n B_n(\frac{x}{\omega})$ is a solution of the more general difference equation

$$\frac{f(x+\omega) - f(x)}{\omega} = \phi(x), \tag{1.17}$$

with $\phi(x) = nx^{n-1}$. When we want to solve this equation for general $\phi(x)$, we may call

$$f(x) = A - \omega \sum_{n=0}^{\infty} \phi(x + n\omega)$$

a formal solution of the difference equation (1.17), where A is independent of x. For example, when $\phi(x) = \exp(-x)$, we obtain

$$f(x) = A - \omega \sum_{n=0}^{\infty} e^{-x-n\omega} = A - \frac{\omega e^{-x}}{1 - e^{-\omega}},$$

which indeed is a solution of (1.17). The series in this example is convergent, but in general this condition is not satisfied. Several methods are available to use a modified form of the formal solution, from which well-defined solutions can be obtained. For instance, we can take $A = \int_c^N \phi(x)\, dx$, with $c \geq 0$ and N a large integer, and we define

$$f_N(x) = \int_c^N \phi(x)\, dx - \sum_{n=0}^{N} \phi(x + n\omega).$$

When the limit of $f_N(x)$ exists as $N \to \infty$, this limit may be a solution. For example, let $c = 1, \omega = 1$ and $\phi(x) = 1/x,\ x > 0$. Then

$$f_N(x) = \left[\ln N - \sum_{n=0}^{N} \frac{1}{n+1}\right] + \left[\sum_{n=0}^{N} \left(\frac{1}{n+1} - \frac{1}{x+n}\right)\right],$$

and each quantity between square brackets tends to a finite limit, as $N \to \infty$; see the next subsection, Example 1.2. From Chapter 3, formula (3.10), we infer that the function $f_N(x)$ tends to a special function, the logarithmic derivative of the gamma function $\psi(x)$, which indeed satisfies the difference equation $f(x+1) - f(x) = 1/x$.

In a second method the function $\phi(x)$ in (1.17) is replaced with $\phi(x, \mu)$ that satisfies $\lim_{\mu\to 0} \phi(x,\mu) = \phi(x)$. For instance, we can take

$$\phi(x,\mu) = \phi(x)e^{-\mu x}, \quad \mu > 0.$$

Let c be a number independent of x, and assume that

$$\int_c^\infty \phi(x,\mu)\,dx, \quad \text{and} \quad \sum_{n=0}^\infty \phi(x+n\omega,\mu)$$

both converge. Then we define as the solution of (1.17) the function $f(x) = \lim_{\mu\to 0} f(x,\mu)$, where

$$f(x,\mu)) = \left[\int_c^\infty \phi(x,\mu)\,dx - \sum_{n=0}^\infty \phi(x+n\omega,\mu)\right], \tag{1.18}$$

provided that this limit exists. It is shown in the classical literature (for instance, in NÖRLUND (1924)) that this $f(x)$ indeed satisfies (1.17), and that this solution is independent of the particular choice of $\phi(x,\mu)$. Other choices are also possible. It is easily verified that for (1.17) with $c = 1, \omega = 1, \phi(x) = 1$, the function $f(x,\mu)$ is given by

$$f(x,\mu) = \frac{e^{-\mu}}{\mu} - \frac{e^{-\mu x}}{1-e^{-\mu}},$$

and that $\lim_{\mu\to 0} f(x,\mu) = x - \frac{1}{2} = B_1(x)$, a Bernoulli polynomial.

Example 1.1. Consider the difference equation

$$f(x+1) - f(x) = nx^{n-1}e^{-\mu x}, \quad \mu > 0, \quad x > 0,$$

which for $\mu = 0$ reduces to the difference equation of the Bernoulli polynomials. We try to find $f(x, \mu)$ of (1.18). Take $c = 0$, then

$$\begin{aligned} f(x,\mu) &= \int_0^\infty n\,t^{n-1}e^{-\mu t}\,dt - \sum_{m=0}^{\infty} n\,(x+m)^{n-1}e^{-\mu(x+m)} \\ &= n\,(-1)^{n-1}\frac{\partial^{n-1}}{\partial\mu^{n-1}}\left[\int_0^\infty e^{-\mu t}\,dt - \sum_{m=0}^{\infty} e^{-\mu(x+m)}\right] \\ &= n\,(-1)^{n-1}\frac{\partial^{n-1}}{\partial\mu^{n-1}}\left[\frac{1}{\mu} + \frac{e^{-\mu x}}{e^{-\mu}-1}\right] \\ &= n\,(-1)^{n-1}\frac{\partial^{n-1}}{\partial\mu^{n-1}}\left[\sum_{m=1}^{\infty}(-\mu)^{m-1}\frac{B_m(x)}{m!}\right] \\ &= n\sum_{m=n}^{\infty}\frac{(-\mu)^{m-n}B_m(x)}{m\,(m-n)!}. \end{aligned}$$

In this derivation we have used the generating function (1.2). When $\mu \to 0$, we have $f(x,\mu) = B_n(x)$, which again shows that $B_n(x)$ satisfies the second relation in (1.9).

1.1.3. Euler's Summation Formula

A striking application of Bernoulli numbers and polynomials is *Euler's summation formula*, that links a finite or infinite series and an integral. This formula yields an efficient method for evaluating some slowly convergent series by means of an integral. Turning it round, by this method also an integral can be approximated by discretization, which leads to the trapezoidal rule. In EULER (1732) the proof of the formula can be found.

Theorem 1.1. *Let the function f: $[0,1] \to \mathbb{C}$ have k continuous derivatives $(k = 0, 1, 2, \ldots)$. Then for $k \geq 1$*

$$f(1) = \int_0^1 f(x)\,dx + \sum_{i=1}^{k}\frac{(-1)^i B_i}{i!}\left[f^{(i-1)}(1) - f^{(i-1)}(0)\right] + R_k,$$

with

$$R_k = \frac{(-1)^{k+1}}{k!}\int_0^1 f^{(k)}(x)B_k(x)\,dx.$$

Proof. The proof runs with induction with respect to k. For $k = 1$ the claim is true, which follows from integrating by parts. Then the property

$$B_m(x) = \frac{1}{m+1}B'_{m+1}(x)$$

is used to go from $k = m \geq 1$ to $k = m + 1$. ∎

With similar conditions for f on the interval $[j-1, j]$ we have

$$f(j) = \int_{j-1}^{j} f(x)\,dx + \sum_{i=1}^{k} \frac{(-1)^i B_i}{i!} \left[f^{(i-1)}(j) - f^{(i-1)}(j-1)\right] + R_k,$$

with

$$R_k = \frac{(-1)^{k+1}}{k!} \int_{j-1}^{j} f^{(k)}(x)\widetilde{B}_k(x)\,dx,$$

where $\widetilde{B}_k(x)$ is the function introduced in (1.16).

The next step joins a number of these intervals:

$$\sum_{i=1}^{n} f(i) = \int_0^n f(x)\,dx + \sum_{i=1}^{k} \frac{(-1)^i B_i}{i!} \left[f^{(i-1)}(n) - f^{(i-1)}(0)\right] + R_k,$$

with

$$R_k = \frac{(-1)^{k+1}}{k!} \int_0^n f^{(k)}(x)\widetilde{B}_k(x)\,dx.$$

For $k = 1$ this gives the formula

$$f(1) + f(2) + \cdots + f(n) = \int_0^n f(x)\,dx + \frac{1}{2}[f(n) - f(0)] + \int_0^n \widetilde{B}_1(x) f'(x)\,dx,$$

with $\widetilde{B}_1(x)$ a sawtooth function on $[0, n]$. This is *Euler's summation formula* in its simplest form. The formula expresses a connection between the sum of the first n terms of a series and the integral of the corresponding function over the interval $[0, n]$.

Example 1.2. Take

$$f(x) = \frac{1}{x+1}$$

and replace in the above formula n with $n-1$. Then we obtain the classical example

$$1 + \frac{1}{2} + \frac{1}{3} + \frac{1}{4} + \cdots + \frac{1}{n} = \ln n + \frac{1}{2n} + \frac{1}{2} - \int_0^{n-1} \widetilde{B}_1(x) \frac{dx}{(1+x)^2}.$$

The integral is convergent when $n \to \infty$. From this we infer that

$$\gamma = \lim_{n\to\infty} \left(1 + \frac{1}{2} + \frac{1}{3} + \frac{1}{4} + \cdots + \frac{1}{n} - \ln n\right)$$

exists as well. The limit $\gamma = 0.5772\ 15664\ 90153...$ is called *Euler's constant.* From this example also follows that

$$\gamma = \frac{1}{2} - \int_0^\infty \widetilde{B}_1(x) \frac{dx}{(1+x)^2} = \frac{1}{2} - \int_1^\infty \widetilde{B}_1(x) \frac{dx}{x^2}.$$

Since

$$B_k(0) = 0, \quad k = 3, 5, 7, \ldots$$

all terms with odd index can be deleted in the summation formula, except the term with index $i = 1$. And at both sides we can add the term $f(0)$. Then the result is

Theorem 1.2. *Let the function f: $[0, n] \to \mathbb{C}$ have $(2k+1)$ continuous derivatives ($k \geq 0, n \geq 1$). Then*

$$\begin{aligned}\sum_{i=0}^{n} f(i) &= \int_0^n f(x)\,dx + \frac{1}{2}[f(n) + f(0)] \\ &\quad + \sum_{i=1}^{k} \frac{B_{2i}}{(2i)!}\left[f^{(2i-1)}(n) - f^{(2i-1)}(0)\right] + R_k,\end{aligned} \tag{1.19}$$

with

$$R_k = \frac{1}{(2k+1)!}\int_0^n f^{(2k+1)}(x)\widetilde{B}_{2k+1}(x)\,dx.$$

The summation formula is usually presented in this form, and is connected with the *trapezoidal rule* (Exercise 1.7).

Example 1.3. We take $f(x) = x^2$. Since $f^{(3)}(x) = 0$ for each x, the contribution of the remainder R_k in (1.19) is zero when $k \geq 1$. For $k = 1$ (1.19) then reads

$$\sum_{i=0}^{n} i^2 = \int_0^n x^2\,dx + \frac{1}{2}n^2 + \frac{1}{2}B_2[f'(n) - f'(0)] = \frac{1}{3}n^3 + \frac{1}{2}n^2 + \frac{1}{6}n.$$

An alternative summation formula for infinite series arises through the intermediate form

$$\begin{aligned}\sum_{i=m}^{n} f(i) &= \int_m^n f(x)\,dx + \frac{1}{2}[f(n) + f(m)] \\ &\quad + \sum_{i=1}^{k} \frac{B_{2i}}{(2i)!}\left[f^{(2i-1)}(n) - f^{(2i-1)}(m)\right] + R_k,\end{aligned} \tag{1.20}$$

with

$$R_k = \frac{1}{(2k+1)!}\int_m^n f^{(2k+1)}(x)\widetilde{B}_{2k+1}(x)\,dx.$$

In this formula we replace n with ∞, which is allowed when the infinite series and the indefinite integrals

$$\sum_{i=m}^{\infty} f(i), \quad \int_m^{\infty} f(x)\,dx, \quad \int_m^{\infty} f^{(2k+1)}(x)\widetilde{B}_{2k+1}(x)\,dx$$

exist. In addition we assume that f and the derivatives occurring in the formula tend to zero when their arguments tend to infinity. The result is

$$\sum_{i=m}^{\infty} f(i) = \int_m^{\infty} f(x)\,dx + \tfrac{1}{2}f(m) - \sum_{i=1}^{k} \frac{B_{2i}}{(2i)!} f^{(2i-1)}(m) + R_k, \tag{1.21}$$

with

$$R_k = \frac{1}{(2k+1)!}\int_m^{\infty} f^{(2k+1)}(x)\widetilde{B}_{2k+1}(x)\,dx.$$

This form of Euler's summation formula can be fruitfully applied in summing infinite series. It is important to have information on the remainder R_k. It is not always necessary to know the integral in R_k exactly. Also, it is not necessary to know whether

$$\lim_{k\to\infty} R_k = 0.$$

In many cases this condition is not fulfilled, or the limit does not even exist. An estimate of the remainder can be obtained through the following theorem.

Theorem 1.3. *Let f and all its derivatives be defined on the interval $[0,\infty)$ on which they should be monotonic and tend to zero when $x \to \infty$. Then R_k of (1.19) satisfies*

$$R_k = \theta_k \frac{B_{2k+2}}{(2k+2)!}\left[f^{(2k+1)}(n) - f^{(2k+1)}(0)\right], \quad \textit{with} \quad 0 \le \theta_k \le 1.$$

Proof. First we remark that

$$f^{(k)}(x), \quad f^{(k+1)}(x), \quad k = 0,1,2,\ldots$$

have fixed and different signs on $[0,\infty)$. Let $f(x) > 0$, $x \ge 0$. Then it is easily verified (consider the graph of the sine function) that the sign of

$$\int_0^1 \sin(2\pi m x) f(x)\,dx, \quad m = 1,2,3,\ldots$$

is also positive. From (1.13) and (1.16) then follows that the sign of

$$\int_0^n \widetilde{B}_{2n+1}(x) f(x)\,dx, \quad n = 1,2,3,\ldots, \quad m = 0,1,2,\ldots$$

equals the sign of $(-1)^{n+1}$. From this we also conclude that the remainder R_k of (1.19) have different signs for subsequent values of k. This implies that R_k and $(R_k - R_{k+1})$ have the same sign and hence that

$$|R_k| \le |R_k - R_{k+1}|.$$

From (1.19) it follows, however, that

$$R_k - R_{k+1} = \frac{B_{2k+2}}{(2k+2)!}\left[f^{(2k+1)}(n) - f^{(2k+1)}(0)\right].$$

This is exactly the 'first neglected term' in Euler's summation formula. The sign of this term equals the sign of R_k and the absolute value of this term at least equals the absolute value of R_k. ∎

A similar result holds for formula (1.21). In this case we have

$$R_k = -\theta_k \frac{B_{2k+2}}{(2k+2)!} f^{(2k+1)}(m), \quad \text{with} \quad 0 \le \theta_k \le 1. \tag{1.22}$$

The theorem says that, with the conditions on f, the error in taking in (1.21) k terms of the series in the right-hand side is smaller than the first neglected $(k+1)$–th term. In practical problems one tries to find this $(k+1)$–th term that falls below the requested accuracy, and one sums the series on the right-hand side of (1.21) as far as the k–th term. In other words, one may sum the series until a particular term falls below the accuracy. This naive criterion, which is very popular in summing infinite series, is fully legitimate here.

Example 1.4. Sum the series

$$\sum_{i=1}^{\infty} \frac{1}{i^3}$$

with an error less than 10^{-9}. First we compute

$$\sum_{i=1}^{9} \frac{1}{i^3} = 1.19653198567\cdots.$$

Next we apply (1.21) with $f(x) = 1/x^3$ and $m = 10$. Namely, (1.21) should not be used with the low value $m = 1$, but with one that makes R_k small enough (for an acceptable value of k). Our f fulfills the conditions of Theorem 1.3. Verify that the third term in the series of the right-hand side of (1.21) equals

$$-\frac{1}{42}\frac{1}{6!}2520 \times 10^{-8} = -\frac{1}{12} \times 10^{-8} = -0.83 \times 10^{-9}.$$

Hence, we apply (1.21) with $k = 2$, and we obtain

$$\sum_{i=1}^{\infty} = \frac{1}{i^3} = \sum_{i=1}^{9} + \frac{1}{i^3} + \sum_{i=10}^{\infty} \frac{1}{i^3} = 1.19653198567\cdots$$
$$+ \int_{10}^{\infty} \frac{dx}{x^3} + \frac{1}{2000} + \frac{1}{40000} - \frac{1}{12000000} = 1.20205690234,$$

with an error that is smaller than 0.83×10^{-9}. The actual error is 0.82×10^{-9}.

From this example we see that the error estimate can be very sharp. Another point is that Euler's summation formula may produce a quite accurate result, with almost no effort. To obtain the same accuracy, straightforward numerical summation of the series $\sum i^{-3}$ requires about 22360 terms.

Not all series can be evaluated by Euler's formula in this favorable way. Although the class of series for which the formula is applicable is quite interesting, Euler's method has its limitations. Alternating series should be tackled through Boole's summation method, which is based on the Euler polynomials (see §1.2.2).

Several other summation formulas have been invented to improve the convergence of *slowly convergent series*. Each method has a favorite class of series for which the method is extremely successful. Monotonicity and regularity of the derivatives of the function f that generates the terms of the series always is a good starting point.

To obtain information on how many terms one needs using (1.22) one may use estimates of the Bernoulli numbers. Since the radius of convergence of the series in (1.1) equals 2π, one can use the rough estimate

$$\frac{B_{2k+2}}{(2k+2)!} = \mathcal{O}\left[(2\pi)^{-(2k+2)}\right], \quad \text{as} \quad k \to \infty.$$

This estimate can be refined by using the first series in (1.13). Since the series assumes values between 1 and 2, we have (see also Exercise 1.2)

$$\frac{2}{(2\pi)^{2n}} < (-1)^{n+1}\frac{B_{2n}}{(2n)!} < 2\frac{2}{(2\pi)^{2n}} \qquad n = 1, 2, 3, \ldots . \tag{1.23}$$

When also estimates of the derivatives of f are known, much information on R_k of (1.22) may become available.

1.2. Euler Numbers and Polynomials

The *Euler numbers* have a less dominant place in mathematics than those of Bernoulli, although the definitions are quite similar. Again definitions are based on generating functions. A short introduction is worthwhile, in particular in connection with the summation of *alternating series*.

1.2.1. Definitions and Properties

The Euler numbers are introduced through the series:

$$\boxed{\frac{1}{\cosh z} = \frac{2e^z}{e^{2z}+1} = \sum_{n=0}^{\infty} \frac{E_n}{n!} z^n, \quad |z| < \frac{1}{2}\pi}$$

The Euler numbers are integers, in contrast to the Bernoulli numbers. The first few are

$$E_0 = 1, \quad E_2 = -1, \quad E_4 = 5, \quad E_6 = -61, \quad E_8 = 1385,$$

while those with odd index are zero:

$$E_{2n+1} = 0, \quad n = 0, 1, 2, \ldots .$$

The *Euler polynomials* are defined by

$$\boxed{\frac{2e^{xz}}{e^z+1} = \sum_{n=0}^{\infty} \frac{E_n(x)}{n!} z^n, \quad |z| < \pi,}$$

with as relation between numbers and polynomials

$$E_n = 2^n E_n(\tfrac{1}{2}). \tag{1.24}$$

The first few Euler polynomials are:

$$\begin{aligned}
E_0(x) &= 1,\\
E_1(x) &= x - \frac{1}{2},\\
E_2(x) &= x^2 - x,\\
E_3(x) &= x^3 - \frac{3}{2}x^2 + \frac{1}{4},\\
E_4(x) &= x^4 - 2x^3 + x.
\end{aligned}$$

Further generalizations give:

$$\boxed{\left(\frac{2e^z}{e^{2z}+1}\right)^m = \sum_{n=0}^{\infty} \frac{E_n^{(m)}}{n!} z^n, \quad |z| < \frac{1}{2}\pi,}$$

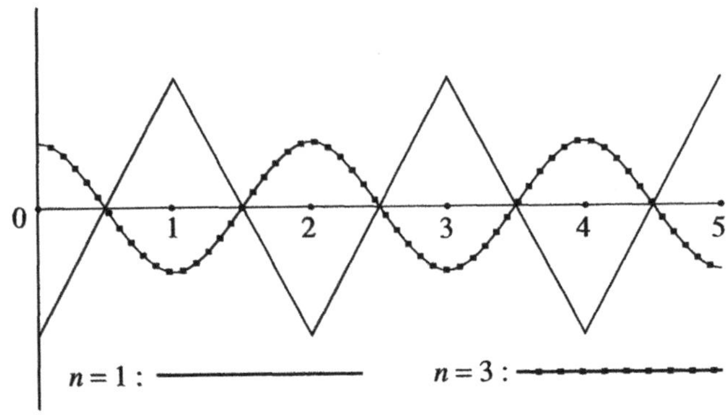

Figure 1.3. The functions $\widetilde{E}_n(x)$, $n = 1$ and $n = 3$.

$$e^{xz}\left(\frac{2}{e^z+1}\right)^m = \sum_{n=0}^{\infty} \frac{E_n^{(m)}(x)}{n!} z^n, \quad |z| < \pi.$$

Note that the system of generalizations is not as clear as in the Bernoulli case. However, the numbers of Bernoulli and Euler do share many properties, such as for example the analogs of the Fourier series in (1.13)

$$E_{2n-1}(x) = 4(-1)^{n+1}(2n-1)! \sum_{m=0}^{\infty} \frac{\cos[(2m+1)\pi x]}{[(2m+1)\pi]^{2n}}$$

$$E_{2n}(x) = 4(-1)^n(2n)! \sum_{m=0}^{\infty} \frac{\sin[(2m+1)\pi x]}{[(2m+1)\pi]^{2n+1}},$$

which hold for $n = 1, 2, 3, \ldots$ and $x \in [0, 1]$. We see that $E'_{2n}(x) = 0$ for $x = \frac{1}{2}$ and with (1.24) it follows that

$$|E_{2n}(x)| \le |E_{2n}(\tfrac{1}{2})| = 2^{-2n} E_{2n}, \quad n = 1, 2, 3, \ldots .$$

We define as in (1.16) the periodic Euler functions $\widetilde{E}_n(x)$ by writing

$$\widetilde{E}_n(x) = E_n(x), \quad 0 \le x < 1,$$

and

$$\widetilde{E}_n(x+1) = \widetilde{E}_n(x), \quad x \in \mathbb{R}.$$

The function $\widetilde{E}_n(x)$ has $n-1$ continuous derivatives.

In Figures 1.3 and 1.4 we show the first four functions $\widetilde{E}_n(x), n = 1, 2, 3, 4$.

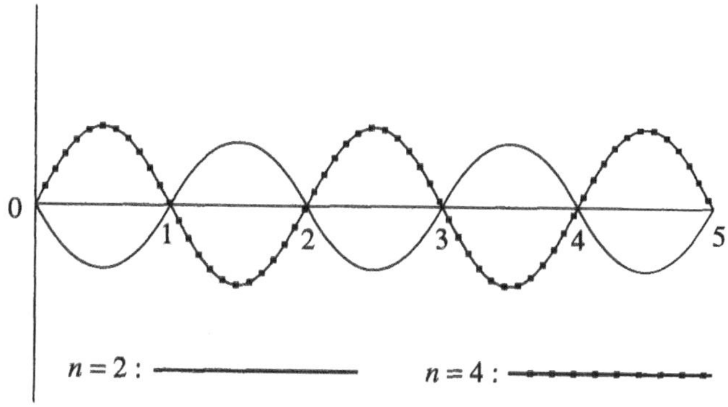

Figure 1.4. The functions $\widetilde{E}_n(x)$, $n = 2$ and $n = 4$.

1.2.2. Boole's Summation Formula

This method is given in BOOLE (1860), but Euler knew the method also. The analog of Theorem 1.1 is given in the form:

Theorem 1.4. *Let the function* f: $[0,1] \to \mathbb{C}$ *have* k *continuous derivatives,* $(k = 0, 1, 2, \ldots)$. *Then, for* $k \geq 1$,

$$f(1) = \frac{1}{2}\sum_{i=0}^{k-1} \frac{E_i(1)}{i!}\left[f^{(i)}(1) + f^{(i)}(0)\right] + R_k,$$

with

$$R_k = \frac{1}{2(k-1)!}\int_0^1 f^{(k)}(x)E_{k-1}(x)\,dx.$$

Proof. Use induction with respect to k. For $k = 1$ the claim is obvious, and for the induction step we can use $(n+1)E_n(x) = E'_{n+1}(x)$. ∎

The theory for *Boole's summation formula* is developed in the same way as in Euler's case. The striking difference is that in the Boole version of Theorem 1.2 the left-hand side of (1.19) now shows an *alternating series*. Moreover an integral of f is missing. We write a final result similar to (1.21), with an extra parameter h, which enables a shift in the argument of f. In this way we obtain Euler numbers in Boole's summation formula (taking $h = \frac{1}{2}$), instead of the quantities $E_n(1)$. The proof of the following theorem is left to the reader.

Theorem 1.5. *Let* f: $[m, \infty) \to \mathbb{R}$ *have* k *continuous derivatives,* $(k = 0, 1, 2, \ldots)$. *Assume that* $f^{(i)}(x) \to 0$ *as* $x \to \infty$ *for each* $i = 0, 1, \ldots, k$. *Let*

$h \in [0,1]$. *Then*

$$\sum_{i=m}^{\infty}(-1)^{i-m}f(h+i) = \frac{1}{2}\sum_{i=0}^{k-1}\frac{E_i(h)}{i!}f^{(i)}(m) + R_k \tag{1.25}$$

with

$$R_k = \frac{1}{2}\int_m^{\infty}\frac{\widetilde{E}_{k-1}(m+h-x)}{(k-1)!}f^{(k)}\,dx.$$

Applying this with $h = \frac{1}{2}, f(x) = 1/x$ we obtain after some algebra

$$2\sum_{k=n}^{\infty}\frac{(-1)^k}{(2k+1)} = (-1)^n\sum_{k=0}^{N}\frac{E_{2k}}{(2n)^{2k+1}} + R,$$

where, on account of earlier given estimates of the Euler polynomials and evaluation of the derivatives of f, it easily follows that

$$|R| \le \frac{|E_{2N}|}{(2n)^{2N+1}}.$$

This result holds for each positive n and N. With $n = 50\,000$ and $N > 1$ we obtain an explanation of the intriguing phenomenon we observed at the beginning of this chapter. The error equals

$$10^{-5} - 10^{-15} + 5\times 10^{-25} - 61\times 10^{-35} + 1385\times 10^{-45}\ldots .$$

Adding this to the computed 50-d approximation of $\frac{1}{2}\pi$, we will see that the result indeed is correct to an accuracy of 50 digits. The striking point is that the digits in the asymptotic correction represent integers (the Euler numbers). That is why the indicated effect is clearly visible. In this example we can also take other values of $n = 0.5\times 10^m$. Only sufficiently large values of m show the effect appropriately.

1.3. Stirling Numbers

James Stirling introduced in the beginning of *Methodus Differentialis* (STIRLING (1730)) certain coefficients that became famous and now bear his name. We define the *Stirling numbers* of the first and second kind, respectively denoted by $S_n^{(m)}$ and $\mathfrak{S}_n^{(m)}$, as the coefficients in the expansions

$$\boxed{x(x-1)\cdots(x-n+1) = \sum_{m=0}^{n} S_n^{(m)} x^m,} \tag{1.26}$$

$$\boxed{x^n = \sum_{m=0}^{n} \mathfrak{S}_n^{(m)} x(x-1)\cdots(x-m+1),} \tag{1.27}$$

where we give the left-hand side of (1.26) the value 1 if $n = 0$; similarly the factors on the right-hand side of (1.27) have the value 1 if $m = 0$. This gives the 'boundary values'

$$S_n^{(n)} = \mathfrak{S}_n^{(n)} = 1,\ n \geq 0, \quad \text{and} \quad S_n^{(0)} = \mathfrak{S}_n^{(0)} = 0,\ n \geq 1.$$

Furthermore it is convenient to agree on $S_n^{(m)} = \mathfrak{S}_n^{(m)} = 0$ if $m > n$.

The Stirling numbers are integer numbers; apart from the above mentioned zero values, the numbers of the second kind are positive, and those of the first kind have the sign of $(-1)^{n+m}$. We have the following recurrence relations

$$S_{n+1}^{(m)} = S_n^{(m-1)} - nS_n^{(m)}, \tag{1.28}$$

$$\mathfrak{S}_{n+1}^{(m)} = m\mathfrak{S}_n^{(m)} + \mathfrak{S}_n^{(m-1)}. \tag{1.29}$$

A proof of (1.28) follows from

$$x(x-1)\cdots(x-n) = \sum_{m=0}^{n+1} S_{n+1}^{(m)} x^m = (x-n)\sum_{m=0}^{n} S_n^{(m)} x^m$$

and comparing corresponding powers of x. A similar proof can be used for (1.29).

The Stirling numbers of the first and second kind satisfy an interesting orthogonality relation. Substitution of (1.26) into (1.27) yields

$$x^n = \sum_{m=0}^{n} \mathfrak{S}_n^{(m)} \sum_{k=0}^{m} S_m^{(k)} x^k = \sum_{k=0}^{n}\sum_{m=k}^{n} \mathfrak{S}_n^{(m)} S_m^{(k)} x^k.$$

Comparison of corresponding powers of x gives

$$\sum_{m=k}^{n} \mathfrak{S}_n^{(m)} S_m^{(k)} = \delta_{k,n}.$$

In a similar way one proves

$$\sum_{m=k}^{n} \mathfrak{S}_m^{(k)} S_n^{(m)} = \delta_{k,n}.$$

These properties lead to a general inversion which can be applied to sequences of numbers. Let

$$a_0, a_1, \ldots \quad b_0, b_1, \ldots$$

be two number sequences, then we have the *Stirling inversion*

$$a_n = \sum_{k=0}^{n} S_n^{(k)} b_k \iff b_n = \sum_{k=0}^{n} \mathfrak{S}_n^{(k)} a_k,$$

with as alternative for infinite sequences (the formal) inversion

$$a_n = \sum_{k=n}^{\infty} S_k^{(n)} b_k \iff b_n = \sum_{k=n}^{\infty} \mathfrak{S}_k^{(n)} a_k.$$

For example, it is not difficult to verify from (1.26) that

$$\sum_{m=0}^{n} (-1)^{n-m} S_n^{(m)} = n!.$$

Then from the inversion relation we obtain

$$\sum_{m=0}^{n} (-1)^{n-m} m! \mathfrak{S}_n^{(m)} = 1.$$

Several other generating functions are available for Stirling numbers. We have

$$\frac{[\ln(x+1)]^m}{m!} = \sum_{n=m}^{\infty} S_n^{(m)} \frac{x^n}{n!}, \tag{1.30}$$

$$\frac{(e^x - 1)^m}{m!} = \sum_{n=m}^{\infty} \mathfrak{S}_n^{(m)} \frac{x^n}{n!}. \tag{1.31}$$

When (1.31) is proved, one can use (with some imagination) the inversion formula for infinite sequences to verify (1.30). To do so, write in (1.31) $z = e^x - 1$, $x = \ln(z+1)$.

We sketch the proof of (1.31). Call $F_m(x)$ the (unknown) left-hand side of (1.31). Then we know that $F_0(x) = 1$ and furthermore (since $\mathfrak{S}_n^{(1)} = 1$) $F_1(x) = e^x - 1$. Using the recurrence relation in (1.29) we arrive at

$$\frac{d}{dx} F_m(x) = m F_m(x) + F_{m-1}(x).$$

Considering the condition at $x = 0$, we obtain a solution of this recursion, which is indeed in the form of the left-hand side of (1.31).

A closed expression for the numbers of the second kind reads

$$\mathfrak{S}_n^{(m)} = \frac{1}{m!} \sum_{k=0}^{m} (-1)^{m-k} \binom{m}{k} k^n. \tag{1.32}$$

A proof follows simply by expanding the left-hand side of (1.31) by Newton's binomial formula and by comparing the power series of the resulting exponential functions on the right-hand side of (1.31). A similar closed expression for the numbers of the first kind is not known.

The Stirling numbers play an important role in difference calculus, combinatorics, and probability theory. An example from *combinatorics* is: $\mathfrak{S}_n^{(m)}$ is the number of ways of partitioning a set of n elements into m non-empty subsets. We find for $m = 2, n = 4$ the value $\mathfrak{S}_4^{(2)} = 7$, since

$$\begin{aligned}\{a,b,c,d\} &= \{a\} \cup \{b,c,d\} = \{b\} \cup \{a,c,d\} \\ &= \{c\} \cup \{a,b,d\} = \{d\} \cup \{a,b,c\} \\ &= \{a,b\} \cup \{c,d\} = \{a,c\} \cup \{b,d\} = \{a,d\} \cup \{b,c\}.\end{aligned}$$

1.4. Remarks and Comments for Further Reading

1.1. An extensive bibliography on Bernoulli numbers (and the other quantities treated in this chapter) is given in DILCHER et al.(1991).

1.2. Further information on classical difference calculus can be found in JORDAN (1947), NÖRLUND (1924), and MILNE-THOMSON (1933), where also the summation methods of Euler and Boole are considered. Euler's method can be found in KNOPP (1946).

1.3. Euler's name is connected with another method for applying transformations on series: the *Euler transformation*, which should not be confused with the summation methods discussed earlier in this chapter. Euler's transformation is a powerful method to improve the convergence of series, in particular of slowly convergent alternating series. Let the given convergent series be $S = \sum_{n=0}^{\infty} (-1)^n a_n$. Then

$$S = \sum_{n=0}^{\infty} \frac{\Delta^n a_0}{2^{n+1}},$$

where Δ is the *forward difference operator* defined by

$$\Delta a_k = a_k - a_{k+1}, \quad \Delta^2 a_k = \Delta a_k - \Delta a_{k+1} = a_k - 2a_{k+1} + a_{k+2},$$

and, in general

$$\Delta^n a_0 = \sum_{k=0}^{n} (-1)^{n-k} \binom{n}{k} a_k.$$

For example, when $a_n = 1/(n+1)$, then $\Delta^n a_0 = 1/(n+1)$, which gives

$$\ln 2 = \sum_{n=0}^{\infty} \frac{(-1)^n}{n+1} = \sum_{n=0}^{\infty} \frac{1}{(n+1)\,2^{n+1}},$$

a startling improvement with respect to convergence; see KNOPP (1946).

1.4. Yet another method for transforming finite and infinite sums into integrals, based on residue calculus, is the *Abel-Plana formula*. See Chapter 8 in OLVER (1974), where also the connection with Euler's summation method is described.

1.5. The remarkable phenomenon observed in summing the series in the introduction of this chapter, and in Exercise 1.9, which is explained in §1.2.2 with the help of Boole's summation formula, seems to be discussed for the first time in BORWEIN, BORWEIN & DILCHER (1989).

1.6. More information on Stirling numbers can be found in JORDAN (1947, Ch. 4) and COMTET (1974). See also Chapter 24 of ABRAMOWITZ & STEGUN (1964), for tables and more formulas. KNUTH (1992) discusses the notation and other interesting historical aspects. Asymptotic expansions have been given by MOSER & WYMAN (1958a, 1958b). They consider several overlapping domains in the n, m-plane with $n \ge m$. See also DINGLE (1973, p. 199). A more recent result is given in TEMME (1993), where approximations are given for Stirling numbers of both kinds, with n large, uniformly with respect to $m, 0 \le m \le n$.

1.5. Exercises and Further Examples

1.1. A number of Taylor series for elementary functions can be derived from the generating functions for the Bernoulli and Euler numbers. A few important series follow from the following results.

a. Show that the function $f(z)$ introduced in connection with (1.1) can be written in the form

$$f(z) = \frac{1}{2} z \coth \frac{1}{2} z - 1.$$

b. Determine via (1.1) the Taylor series of the function $z \coth z$.

c. Show that $\tanh z = 2\coth 2z - \coth z$ and determine the Taylor series of the function $\tanh z$.

d. Show that $2/\sinh 2z = \coth z - \tanh z$ and determine the Taylor series of the function $z/\sinh z$.

e. Integrate $\coth z - 1/z$ and determine the Taylor series of $\ln[\sinh(z)/z]$.
f. Determine the Taylor series of the functions $\ln\cosh z$ and $\ln[\tanh(z)/z]$.
g. Determine with these results corresponding expansions for trigonometric functions.

1.2. Evaluate as for (1.14), the following series:

$$\sum_{m=1}^{\infty} \frac{(-1)^{m+1}}{m^{2n}} = \frac{(-1)^{n+1}(2\pi)^{2n}\left(1-2^{1-2n}\right)B_{2n}}{2\,(2n)!}, \qquad n = 1, 2, 3, \ldots,$$

$$\sum_{m=0}^{\infty} \frac{1}{(2m+1)^{2n}} = \frac{(-1)^{n+1}(2\pi)^{2n}\left(1-2^{-2n}\right)B_{2n}}{2\,(2n)!}, \qquad n = 1, 2, 3, \ldots.$$

For the first formula you will need the result

$$B_n(\tfrac{1}{2}) = -\left(1-2^{1-n}\right)B_n, \quad n = 0, 1, 2, \ldots.$$

With these results improve the inequalities in (1.23) to obtain the rather accurate estimates (at least, when n is large)

$$\frac{2}{(2\pi)^{2n}}\frac{1}{1-2^{-2n}} < (-1)^{n+1}\frac{B_{2n}}{(2n)!} < \frac{2}{(2\pi)^{2n}}\frac{1}{1-2^{1-2n}}, \quad n = 1, 2, 3, \ldots.$$

1.3. Show that from both equations in (1.9) follows:

$$\int_x^y B_p(t)\,dt = \frac{1}{p+1}\left[B_{p+1}(y) - B_{p+1}(x)\right], \qquad \int_x^{x+1} B_p(t)\,dt = x^p,$$

$p = 0, 1, 2, \ldots$, and that, hence, for $n = 1, 2, 3, \ldots$

$$\sum_{i=0}^{n-1} i^p = \sum_{i=0}^{n-1}\int_i^{i+1} B_p(t)\,dt = \int_0^n B_p(t)\,dt = \frac{1}{p+1}\left[B_{p+1}(n) - B_{p+1}\right].$$

Expand in powers of n:

$$\sum_{i=0}^{n-1} i^p = \frac{1}{p+1}\sum_{k=0}^{p}\binom{p+1}{k} B_k n^{p+1-k}.$$

This formula holds for $p = 0$ when we put $0^0 = 1$.

1.4. The series

$$\sum_{i=1}^{\infty}\frac{1}{i^2} = \frac{\pi^2}{6}$$

can be computed with Euler's formula. Determine a value of m that satisfies

$$\left|\frac{B_6}{6!} f^{(5)}(m)\right| < 10^{-10}, \quad \text{with} \quad f(x) = \frac{1}{x^2}.$$

1.5. Compute *Euler's constant*

$$\boxed{\gamma = \lim_{n\to\infty}\left[1 + \frac{1}{2} + \frac{1}{3} + \frac{1}{4} + \cdots + \frac{1}{n} - \ln(n+1)\right]}$$

by treating the series

$$\sum_{n=1}^{\infty} u_n, \quad \text{with} \quad u_n = \frac{1}{n} + \ln n - \ln(n+1)$$

via Euler's summation formula. Show that the series converges and that the corresponding function f satisfies the conditions of Theorem 1.4. If you find it difficult to prove that the derivatives of $f(x) = 1/x + \ln x - \ln(x+1)$ are monotonic, you may use the representation

$$f(x) = \int_0^1 \left(\frac{1}{x} - \frac{1}{t+x}\right) dt.$$

Now it easily shown that all derivatives have fixed signs on $(0, \infty)$ and that $f^{(k)}$ and $f^{(k+1)}$ have different signs.

1.6. Prove *Stirling's formula* (STIRLING (1730, page 135))

$$\boxed{n! \sim \sqrt{2\pi n}\, n^n e^{-n}, \quad \text{as} \quad n \to \infty}$$

with the help of Euler's summation formula. Prove first *Wallis' product*

$$\lim_{n\to\infty} \frac{2^2\, 4^2\, 6^2 \cdots (2n)^2}{1^2\, 3^2\, 5^2 \cdots (2n-1)^2\, (2n+1)} = \frac{\pi}{2}.$$

Consider for this the integral

$$I_n = \int_0^{\pi/2} \sin^n x\, dx, \quad n = 0, 1, 2, \ldots.$$

Integrating by parts we can show that the recursion $I_n = [(n-1)/n]\, I_{n-2}$ holds, with initial values $I_0 = \pi/2$, $I_1 = 1$. From $I_{n+1} < I_n < I_{n-1}$ it then follows that the ratio I_{2n+1}/I_{2n} tends to 1 as n tends to infinity. This proves Wallis' product; a more concise notation is

$$\lim_{n\to\infty} \frac{(n!)^2\, 2^{2n}}{(2n)!\, \sqrt{n}} = \sqrt{\pi}.$$

Now apply (1.20) with $f(x) = \ln x$, $n = 2m$, $k = 0$. The result is

$$\ln \frac{(2m)!}{(m-1)!} = 2m \ln 2 + (m+1)\ln m - m + \frac{1}{2}\ln 2 + R_0,$$

with

$$R_0 = \int_m^{2m} \frac{\widetilde{B}_1(x)}{x}\,dx = \left[\frac{\widetilde{B}_2(x)}{2x}\right]_m^{2m} + \int_m^{2m} \frac{\widetilde{B}_2(x)}{2x^2}\,dx.$$

Since $\widetilde{B}_2(m) = \widetilde{B}_2(2m) = \frac{1}{6}$ the integrated term equals $-\frac{1}{24m}$. Because of

$$-\frac{1}{12} \le \widetilde{B}_2(x) \le \frac{1}{6} \quad \text{and} \quad 2x^2 \ge 2m^2,$$

we see that the integral with $\widetilde{B}_2(x)$ lies between $-\frac{1}{24m}$ and $\frac{1}{12m}$. So,

$$-\frac{1}{12m} \le R_0 \le \frac{1}{24m}.$$

Application of Wallis' product finally gives

$$\lim_{m\to\infty} \frac{m!}{m^m\, e^{-m}\sqrt{2\pi m}} = 1,$$

a different way of writing Stirling's formula.

1.7. Apply (1.19) with the function $f(x) = g(x/n)$. The result is the *extended trapezoidal rule* on the interval $[0,1]$ in the form

$$\begin{aligned}\int_0^1 g(t)\,dt = {}& h\left[\frac{1}{2}g(0) + g(h) + g(2h) + \cdots + g(1-h) + \frac{1}{2}g(1)\right]\\ &+ \sum_{i=1}^{k} \frac{B_{2i}}{(2i)!} h^{2i}\left[g^{(2i-1)}(0) - g^{(2i-1)}(1)\right]\\ &- \frac{h^{2k+1}}{(2k+1)!}\int_0^1 g^{(2k+1)}(t)\widetilde{B}_{2k+1}(nt)\,dt.\end{aligned}$$

Here, $h = 1/n$. Notice that as more derivatives of g vanish at 0 and 1 the trapezoidal rule progressively improves in accuracy. For a C^∞–function with compact support in $(0,1)$, all terms in the sum disappear, whatever i and k. Only the integral remains on the right-hand side , in which we may take k as large as we please.

1.8. Verify the following relations for the Euler polynomials:

$$\begin{aligned}&E'_{n+1}(x) = (n+1)E_n(x),\\ &E_n(x+1) + E(x) = 2x^n,\\ &E_n(1-x) = (-1)^n E_n(x),\\ &E_{2n} = 2\,(-1)^n\,(2n)!\left(\frac{2}{\pi}\right)^{2n+1}\sum_{m=0}^{\infty}\frac{(-1)^m}{(2m+1)^{2n+1}},\end{aligned}$$

for $n = 0, 1, 2, \ldots$, and

$$\sum_{k=1}^{m} (-1)^{m-k} k^n = \frac{1}{2} \left[E_n(m+1) + (-1)^m E_n(0) \right], \quad m, n \geq 1.$$

1.9. Summing the first 50 000 terms in the series $\ln 2 = \sum_{n=1}^{\infty} \frac{(-1)^{n+1}}{n}$ gives

$$\sum_{n=1}^{50\,000} \frac{(-1)^{n+1}}{n}$$
$$= 0.69313\ 71806\ 59945\ 30939\ 72321\ 21474\ 17656\ 80483\ 00134\ 43962.$$

Show that the error (for obtaining a correct 50-d result) equals

$$10^{-5} - 10^{-10} + 2 \times 10^{-20} - 16 \times 10^{-30} + 272 \times 10^{-40} - 7936 \times 10^{-50}.$$

Apply Theorem 1.5 with $h = 1$. The integer numbers in the asymptotic expansion of the error are the tangent numbers introduced in (1.12). Namely,

$$T_n = (-1)^n 2^n E_n(1),$$

which relation follows from

$$\tan z = i - \frac{2ie^{2iz}}{e^{2iz} + 1}.$$

1.10. Verify the following relations for the Stirling numbers:

$$S_n^{(1)} = (-1)^{n-1}(n-1)!, \quad \mathfrak{S}_n^{(1)} = 1, \quad -S_n^{(n-1)} = \mathfrak{S}_n^{(n-1)} = \binom{n}{2}$$

$$\sum_{m=1}^{n} S_n^{(m)} = 0, \quad \lim_{n \to \infty} m^{-n} \mathfrak{S}_n^{(m)} = \frac{1}{m!}.$$

For the final relation you may use (1.32).

1.11. Show, with the help of (1.31) that the =Stirling numbers of the second kind, in fact, are special cases of the generalized Bernoulli numbers defined in (1.3) with $x = 0$, and show that a similar relation exists for the Stirling numbers of the first kind. That is,

$$S_n^{(m)} = \binom{n-1}{m-1} B_{n-m}^{n}, \quad \mathfrak{S}_n^{(m)} = \binom{n}{m} B_{n-m}^{-m}.$$

1.12. Verify that, for $n > 1$, the Fourier series of $B_n(x)$ in (1.13) can be obtained by the calculus of residues. Consider $\int_C f(z)\,dz$ with

$$f(z) = z^{-n} e^{xz} (e^z - 1)^{-1}, \quad n \text{ an integer} > 1,$$

the contour C being a (large) circle with radius $(2N+1)\pi$ (N an integer), center at the origin. The poles of the integrand are $z_r = 2\pi i r$, $(r = 0, \pm 1, \pm 2, \ldots)$ and the residues of $f(z)$ at these poles are $(2\pi i r)^{-n} \exp(2\pi i r x)$; the residue at $z = 0$ is $B_n(x)/n!$. The integral around the circle C tends to zero as $N \to \infty$ provided $0 \le x \le 1$. Verify that, by the theorem of residues,

$$B_n(x) = -n! \sum_{r=\infty}^{\infty}{}' (2\pi i r)^{-n} e^{2\pi i r x}, \quad n > 1, \quad 0 \le x \le 1,$$

where the prime indicates that the term corresponding to $r = 0$ must be omitted. This gives the expansions in (1.13).

2

Useful Methods and Techniques

When manipulating series and integrals we often encounter problems connected with

- interchanging summation and integration,
- interchanging the order of integration, and
- differentiation of integrals with respect to a parameter.

In this chapter we quote some tools that are frequently used in advanced calculus and analysis. The first two topics can be considered from the point of view of Lebesgue or Stieltjes integrals, for which Fubini's theorem or Lebesgue's dominated convergence theorem can be applied. For readers more familiar with Riemann integration it is useful to reformulate these theorems.

We also give an introduction to asymptotic analysis of integrals, which is also a basic tool in the theory of special functions. In particular we consider Watson's lemma and the saddle point method. In subsequent chapters asymptotic methods will be applied to obtain asymptotic expansions for the gamma function, Bessel functions, and so forth.

2.1. Some Theorems from Analysis

The first theorem gives the conditions for justification of what is usually named *interchanging summation and integration*, a technique that is used very frequently in applied analysis and in the field of special functions. It is called the *theorem of dominated convergence of Lebesgue* in the setting of Riemann integrals. A proof can be found in the standard works of classical analysis, for instance in BROMWICH (1926, §§175 and 176) or TITCHMARSH (1939, §1.77).

Theorem 2.1. *Let (a, b) be a given finite or infinite interval, and let $\{U_n(t)\}$ be a sequence of real or complex valued continuous functions, which satisfy the following conditions:*

(1) $\sum_{n=0}^{\infty} U_n(t)$ *converges uniformly on any compact interval in (a, b).*

(2) *At least one of the following two quantities is finite:*

$$\int_a^b \sum_{n=0}^{\infty} |U_n(t)|\, dt, \quad \sum_{n=0}^{\infty} \int_a^b |U_n(t)|\, dt.$$

Then we have

$$\int_a^b \sum_{n=0}^{\infty} U_n(t)\, dt = \sum_{n=0}^{\infty} \int_a^b U_n(t)\, dt.$$

The second theorem treats the interchanging of the order of integration for improper Riemann integrals. For the principles of Lebesgue theory, measurable functions, and so on, we refer to RUDIN (1976). We have

Theorem 2.2. *If $f(x, y)$ is measurable (in particular continuous) on the open quadrant $(0, \infty) \times (0, \infty)$, and the repeated improper Riemann integrals*

$$\int_0^\infty dx \int_0^\infty f(x, y)\, dy \qquad \int_0^\infty dy \int_0^\infty f(x, y)\, dx$$

both exist and are both absolutely convergent, then these integrals are equal.

This theorem is given in LOVE (1970), *where also instructive comments on this theorem are given.*

The third theorem is an extension to complex variables of a standard theorem concerning differentiation of an integral over an infinite contour with respect to a parameter; for proofs see, for example, LEVINSON & REDHEFFER (1970, Ch. 6) or COPSON (1935, §5.51).

Theorem 2.3. *Let t be a real variable ranging over a finite or infinite interval (a, b) and z a complex variable ranging over a domain Ω. Assume that the function $f: (\Omega \times (a, b)) \to \mathbb{C}$ satisfies the following conditions:*

(1) *f is a continuous function of both variables.*

(2) *For each fixed value of t, $f(., t)$ is a holomorphic function of the first variable.*

(3) *The integral*

$$F(z) = \int_a^b f(z, t)\, dt, \qquad z \in \Omega$$

converges uniformly at both limits in any compact set in Ω.

Then $F(z)$ is holomorphic in Ω, and its derivatives of all orders may be found by differentiating under the sign of integration.

Often, the theorem will be applied to a contour integral in the complex plane, for which the contour of integration can be parameterized by a real parameter t.

2.2. Asymptotic Expansions of Integrals

The next topic is from the theory of asymptotics for Laplace integrals. We mention a very useful result known as *Watson's Lemma* and we discuss the basic elements of the method of saddle points. First we give a definition of an asymptotic expansion.

Definition 2.1. Let F be function of a real or complex variable z; let $\sum_{n=0}^{\infty} a_n z^{-n}$ denote a (convergent or divergent) formal power series, of which the sum of the first n terms is denoted by $S_n(z)$; let

$$R_n(z) = F(z) - S_n(z).$$

That is,

$$F(z) = a_0 + \frac{a_1}{z} + \frac{a_2}{z^2} + \cdots + \frac{a_{n-1}}{z^{n-1}} + R_n(z), \quad n = 0, 1, 2 \ldots, \tag{2.1}$$

where we assume that when $n = 0$ we have $F(z) = R_0(z)$. Next, assume that for each $n = 0, 1, 2, \ldots$ the following relation holds

$$R_n(z) = \mathcal{O}\left(z^{-n}\right), \quad \text{as} \quad z \to \infty \tag{2.2}$$

in some unbounded domain Δ. Then $\sum_{n=0}^{\infty} a_n z^{-n}$ is called an *asymptotic expansion* of the function $F(z)$ and we denote this by

$$F(z) \sim \sum_{n=0}^{\infty} a_n z^{-n}, \quad z \to \infty, \quad z \in \Delta. \tag{2.3}$$

This definition is due to Poincaré (1886). Analogous definitions can be given for $z \to 0$, and so on.

Observe that we do not assume that the infinite series $\sum_{n=0}^{\infty} a_n z^{-n}$ converges for certain z–values. This is not relevant in asymptotics; in the definition only a property of $R_n(z)$ is requested, with n fixed.

Example 2.1. The classical example is the so-called *exponential integral*, that is,

$$F(x) = x \int_x^{\infty} t^{-1} e^{x-t}\, dt = x\, e^x E_1(x),$$

(see §7.3.5) where x is real and positive. Repeatedly using integration by parts, we obtain

$$F(x) = 1 - \frac{1}{x} + \frac{2!}{x^2} - \cdots + \frac{(-1)^{n-1}(n-1)!}{x^{n-1}} + (-1)^n n!\, x \int_x^\infty \frac{e^{x-t}}{t^{n+1}}\, dt.$$

In this case we have, since $t > x$,

$$(-1)^n R_n(x) = n!\, x \int_x^\infty \frac{e^{x-t}}{t^{n+1}}\, dt \le \frac{n!}{x^n} \int_x^\infty e^{x-t}\, dt = \frac{n!}{x^n}.$$

Indeed, $R_n(x) = \mathcal{O}(x^{-n})$ as $x \to \infty$. Hence

$$x \int_x^\infty t^{-1} e^{x-t}\, dt \sim \sum_{n=0}^{\infty} (-1)^n \frac{n!}{x^n}, \qquad x \to \infty.$$

This series is divergent for any finite value of x. However, when x is sufficiently large and n is fixed, the finite part of the series $S_n(x)$ given by

$$S_n(x) = F(x) - R_n(x),$$

approximates the function $F(x)$ with any desired accuracy.

In this example we can derive the asymptotic expansion in a different way. We write (using the transformation $t = x(1+u)$) $F(x)$ as a Laplace integral

$$F(x) = x \int_0^\infty e^{-xu} f(u)\, du, \quad f(u) = 1/(1+u). \tag{2.4}$$

We now write

$$f(u) = 1 - u + u^2 - \cdots + (-1)^{n-1} u^{n-1} + (-1)^n u^n/(1+u),$$

and we obtain exactly the same expansion, with the same expression and upper bound for $|R_n(x)|$.

2.2.1. Watson's Lemma

The second approach above gives the set-up for the following result (the Euler gamma function appearing in (2.7) is treated in the next chapter).

Theorem 2.4. *(Watson's lemma). Assume that:*

(i) *$f(t)$ is a real or complex function of the positive real variable t with a finite number of discontinuities and infinities.*

(ii) *As $t \to 0+$*

$$f(t) \sim t^{\lambda-1} \sum_{n=0}^{\infty} a_n t^n, \quad \Re\lambda > 0. \tag{2.5}$$

(iii) *The integral*

$$F(z) = \int_0^\infty f(t)e^{-zt}\,dt \tag{2.6}$$

is convergent for sufficiently large values of $\Re z$.

Then

$$F(z) \sim \sum_{n=0}^{\infty} \Gamma(n+\lambda)\frac{a_n}{z^{n+\lambda}}, \quad z \to \infty \tag{2.7}$$

in the sector $|\arg z| \le \frac{1}{2}\pi - \delta(< \frac{1}{2}\pi)$, *where* $z^{n+\lambda}$ *has its principal value.*

A larger z–sector can be obtained when we know that f is analytic in a certain domain of the complex plane. For example, when f is analytic in the sector $|\arg z| < \pi/2$ and $f(t) = \mathcal{O}[\exp(\sigma|t|)]$ in that sector, for some number σ, then the asymptotic expansion in Watson's lemma holds in the sector $|\arg z| \le \pi - \delta(< \pi)$.

For a proof we refer to OLVER (1974, p. 113), where a more general condition (ii) is assumed.

When applying Watson's lemma in the theory of special functions, condition (i) often holds, since the function $f(t)$ is, up to the factor $t^{\lambda-1}$, usually an analytic function in a domain containing $[0, \infty)$. Compare Example 2.1 for the exponential integral with $f(t) = 1/(1+t)$. In that case $f(t)$ is analytic in the sector $|\arg t| < \pi$.

Next we formulate a second theorem in which a much larger domain than in the previous theorem for the phase of the large parameter z is possible. For a proof we refer to OLVER (1974, p. 114).

Theorem 2.5. *Assume that:*

(i) *$f(t)$ is analytic inside a sector Ω: $\alpha_1 < \arg t < \alpha_2$, where $\alpha_1 < 0$ and $\alpha_2 > 0$.*

(ii) *For each $\delta \in (0, \frac{1}{2}\alpha_2 - \frac{1}{2}\alpha_1)$ (2.5) holds as $t \to 0$ in the sector*

$$\Omega_\delta : \alpha_1 + \delta < \arg t < \alpha_2 - \delta;$$

for λ we again assume that $\Re\lambda > 0$.

(iii) *There is a real number σ such that $f(t) = \mathcal{O}(e^{\sigma|t|})$ as $t \to \infty$ in Ω_δ.*

Then the integral (2.6), or its analytic continuation, has the asymptotic expansion (2.7) in the sector

$$-\alpha_2 - \frac{1}{2}\pi + \delta \le \arg z \le -\alpha_1 + \frac{1}{2}\pi - \delta. \tag{2.8}$$

In this result the many-valued functions $t^{n+\lambda}$, $z^{n+\lambda}$ have their principal values on the positive real axis and are defined by continuity elsewhere.

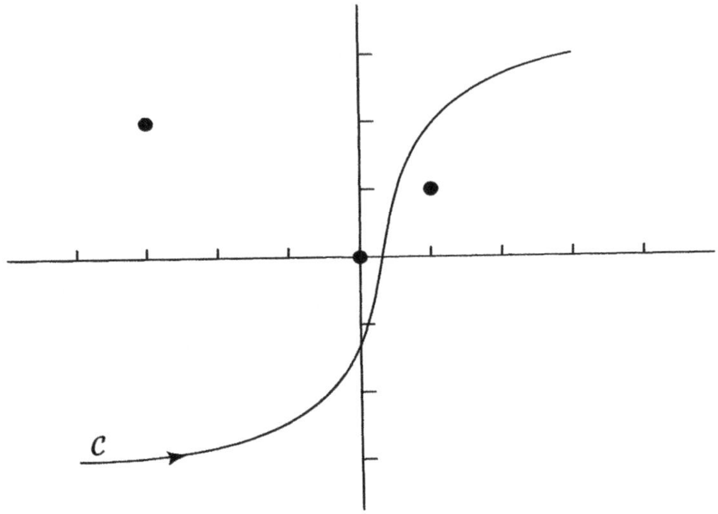

Figure 2.1. Contour C in the complex plane; the dots indicate singularities of ϕ or ψ or saddle points of ϕ.

To explain how the bounds in (2.8) arise, we write $\arg t = \tau$ and $\arg z = \theta$, where $\alpha_1 < \tau < \alpha_2$. The condition for convergence in (2.6) is $\cos(\tau + \theta) > 0$, that is, $-\frac{1}{2}\pi < \tau + \theta < \frac{1}{2}\pi$. Combining this with the bounds for τ we obtain the bounds for θ in (2.8).

For the exponential integral in (2.4) we can take $\alpha_1 = -\pi$, $\alpha_2 = \pi$. Hence, the asymptotic expansion given in Example 2.1 holds in the sector $|\arg z| \le \frac{3}{2}\pi - \delta$. This range is much larger than the usual domain of definition for the exponential integral, which reads: $|\arg z| < \pi$. The phrase *or its analytic continuation* is indeed important in this theorem.

2.2.2. The Saddle Point Method

The saddle point method is usually applied to contour integrals in the complex plane:

$$F(\omega) = \int_C e^{-\omega\phi(z)}\psi(z)\,dz$$

where ϕ, ψ are functions of the complex variable z and which are analytic in a domain $\mathcal{D}$ of the complex plane. The integral is taken along a path C in the $z-$plane, as shown in Figure 2.1, and avoids the singularities of the integrand; ω is a real or complex large parameter. Integrals of this type arise naturally in the context of linear wave propagation.

The first ideas of the saddle point method have been sketched by Riemann (1863). Debye (1909) has used the method for Bessel functions of large orders. Consider this problem from the view point of numerical quadrature of the above integral. Assume that ω is real. Separating ϕ into its real and imaginary parts, writing

$$z = x + iy, \quad \phi(z) = R(x,y) + iI(x,y),$$

we know that, when ω is large, the evaluation of the integral is hampered by the strong oscillatory behavior of the integrand, caused by the expression $\exp[i\omega I(x,y)]$. Usually we have much freedom in choosing the path C in the complex plane (by invoking Cauchy's theorem). When the contour C can be chosen such that

$$I(x,y) = I_0 = constant \quad \text{for} \quad z = x + iy \in C,$$

we can write

$$F(\omega) = e^{-i\omega I_0} \int_C e^{-\omega \Re\phi(z)} \psi(z)\, dz,$$

where the dominant part $\exp[-\omega R(x,y)]$ of the integral is non-oscillating (in some cases the new path C is split up in more branches, each branch being defined by a different I_0, resulting in a sum of integrals of the above type). From a numerical point of view the new representation of $\phi(\omega)$ is very attractive. The question that remains is: which constant I_0 should be used? Luckily, *there is not much choice.*

Considering the real part of the phase function $R(x,y)$, and the landscape of mountains and valleys defined by $\exp[-\omega R(x,y)]$, we may assume that, if the original contour C extends to infinity, C will certainly extend to infinity by descending into one of the valleys. Recall from complex function theory that $R(x,y)$ is a harmonic function, $\Delta R(x,y) = 0$ and that, hence, $R(x,y)$ cannot have local maxima. When the path C runs in two different valleys, the path should pass the saddle point between these two valleys.

This can be seen as follows. When C is defined by $I(x,y) = constant$, we have $I_x dx + I_y dy = 0$. Using the Cauchy-Riemann equations $R_x = I_y, R_y = -I_x$ this gives $R_y dx - R_x dy = 0$. Hence the path C runs orthogonal to the level curves $R(x,y) = constant$; in other words, C is a *steepest descent path.* Such a path, joining two valleys, should pass through the saddle point. The constant I_0 used above should be equal to $I(x_0, y_0)$ where $z_0 = x_0 + iy_0$ is a saddle point: $\phi'(z_0) = 0$.

Summarizing, one tries to deform the contour C through one or more points where the dominant part of the integrand locally behaves like a Gaussian curve. These points are found at the saddle points of the integrand. The method is best learned from clear examples.

Example 2.2. It is instructive to see what happens in a simple example. Let

$$\phi(z) = \frac{1}{2}z^2 - iz \quad \text{and} \quad C = (-\infty, +\infty).$$

We have

$$I(x,y) = c \quad \text{when} \quad x(y-1) = c.$$

The path C can be deformed into the path $y = 1$, corresponding with $c = 0$. Otherwise, when $c \neq 0$, the path C will be such that the integral is divergent. Observe that the saddle point is $z = i$, at which point $I(x,y) = 0$. In this case we have, integrating along the path $y = 1$ and substituting $z = x + i$:

$$F(\omega) = \int_{-\infty+i}^{+\infty+i} e^{-\omega(\frac{1}{2}z^2 - iz)}\, dz = e^{-\frac{1}{2}\omega} \int_{-\infty}^{\infty} e^{-\frac{1}{2}\omega x^2}\, dx = \sqrt{\frac{2\pi}{\omega}}\, e^{-\frac{1}{2}\omega}.$$

In the following case

$$F_0(\omega) = \int_0^{\infty} e^{-\omega(\frac{1}{2}z^2 - iz)}\, dz$$

the best contour of integration is constituted by two parts: one part from 0 to i and a second part from i to $i + \infty$. In Exercise 2.1 we consider this example again.

In §3.6.3 of Chapter 3 we give a detailed example of the saddle point method for the reciprocal gamma function. In the present subsection we give another example, involving Hankel functions with large argument and fixed order.

Example 2.3. Consider the *Hankel function* defined by

$$H_\nu^{(1)}(z) = \frac{1}{\pi i} \int_{-\infty-\pi i}^{\infty} e^{-z \sinh w + \nu w}\, dw, \tag{2.9}$$

with ν a fixed real number and z a large positive number. This function will be considered in Chapter 9 on Bessel functions. The dominant part of the integrand is $\exp(-z \sinh w)$, which has saddle points at the zeros of its derivative; that is at the zeros of $\cosh w$. The saddle points are

$$w_k = -\frac{1}{2}\pi i + k\pi i, \quad k \in \mathbb{Z}.$$

We concentrate on the point w_0; we have $\exp(-z \sinh w_0) = \exp(iz)$. We try to choose the contour of integration such that along this contour $\Im \sinh w = \Im \sinh w_0 = -1$. When this is satisfied, the dominant part of the integrand has a constant phase; this gives a very convenient representation, as we will

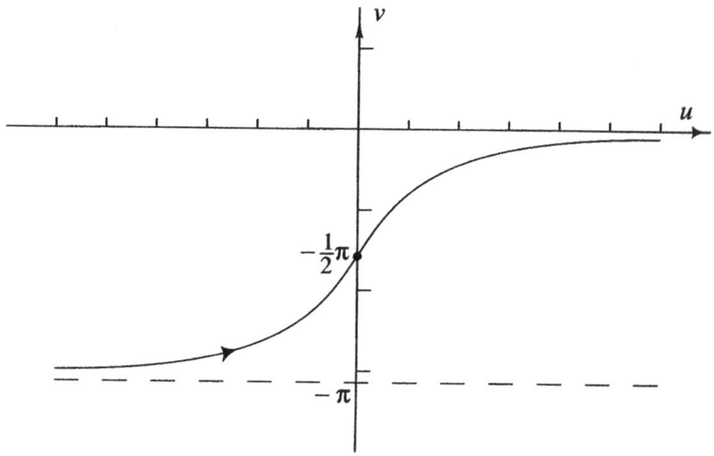

Figure 2.2. Contour for the integral in (2.9), $w = u + iv$.

shortly see. We note, writing $w = u + iv$, that equation $\Im \sinh w = -1$ is equivalent to $\cosh u \sin v = -1$. A suitable solution for the present problem is

$$v(u) = -\frac{1}{2}\pi + \arctan \sinh u, \ u \in \mathbb{R}. \tag{2.10}$$

We integrate (2.9) on the path defined by this relation; see Figure 2.2. It is straightforward to verify that, on this path,

$$\Re(-z \sinh w) = -z \sinh^2 u / \cosh u.$$

It follows that, when we integrate with respect to u, (2.9) can be written as

$$H_\nu^{(1)}(z) = \frac{e^{iz}}{\pi i} \int_{-\infty}^{\infty} e^{-z \sinh^2 u/\cosh u + \nu(u+iv)} f(u)\, du, \tag{2.11}$$

where

$$f(u) = \frac{dw}{du} = 1 + i\frac{dv}{du} = 1 + \frac{i}{\cosh u}$$

and with $v = v(u)$ given in (2.10). It is also possible to write (2.11) as an integral with respect to v over the finite interval $(-\pi, 0)$.

The dominant part of the integrand in (2.11) has the desired Gaussian shape. The integral is a suitable starting point for numerical quadrature. The oscillations are only caused by $\exp(i\nu v)$; this factor is quite harmless with ν fixed and v in the bounded interval $(-\pi, 0)$.

A few manipulations then give

$$H_\nu^{(1)}(z) = \frac{2^{3/2} e^{i\chi}}{\pi} \int_0^\infty e^{-z \sinh^2 u / \cosh u} g(u)\, du, \tag{2.12}$$

where

$$g(u) = \frac{1}{\sqrt{2}} \cosh \nu(u + i \arctan \sinh u) \left(1 + \frac{i}{\cosh u}\right) e^{-\frac{1}{4}\pi i}, \quad g(0) = 1,$$

and $\chi = z - (\frac{1}{2}\nu + \frac{1}{4})\pi$. A further transformation $\sinh^2 u / \cosh u = 2t$, or $\cosh u = t + \sqrt{1+t^2}$, brings this integral in the form of a Laplace integral, on which Watson's lemma can be applied. This eventually leads to a well-known asymptotic representation of the Hankel function. We do not give details on the computation of the coefficients, because in Chapter 10 on Bessel functions the coefficients are obtained rather straightforwardly.

Remark 2.1. The choice of saddle point t_0 is rather obvious in this case. In more complicated cases it is not always clear which saddle point should be chosen. In the present case t_0 is the only saddle point through which the contour $\mathcal{L}_1$ can pass such that the contour ends at $-\infty - \pi i$ and $+\infty$, and on which $\Im \sinh t$ is constant.

Several monographs give extensive treatments of the saddle point method. See, for instance, De Bruijn (1961), Dingle (1973), Bleistein & Handelsman (1975), Lauwerier (1974), Olver (1974), and Wong (1989).

2.2.3. Other Asymptotic Methods

Other standard methods in this area of asymptotics of integrals, such as the method of stationary phase, Laplace's method, can be found in the literature.

- Olver (1974) treats all topics and is very thorough and rigorous. Many results for special functions are derived from their differential equations. It is an important book for uniform asymptotics.
- Erdélyi (1956) is a very concise book, and of interest for the method of stationary phase.
- De Bruijn (1961) gives several very interesting topics, in particular the saddle point method, asymptotic inversion and asymptotic iteration.
- Dingle (1973) has developed its own terminology and has recently received much interest in connection with the study and re-expansion of remainders in asymptotic expansions.
- Bleistein & Handelsman (1975) gives a nice treatment of the saddle point method; this book is also important because of the extensive discussion of Mellin transformation techniques with applications to asymptotics.

- LAUWERIER (1974) is an attractive practical introduction to several methods in asymptotics of integrals.
- WONG (1989) is a sequel to Olver's book, with a sound mathematical approach; a distinctive and important feature is the application of distributional and summation methods; no other book on asymptotics pays attention to this. Wong also treats uniform asymptotics and asymptotics for multi-dimensional integrals.

New interest in several aspects of asymptotics arose recently, starting with the paper BERRY (1989), in which the so-called *Stokes phenomenon* has been given a new interpretation. We return to these new developments in §11.1, when we discuss properties of the error function.

2.3. Exercises and Further Examples

2.1. Let $z_0 = x_0 + iy_0$ be any point in the finite plane. Describe the path such that the integral

$$F_{z_0}(\omega) = \int_{z_0}^{+\infty} \exp -\omega[\phi(z)]\, dz, \quad \phi(z) = \frac{1}{2}z^2 - iz$$

converges, and on which path the integrand does not oscillate.

2.2. Consider the Hankel function defined in (2.9) for the case that order and argument are equal; that is, $z = \nu > 0$. Describe the path of steepest descent for this case.

3

The Gamma Function

3.1. Introduction

The first occurrence of the *gamma function* happens in 1729 in a correspondence between Euler and Goldbach. We take as the definition of the gamma function the integral

$$\boxed{\Gamma(z) = \int_0^\infty t^{z-1} e^{-t}\, dt, \quad \Re z > 0.} \tag{3.1}$$

Closely connected with this is the *beta integral* (EULER (1772)) with definition

$$\boxed{B(p, q) = \int_0^1 t^{p-1}(1-t)^{q-1}\, dt, \quad \Re p > 0, \quad \Re q > 0.} \tag{3.2}$$

The latter is called the *Euler integral of the first kind*, and (3.1) the *Euler integral of the second kind.*

The gamma function is the most obvious generalization of the factorial. Euler was confronted with this matter when he was offered a seemingly simple problem. It was expected that $n!$ (this notation was not used at that time) could be expressed in terms of simple algebraic quantities. This was possible for the so-called *triangular numbers* $T_n = 1 + 2 + 3 + \cdots + n$, which can be written as $T_n = \frac{1}{2}n(n+1)$. In Euler's days one paid much attention to problems of this kind. First, because a formula like the one for T_n gives the opportunity for quick computations. Second, because of the possibility of interpolating. The formula $T_n = \frac{1}{2}n(n+1)$ also can be interpreted for non-integer values of n.

In 1729 Euler proved that for $n!$ such a simple formula does not exist. That is, there is no formula with a finite number of algebraic terms available for $n!$, and he derived the formula

$$n! = \int_0^1 (-\ln x)^n \, dx.$$

The right-hand side is indeed defined for non-negative real values of n. Nowadays Euler's integral is usually written as (3.1), a notation, in fact, due to LEGENDRE (1809a), who also introduced the name gamma function.

3.1.1. The Fundamental Recursion Property

Integrating by parts in (3.1), we obtain

$$\Gamma(z+1) = -\int_0^\infty t^z \, de^{-t} = z\int_0^\infty t^{z-1} e^{-t} \, dt = z\Gamma(z),$$

and this immediately gives the fundamental property $\Gamma(z+1) = z\Gamma(z)$ of the factorial. From $\Gamma(1) = 1$ then we have $\Gamma(n+1) = n!$.

This does not make clear why Euler's choice of generalization is the best one. But, afterwards, this became abundantly clear. Time and again, the Euler gamma function shows up in a very natural way in all kinds of problems. Moreover, the function has a number of interesting properties.

One of the striking properties of the gamma function is that it cannot satisfy a differential equation with algebraic coefficients (this result is due to HÖLDER (1887)). This property makes the gamma function completely different from other well-known transcendental functions, such as the exponential and trigonometric functions. While the functional relation $\Gamma(z+1) = z\Gamma(z)$ is so simple! From a certain point of view, the gamma function is at the basis of a theory of a class of difference equations, just as the exponential function, the solution of the simple equation $y' = y$, plays a principal role in the theory of differential equations.

3.1.2. Another Look at the Gamma Function

Apart from (3.1), various other starting points are available for defining the generalization of the factorial. For instance, for positive values of the argument we have a definition in the form:
The gamma function Γ: $(0, \infty) \to (0, \infty)$ is the function f with $f(1) = 1$ that, for $x > 0$, satisfies the following three conditions

- $f(x) > 0$,
- $f(x+1) = xf(x)$,
- f is log-convex (this means, $\ln f$ is convex).

In this way the gamma function is introduced in the legendary works of Bourbaki. In BOHR & MOLLERUP (1922) it is shown that these three properties characterize $\Gamma(x)$ completely. The interested reader may verify the equivalence of this definition with the other ones, say (3.1), by consulting RUDIN (1976). There it is also observed that the log-convexity of $\Gamma(x)$ on $(0,\infty)$ follows from

$$\Gamma\left(\frac{x}{p}+\frac{y}{q}\right) \le \Gamma(x)^{1/p}\,\Gamma(y)^{1/q},$$

if $1 < p < \infty$ and $(1/p) + (1/q) = 1$. This inequality is easily derived from Euler's integral (3.1) and *Hölder's inequality*:

$$\left|\int_a^b f(x)g(x)\,dx\right| \le \left\{\int_a^b |f(x)|^p\,dx\right\}^{1/p} \left\{\int_a^b |g(x)|^q\,dx\right\}^{1/q},$$

where f, g are (complex) integrable functions on $[a, b]$

3.2. Important Properties

The gamma function is defined via (3.1) in the half-plane $\Re z > 0$; in this domain the function is analytic. This follows from a well-known theorem of complex function theory for integrals that depend on a parameter; see Theorem 2.3. Using the recursion $\Gamma(z) = \Gamma(z+1)/z$ we can enter the left half-plane step by step. The first step, and the principal of analytic continuation, shows that the gamma function is analytic in the strip $-1 < \Re z < 0$. The question about the nature of the singularities is not answered in this way. From $\Gamma(z) = \Gamma(z+1)/z$ and $\Gamma(1) = 1$, we expect the origin to be a pole of first order.

3.2.1. Prym's Decomposition

More insight gives *Prym's decomposition* (PRYM (1877)):

$$\Gamma(z) = \int_0^1 t^{z-1}e^{-t}\,dt + \int_1^\infty t^{z-1}e^{-t}\,dt.$$

The second integral represents an entire function of z. In the first integral we substitute the series of the exponential function. Since this series converges uniformly, we can interchange summation and integration when z is in the domain $0 < \delta \le \Re z \le A$, where δ and A are arbitrary numbers. For these $z-$values we obtain the expansion due to Mittag-Leffler

$$\Gamma(z) = \sum_{n=0}^{\infty} \frac{(-1)^n}{n!\,(n+z)} + \int_1^\infty t^{z-1}e^{-t}\,dt,$$

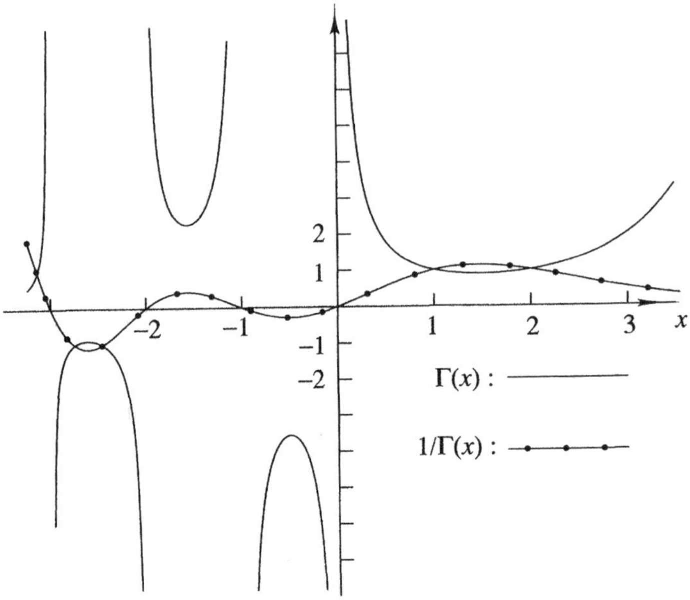

Figure 3.1. Graphs of $\Gamma(x)$ and $1/\Gamma(x)$, x real.

which holds for all $z \neq -n, n = 0, 1, 2, \ldots$. From this analysis it follows that the gamma function has simple poles in the left half-plane at each non-positive integer value of z. Furthermore,

$$\lim_{z \to -n} (z+n)\Gamma(z) = \frac{(-1)^n}{n!}.$$

In other words, the residue at the pole $-n$ equals $(-1)^n/n!$.

In Figure 3.1 we show the graphs of the gamma function and the reciprocal gamma function.

3.2.2. The Cauchy-Saalschütz Representation

There is another way to consider the analytic continuation of the gamma function. In Cauchy (1827) and Saalschütz (1887-1888) it has been shown that for (3.1) an analogous representation exists for negative values of $\Re z$. Consider the integral

$$\int_0^\infty t^{z-1}\left(e^{-t}-1\right) dt, \quad -1 < \Re z < 0.$$

Integration by parts gives (the integrated terms vanish at both limits)

$$\int_0^\infty t^{z-1}\left(e^{-t}-1\right) dt = \frac{1}{z}\int_0^\infty t^z e^{-t}\, dt = \frac{1}{z}\Gamma(z+1) = \Gamma(z).$$

It follows that the left-hand side defines the gamma function in the strip $-1 < \Re z < 0$. In general we can write the *Cauchy – Saalschütz representation*, which holds for any integer $n \geq 0$ with $-(n+1) < \Re z < -n$,

$$\Gamma(z) = \int_0^\infty t^{z-1} \left[e^{-t} - 1 + t - \frac{1}{2!}t^2 + \ldots + (-1)^{n+1} \frac{1}{n!} t^n \right] dt.$$

3.2.3. The Beta Integral

The beta integral defined in (3.2) can be expressed in terms of the gamma function:

$$\boxed{B(p,q) = \frac{\Gamma(p)\Gamma(q)}{\Gamma(p+q)}.} \tag{3.3}$$

This will be shown by computing the double integral

$$I(p,q) = \int_0^\infty \int_0^\infty x^{2p-1} y^{2q-1} e^{-(x^2+y^2)}\, dx\, dy$$

in two ways. On the one hand we know that $I(p,q) = I(p)I(q)$, with $I(p) = \frac{1}{2}\Gamma(p$; see Exercise 3.2. On the other hand we can introduce polar coordinates

$$x = r\cos\theta, \quad y = r\sin\theta$$

in $I(p,q)$, with the result

$$I(p,q) = \int_0^\infty r^{2p+2q-1} e^{-r^2}\, dr \int_0^{\pi/2} \cos^{2p-1}\theta \sin^{2q-1}\theta\, d\theta.$$

Using the substitution $t = \cos^2\theta$ we then arrive at the required result (3.3).

An alternative proof is based on a simple theorem from Laplace transforms. Consider

$$f(t) = \int_0^t \tau^{p-1}(t-\tau)^{q-1}\, d\tau = B(p,q)t^{p+q-1}.$$

We compute, via (3.1), the Laplace transform of f, that is,

$$F(s) = \int_0^\infty f(t)e^{-st}\, dt = B(p,q)\Gamma(p+q)s^{-p-q}.$$

On the other hand, we can also recognize the integral representation of f as a convolution of two simple functions, and use the *convolution theorem for Laplace transformations*. This reads: If f can be written in the form

$$f(t) = \int_0^t f_1(\tau) f_2(t-\tau)\, d\tau,$$

where the following Laplace transforms

$$F_j(s) = \int_0^\infty f_j(t)e^{-st}\,dt, \quad j = 1, 2$$

exist, then, under mild conditions on f_j, we have $F(s) = F_1(s)F_2(s)$. Applying this to the function f just introduced we again arrive at (3.3).

3.2.4. The Multiplication Formula

Next we prove *Legendre's multiplication formula for the gamma function* (Legendre (1809b))

$$\boxed{2^{2z-1}\,\Gamma(z)\Gamma\left(z+\tfrac{1}{2}\right) = \Gamma\left(\tfrac{1}{2}\right)\Gamma(2z).} \tag{3.4}$$

The proof follows from

$$B(z,z) = \int_0^1 [t(1-t)]^{z-1}\,dt = 2\int_0^{\frac{1}{2}} [t(1-t)]^{z-1}\,dt.$$

Substitute $s = 4t(1-t)$; then we see that

$$B(z,z) = 2^{1-2z}B(z, \tfrac{1}{2}),$$

which is equivalent with (3.4). A generalization of the multiplication formula is given in §3.3.1.

3.2.5. The Reflection Formula

Another interesting formula is the *reflection formula for the gamma function* (Euler (1771, page 136)):

$$\boxed{\Gamma(z)\Gamma(1-z) = \frac{\pi}{\sin \pi z}, \qquad z \notin \mathbb{Z},} \tag{3.5}$$

from which immediately follows

$$\boxed{\Gamma\left(\tfrac{1}{2}\right) = \sqrt{\pi}\,.}$$

With the functional equation (3.5) the gamma function can easily be computed for negative values of $\Re z$. For a proof we first take $0 < \Re z < 1$ and write

$$B(z, 1-z) = \int_0^1 \left(\frac{t}{1-t}\right)^{z-1} \frac{dt}{1-t} = \int_0^\infty \frac{s^{z-1}}{1+s}\,ds,$$

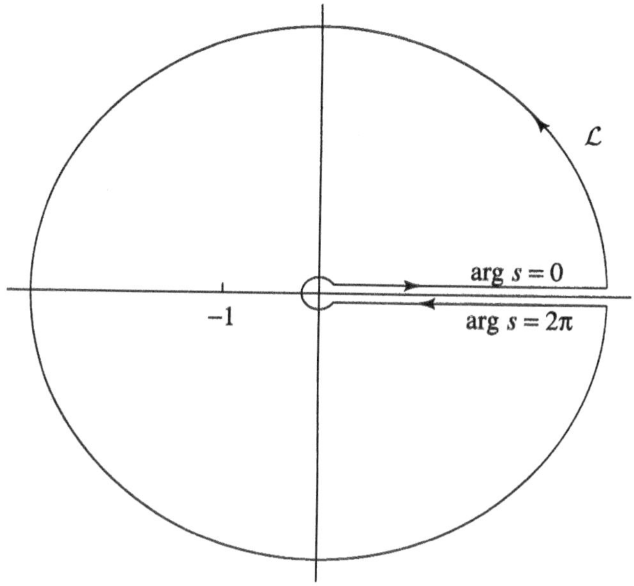

Figure 3.2. Contour for the proof of (3.5).

where we have substituted $s = t/(1-t)$. The final integral can be treated with a standard method from complex function theory. The function

$$f(s) = s^{z-1}/(1+s), \quad 0 < \arg s < 2\pi$$

has a pole at $s = e^{i\pi}$, which is inside the contour $\mathcal{L}$ of Figure 3.2. Hence,

$$\int_{\mathcal{L}} f(s)\,ds = 2\pi i \operatorname*{Res}_{s=e^{i\pi}} f(s) = -2\pi i\, e^{i\pi z}.$$

On the other hand, we can integrate along $\mathcal{L}$. Since $0 < \Re z < 1$, the contributions from the small and the large circles tend to zero when their radii become smaller and larger, respectively. The contributions from the upper and lower sides of the branch cut $(0, \infty)$ give together

$$(1 - e^{2\pi i z}) \int_0^\infty \frac{s^{z-1}}{1+s}\,ds,$$

from which follows

$$B(z, 1-z)\left(1 - e^{2\pi i z}\right) = -2\pi i\, e^{i\pi z}.$$

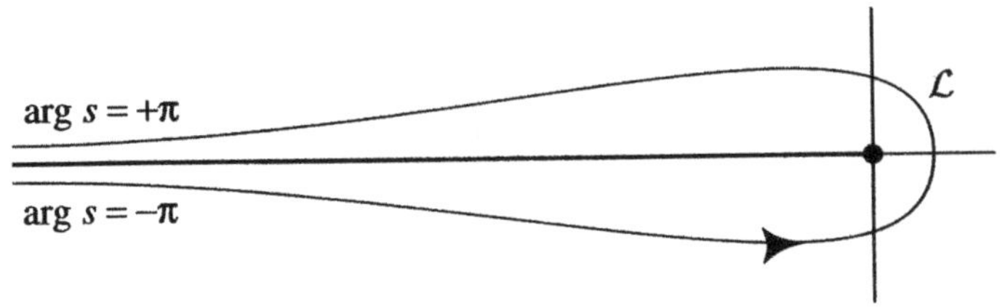

Figure 3.3. Hankel contour for the proof of (3.6).

This is equivalent to (3.5). For the remaining z–values ($z \notin \mathbb{Z}$) we use the principle of analytic continuation.

A symmetric version of the reflection formula (3.5) reads

$$\Gamma(1+z)\Gamma(1-z) = \frac{\pi z}{\sin \pi z},$$

with generalization

$$\Gamma(n+z)\Gamma(n-z) = [(n-1)!]^2 \frac{\pi z}{\sin \pi z} \prod_{m=1}^{n-1} \left(1 - \frac{z^2}{m^2}\right), \qquad n = 1, 2, 3, \ldots,$$

where, in the case of $n = 1$, 0! and the empty product are equal to unity.

Other forms of the reflection formula are given in Exercises 3.7 and 3.8. For instance,

$$\Gamma\left(\tfrac{1}{2} - z\right)\Gamma\left(\tfrac{1}{2} + z\right) = \frac{\pi}{\cos \pi z}, \qquad z - \tfrac{1}{2} \notin \mathbb{Z}.$$

3.2.6. The Reciprocal Gamma Function

Hankel's contour integral (Hankel (1863)) is one of the beautiful representations of the gamma function. In fact it is an integral for the reciprocal gamma function:

$$\boxed{\frac{1}{\Gamma(z)} = \frac{1}{2\pi i} \int_{\mathcal{L}} s^{-z} e^s \, ds, \qquad z \in \mathbb{C}.} \tag{3.6}$$

The contour of integration $\mathcal{L}$ is the *Hankel contour* that runs (see Figure 3.3) from $-\infty, \arg s = -\pi$, encircles the origin in positive direction (that is, counter-clockwise) terminates at $-\infty$, now with $\arg s = +\pi$. For this we also use the notation $\int_{-\infty}^{(0+)}$ instead of= $\int_{\mathcal{L}}$. The many-valued function s^{-z} is assumed to be real for real values of z and s, $s > 0$.

A proof of (3.6) follows immediately from the theory of Laplace transforms: from the well-known integral

$$\frac{\Gamma(z)}{s^z} = \int_0^\infty t^{z-1} e^{-st}\, dt$$

(3.6) follows as a special case of the inversion formula. A direct proof follows from a special choice of $\mathcal{L}$: the negative real axis. This is only possible when $\Re z < 1$. Under this condition, the contribution from a small circle around the origin, with radius tending to zero, can be neglected. Thus we obtain for the right-hand side of (3.6):

$$\frac{1}{2\pi i}\left[-\int_\infty^0 (se^{-i\pi})^{-z} e^{-s}\, ds - \int_0^\infty (se^{+i\pi})^{-z} e^{-s}\, ds\right] = \frac{1}{\pi} \sin \pi z\, \Gamma(1-z).$$

Using (3.5) we infer that this indeed equals the left-hand side of (3.6). In a final step the principle of analytic continuation is used to show that (3.6) holds for all finite complex values of z. Namely, both the left-hand side and the right-hand side of (3.6) are entire functions of z.

Another form of (3.6) is

$$\Gamma(z) = \frac{1}{2i \sin \pi z} \int_{\mathcal{L}} s^{z-1} e^s\, ds.$$

The substitution $s = -t$ gives an integrand as in the starting point (3.1). The message is, that the many-valued function t^{z-1} in (3.1) can be used to "open" the original contour along $[0, \infty)$, and to obtain a representation that is valid in a larger domain of the parameter z. This approach can be useful with other special functions.

3.2.7. A Complex Contour for the Beta Integral

We give here another demonstration of complex contours. Consider the integral

$$I_{p,q} = \frac{1}{2\pi i} \int_0^{(1+)} w^{p-1} (w-1)^{q-1}\, dw,$$

with $\Re p > 0$ and $q \in \mathbb{C}$. The contour starts and ends at the origin, and encircles the point 1 in positive direction; see Figure 3.4. The phase of $w - 1$ is zero at positive points larger than 1. When $\Re q > 0$ we can deform the contour along $(0, 1)$. Then we obtain $I_{p,q} = B(p, q) \sin(\pi q)/q$. It follows that

$$B(p, q) = \frac{q}{\sin \pi q} \frac{1}{2\pi i} \int_0^{(1+)} w^{p-1} (w-1)^{q-1}\, dw.$$

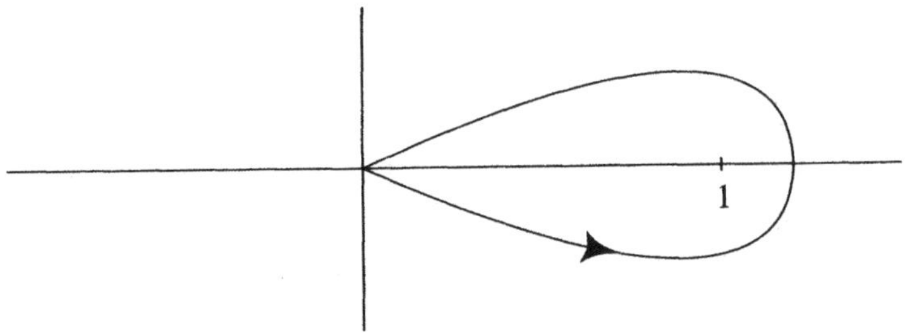

Figure 3.4. Complex contour for $B(p, q)$.

The integral is defined for any complex value of q. For $q = 1, 2, \ldots$ the integral vanishes; this absorbs the infinite values of the term in front of the integral at these points.

3.3. Infinite Products

An alternative approach for introducing the gamma function goes via infinite products. WEIERSTRASS (1856) defined the gamma function in the form

$$\boxed{\frac{1}{\Gamma(z)} = z e^{\gamma z} \prod_{n=1}^{\infty} \left[\left(1 + \frac{z}{n}\right) e^{-z/n}\right].} \tag{3.7}$$

We say that an infinite product $\prod_{k=1}^{\infty}(1 + u_k)$ is *convergent* if there exists a non-zero limit of the sequence of partial products $p_n = \prod_{k=1}^{n}(1 + u_k)$ as $n \to \infty$. The value of the infinite product is the limit $\lim_n p_n = P$, and one writes $\prod_{k=1}^{\infty}(1 + u_k) = P$. The infinite product $\prod_{k=1}^{\infty}(1 + u_k)$ converges if and only if the series $\sum_{k=1}^{\infty} \ln(1 + u_k)$ is convergent. (Requiring a non-zero limit of the sequence $\{p_n\}$ we exclude the special cases $u_k = -1$ for one or more values of k or $u_k \to -1$ if $k \to \infty$, although the sequence $\{p_n\}$ may then converge.)

In the above product we have

$$\ln(1 + u_n) = \ln\left(1 + \frac{z}{n}\right) - \frac{z}{n} = \mathcal{O}\left(n^{-2}\right),$$

as $n \to \infty$, with z fixed, from which follows that the product in (3.7) converges. In (3.7), γ is Euler's constant, which is defined by

$$\gamma = \lim_{n\to\infty} \sum_{k=1}^{n} \frac{1}{k} - \ln(n+1) = 0.5772157\cdots \tag{3.8}$$

and which has already appeared in Chapter 1, Exercise 1.5.

We now derive a representation of the gamma function in the form of a limit of products. This result is due to Euler. From (3.7) it follows that

$$\begin{aligned}\frac{1}{\Gamma(z)} &= z \lim_{m\to\infty} \left\{ e^{(1+\frac{1}{2}+\cdots+\frac{1}{m}\, -\, \ln m)z} \prod_{n=1}^{m} \left[\left(1+\frac{z}{n}\right) e^{-z/n}\right]\right\}\\ &= z \lim_{m\to\infty} \left[m^{-z} \prod_{n=1}^{m} \left(1+\frac{z}{n}\right)\right].\end{aligned}$$

From this we derive after some algebra

$$\boxed{\Gamma(z) = \lim_{n\to\infty} \frac{n!n^z}{z(z+1)\cdots(z+n)},} \tag{3.9}$$

which Euler used as the definition EULER (1729)). We shall verify

Theorem 3.1. *If $\Re z > 0$, then (3.9) and (3.1) represent the same function.*

Proof. Consider for this purpose

$$\Pi(z,n) = \int_0^n \left(1-\frac{t}{n}\right)^n t^{z-1}\, dt = n^z \int_0^1 (1-u)^n u^{z-1}\, du.$$

With the help of the beta integral (3.2) or by integrating by parts it can be verified that $\Pi(z,n)$ equals the fraction on the right-hand side of (3.9). On the other hand, it follows from the second integral, via the substitution $u = 1 - e^{-v}$, that

$$\Pi(z,n) = n^z \int_0^\infty v^{z-1} e^{-nv} f(v)\, dv, \quad f(v) = \left(\frac{1-e^{-v}}{v}\right)^{z-1} e^{-v}.$$

Watson's lemma (see Theorem 2.4) now gives the result (notice that $f(0) = 1$)

$$\Pi(z,n) = n^z f(0) \int_0^\infty v^{z-1} e^{-nv}\, dv \left[1 + \mathcal{O}\left(n^{-1}\right)\right] = \Gamma(z)\left[1 + \mathcal{O}\left(n^{-1}\right)\right],$$

as $n \to \infty$. This shows that when $\Re z > 0$ (3.1) and (3.9) are equivalent. ■

3.3.1. Gauss' Multiplication Formula

The infinite product can be used to prove a generalization of (3.4):

$$\prod_{k=0}^{m-1} \Gamma\left(z+\frac{k}{m}\right) = (2\pi)^{\frac{1}{2}(m-1)} m^{\frac{1}{2}-mz}\Gamma(mz), \quad m = 2,3,4,\ldots .$$

This formula is called *Gauss' multiplication formula* (GAUSS (1812)). The proof is not so straightforward. We denote the left-hand side by $G(z)$. Then by the use of (3.9) we have

$$\begin{aligned}\frac{1}{G(z)} &= \lim_{n\to\infty} \prod_{k=0}^{m-1} \frac{(z+k/m)(z+k/m+1)\cdots(z+k/m+n)}{n!\, n^{z+k/m}} \\ &= \lim_{n\to\infty} \prod_{k=0}^{nm+m-1} \frac{(z+k/m)}{(n!)^m\, n^{(mz+\sum_{k=0}^{m-1} k/m)}} \\ &= \lim_{n\to\infty} \prod_{k=0}^{nm+m-1} \frac{(mz+k)}{(n!)^m\, n^{mz+(m-1)/2} m^{m(n+1)}}\end{aligned}$$

By a slight modification of (3.9) we also have

$$\Gamma(mz) = \lim_{n\to\infty} \frac{(mn)!\,(mn)^{mz}}{\prod_{k=0}^{mn}(mz+k)}$$

so that

$$\frac{\Gamma(mz)}{G(z)} = \lim_{n\to\infty} \frac{(mn)!\,(mn)^{mz}\, n^{m-1} \prod_{k=nm+1}^{nm+m-1}[(mz+k)/n]}{(n!)^m\, n^{mz+(m-1)/2}\, m^{m(n+1)}}.$$

Since

$$\lim_{n\to\infty} \prod_{k=nm+1}^{nm+m-1} \left[\frac{mz+k}{n}\right] = m^{m-1},$$

we obtain

$$\frac{\Gamma(mz)}{m^{mz}\, G(z)} = \lim_{n\to\infty} \frac{(mn)!\, n^{(m-1)/2}}{(n!)^m\, m^{mn+1}},$$

which is independent of z. The right-hand side can be evaluated by using Stirling's formula (see Exercise 1.6 or §3.6). We have

$$\frac{(mn)!\, n^{(m-1)/2}}{(n!)^m\, m^{mn+1}} \sim \frac{\sqrt{2\pi mn}\,(mn/e)^{mn}\, n^{(m-1)/2}}{(2\pi n)^{m/2}\,(n/e)^{mn}\, m^{mn+1}} = (2\pi)^{-(m-1)/2}\, m^{-1/2}$$

so that finally

$$G(z) = (2\pi)^{\frac{1}{2}(m-1)} m^{\frac{1}{2}-mz} \Gamma(mz).$$

There is another method to compute the constant mentioned in the final step of the proof. We can try to evaluate $\Gamma(mz)/[m^{mz}\, G(z)]$, which is independent of z, for a particular value of z, say $z = 1/m$. Then

$$\frac{\Gamma(mz)}{m^{mz}\, G(z)} = \frac{1}{m\Gamma(1/m)\Gamma(2/m)\cdots\Gamma([(m-1)/m)}.$$

Euler simplified this by observing that

$$\frac{1}{G^2(1/m)} = \prod_{k=1}^{m-1} \frac{1}{\Gamma(k/m)\Gamma(1-k/m)} = \prod_{k=1}^{m-1} \frac{\sin \pi k/m}{\pi}$$

(see (3.5)) and by simplifying the product with the sine functions. Let $\varepsilon = \exp(i\pi/m)$ so that $\sin \pi k/m = (\varepsilon^k - \varepsilon^{-k})/2i$. Note that

$$\varepsilon^{1+2+\cdots+m-1} = \varepsilon^{m(m-1)/2} = e^{i\pi(m-1)/2} = i^{m-1}.$$

Then we have after straightforward manipulations

$$\frac{1}{G^2(1/m)} = (2\pi)^{1-m} \prod_{k=1}^{m-1} \left(1 - \varepsilon^{2k}\right).$$

Because the ε^k are certain roots of unity we can write

$$x^{2m} - 1 = \prod_{k=0}^{m-1} \left(x^2 - \varepsilon^{2k}\right),$$

so that

$$\frac{1}{G^2(1/m)} = (2\pi)^{1-m} \lim_{x\to 1} \frac{x^{2m}-1}{x^2-1} = m\,(2\pi)^{1-m}.$$

This again yields the multiplication formula.

3.4. Logarithmic Derivative of the Gamma Function

The derivative the gamma function itself does not play an important role in the theory of special functions. It is not a very manageable function. Much more interesting is the *logarithmic derivative of the gamma function*:

$$\boxed{\psi(z) = \frac{d}{dz} \ln \Gamma(z) = \frac{\Gamma'(z)}{\Gamma(z)}.}$$

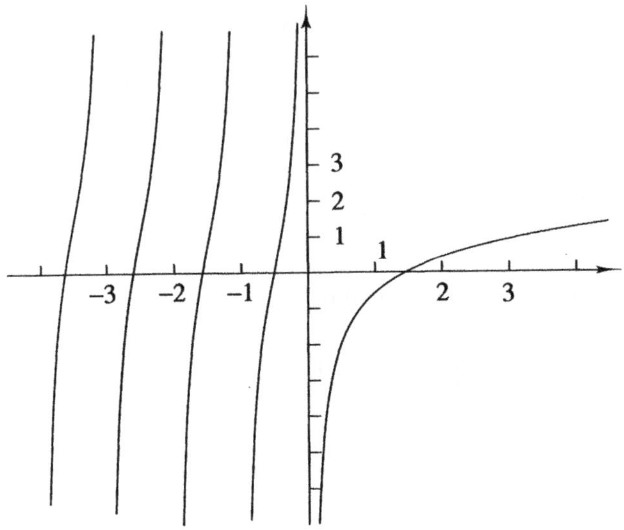

Figure 3.5. Graph of $\psi(x)$, x real; ψ is a meromorphic function with poles at $x = 0, -1, -2, \ldots$

A graph of this function is shown in Figure 3.5. By using the product (3.7) it follows that

$$\psi(z) = -\gamma + \sum_{n=0}^{\infty} \left(\frac{1}{n+1} - \frac{1}{z+n} \right), \quad z \neq 0, -1, -2, \ldots . \tag{3.10}$$

The ψ–function possesses simple poles at all non-positive integers.

We have the recursion relation

$$\boxed{\psi(z+1) = \psi(z) + \frac{1}{z}.}$$

Special values at positive integers at once follow from the series in (3.10):

$$\psi(1) = -\gamma, \quad \psi(k+1) = -\gamma + 1 + \frac{1}{2} + \frac{1}{3} + \cdots + \frac{1}{k}, \quad k \in \mathbb{N}.$$

The derivative of $\psi(z)$ is also a meromorphic function and has double poles. This follows, for example, from

$$\psi'(z) = \sum_{n=0}^{\infty} \frac{1}{(z+n)^2}.$$

Observe that the right-hand side is positive on $(0,\infty)$ and that $\psi'(z)$ is the second derivative of $\ln\Gamma(z)$. This again shows that $\Gamma(x)$ is log-convex on $(0,\infty)$.

The function $\psi(z)$ has the integral representation

$$\psi(z+1) = -\gamma + \int_0^1 \frac{1-t^z}{1-t}\,dt, \quad \Re z > -1. \tag{3.11}$$

The proof follows from

$$\frac{1}{1-t} = \sum_{n=0}^{N-1} t^n + \frac{t^N}{1-t}.$$

Since $(1-t^z)/(1-t)$ is bounded on $t \in [0,1]$, we have

$$\lim_{N\to\infty} \int_0^1 \frac{1-t^z}{1-t} t^N\,dt = 0.$$

Hence, formula (3.10) is obtained by integrating in (3.11) each term of the series $\sum(1-t^z)t^n$.

Of great interest is also (BINET (1839))

$$\psi(z+1) = \ln z + \frac{1}{2z} - \int_0^\infty t\,\beta(t)e^{-zt}\,dt, \quad \Re z > 0, \tag{3.12}$$

with

$$\beta(t) = \frac{1}{t}\left(\frac{1}{e^t-1} - \frac{1}{t} + \frac{1}{2}\right) = \frac{1}{12} - \frac{1}{720}t^2 + \frac{1}{30240}t^4 + \dots .$$

We have encountered a similar function when introducing the Bernoulli numbers, namely,

$$\beta(t) = \sum_{n=1}^{\infty} \frac{B_{2n}}{(2n)!} t^{2n-2}, \quad |t| < 2\pi. \tag{3.13}$$

A proof of (3.12) runs as follows. From (3.11) one has when $\varepsilon \to 0$:

$$\begin{aligned}\psi(z+1) &= -\gamma - \ln\varepsilon - \int_0^{1-\varepsilon} \frac{t^z}{1-t}\,dt + o(1)\\ &= -\gamma - \ln\varepsilon - \int_\varepsilon^\infty \frac{e^{-zt}}{e^t-1}\,dt + o(1),\end{aligned}$$

where $o(1)$ represents quantities which vanish when $\varepsilon \to 0$. Since

$$\Gamma'(1) = -\gamma = \int_0^\infty \ln t\, e^{-t}\,dt,$$

which can be verified by using (3.10) and (3.1), we have similarly (if $\Re z > 0$):

$$\begin{aligned}-\gamma &= \ln(\varepsilon z) + \int_{\varepsilon z}^{\infty} e^{-t}\frac{dt}{t} + o(1)\\ &= \ln\varepsilon + \ln z + \int_{\varepsilon}^{\infty} e^{-zt}\frac{dt}{t} + o(1),\end{aligned}$$

again as $\varepsilon \to 0$. Combining these two results we obtain

$$\psi(z+1) = \ln z + \int_0^{\infty} e^{-zt}\left(\frac{1}{t} - \frac{1}{e^t-1}\right)dt,$$

which is equivalent to (3.12).

With the help of (3.10) we can evaluate series with more complicated rational terms.

Example 3.1. Consider the series

$$S = \sum_{n=1}^{\infty} \frac{1}{(n+1)(2n+1)(4n+1)} = \sum_{n=1}^{\infty} u_n,$$

with

$$\begin{aligned}u_n &= \frac{\frac{1}{3}}{n+1} - \frac{1}{n+\frac{1}{2}} + \frac{\frac{2}{3}}{n+\frac{1}{4}}\\ &= \frac{1}{3}\left(\frac{1}{n+1} - \frac{1}{n}\right) - \left(\frac{1}{n+\frac{1}{2}} - \frac{1}{n}\right) + \frac{2}{3}\left(\frac{1}{n+\frac{1}{4}} - \frac{1}{n}\right).\end{aligned}$$

Application of (3.10) then gives

$$S = -\tfrac{1}{3}\psi(2) + \psi\left(\tfrac{3}{2}\right) - \tfrac{2}{3}\psi\left(\tfrac{5}{4}\right) = \tfrac{1}{3}\pi - 1.$$

These values follow from relations of the ψ–function given in Exercise 3.17.

Alternating series can be evaluated by using

$$\sum_{n=0}^{\infty} \frac{(-1)^n}{n+z} = \frac{1}{2}\psi\left(\frac{1+z}{2}\right) - \frac{1}{2}\psi\left(\frac{z}{2}\right). \qquad (3.14)$$

This formula can be verified with the help of (3.11). The reader can prove that the right-hand side of (3.14) can be brought into the form

$$\int_0^1 \frac{t^{z-1}}{t+1}\,dt.$$

Expansion of $1/(t+1)$ in powers of t concludes the proof of (3.14).

3.5. Riemann's Zeta Function

The *Riemann zeta function* is defined by the series

$$\boxed{\zeta(z) = \sum_{n=1}^{\infty} \frac{1}{n^z}} \tag{3.15}$$

under the condition $\Re z > 1$. The function was known to Euler, but its main properties were discovered by RIEMANN (1859) (an English translation of this paper is given in EDWARDS (1974)). The series converges absolutely and uniformly in each compact domain inside the half-plane $\Re z > 1$, and hence defines there an analytic function.

An integral representation for $\zeta(z)$ is obtained by using an integral for the gamma function in the form

$$\frac{1}{n^z} = \frac{1}{\Gamma(z)} \int_0^\infty t^{z-1} e^{-nt}\, dt, \quad \Re z > 0,$$

and by substituting this in (3.15). With the help of Theorem 2.1 we can verify that, if $\Re z > 1$, summation and integration may be interchanged. The result is

$$\zeta(z) = \frac{1}{\Gamma(z)} \int_0^\infty \frac{t^{z-1}}{e^t - 1}\, dt, \quad \Re z > 1. \tag{3.16}$$

A direct proof easily follows from

$$\lim_{N\to\infty} \sum_{n=1}^{N} \frac{1}{n^z} = \lim_{N\to\infty} \frac{1}{\Gamma(z)} \int_0^\infty t^{z-1} \frac{1 - e^{-(N+1)t}}{e^t - 1}\, dt.$$

Again, when $\Re z > 1$, the limit on the right-hand side can be evaluated.

The analytic continuation of the zeta function into the left half-plane is obtained through a different integral. As with the gamma function we consider the contour integral:

$$I(z) = \int_{-\infty}^{(0+)} \frac{t^{z-1}}{e^{-t} - 1}\, dt,$$

where the contour does not enclose the points $\pm 2\pi i, \pm 4\pi i, \ldots$. This integral is an entire function of z (Theorem 2.3). When $\Re z > 1$ we can deform the contour along the negative real axis. As in the proof of (3.6) we have

$$I(z) = 2i \sin \pi z \int_0^\infty \frac{s^{z-1}}{e^s - 1}\, ds = 2i \sin \pi z\; \Gamma(z)\; \zeta(z).$$

Using the reflection formula (3.5) we get

$$\zeta(z) = \frac{\Gamma(1-z)}{2\pi i}\int_{-\infty}^{(0+)} \frac{t^{z-1}}{e^{-t}-1}\,dt. \tag{3.17}$$

As remarked earlier, the integral is an entire function of z. The only singularities in the above representation are produced by the gamma function, and occur at $z = 1, 2, \ldots$. But we know already that $\zeta(z)$ is analytic when $\Re z > 1$. Hence, the points $z = 2, 3, \ldots$ must be removable singularities. When $z = 1$ the integral in (3.17) gives

$$\int_{-\infty}^{(0+)} \frac{1}{e^{-t}-1}\,dt = -2\pi i.$$

From this we infer that $z = 1$ is the only singularity of $\zeta(z)$; it is a simple pole with residue 1. That is, $(z-1)\zeta(z)$ is an entire function and

$$\lim_{z\to 1}(z-1)\,\zeta(z) = 1.$$

The zeta function satisfies, just as the gamma function, a functional equation in the form of a reflection formula. This arises when we deform the contour in (3.17) into a vertical line in the right half-plane. In doing so we pass singularities at $t = \pm 2n\pi i$. At these points the residues of $t^{z-1}/(e^{-t}-1)$ equal $-(\pm 2n\pi i)^{z-1}$. Since the integral over the vertical line converges only if $\Re z < 0$, we take this condition for the time being. Also, this condition is needed to be able to neglect the contributions at infinity. The result of this operation is

$$\zeta(z) = \Gamma(1-z)\left[\sum_{n=1}^{\infty}(2ni\pi)^{z-1} + \sum_{n=1}^{\infty}(-2ni\pi)^{z-1}\right]$$

plus an integral over the vertical line in the right half-plane. When $\Re z < 0$ this vertical line can be shifted to the right as far as we please, without passing singularities. It is easily verified that the contribution from this vertical line equals zero. The two series in the above result can be expressed in terms of $\zeta(1-z)$, giving

$$\zeta(z) = 2(2\pi)^{z-1}\Gamma(1-z)\,\zeta(1-z)\,\sin\frac{1}{2}\pi z, \quad \Re z < 0.$$

Because the product $\zeta(1-z)\,\sin\frac{1}{2}\pi z$ can be interpreted as an entire function, the right-hand side has a removable singularity at $z = 0$. Hence, invoking the principle of analytic continuation, we infer that in the above relation the condition $\Re z < 0$ can be replaced by $z \neq 1$. Changing $z \to 1 - z$ we obtain the *reflection formula for the Riemann zeta function*:

$$\boxed{\zeta(1-z) = 2\,(2\pi)^{-z}\Gamma(z)\cos\frac{1}{2}\pi z\;\zeta(z), \quad z \neq 0.} \tag{3.18}$$

The restriction $z \neq 0$ is now given because both sides have a simple pole at the origin. By multiplying both sides of this equation by z, both sides become entire functions. This gives, with the definition in (3.15), a complete description of the Riemann zeta function in the complex plane.

By using (3.4) and Exercise 3.7 the reflection formula can be written in the more symmetric form

$$\pi^{-z/2}\Gamma\left(\tfrac{1}{2}z\right)\zeta(z) = \pi^{-(1-z)/2}\Gamma\left(\tfrac{1}{2} - \tfrac{1}{2}z\right)\zeta(1-z).$$

The function on the left side is unchanged when we replace z by $1 - z$. Also, it is meromorphic with simple poles at $z = 0$ and $z = 1$.

The above residue method can be approached more rigorously by considering the integral

$$\int_{C_N} \frac{t^{z-1}}{e^{-t} - 1}\,dt,$$

where C_N is a contour described by a rectangle with corners at $\pm N \pm (2N-1)\pi i$ and a loop around the origin (see Figure 3.6), with N a large positive integer. Since for $t \in C_N$ we have $|e^{-t} - 1| \geq 1 - e^{-N}$, the above integral (except the contribution from the loop) tends to zero as $N \to \infty$, again under the condition $\Re z < 0$. The residue method again produces two series, and finally (3.18) is obtained.

Special values of the zeta function follow for example from (3.18). When $z = 1$ we use $\lim_{z\to 1}(z-1)\zeta(z) = 1$. It follows that $\zeta(0) = -\frac{1}{2}$. It is also easily verified that $\zeta(-2m) = 0, m \geq 1$ and

$$\zeta(1-2m) = (-1)^m 2^{1-2m}\pi^{-2m}\,(2m-1)!\;\zeta(2m), \quad m = 1, 2, \ldots .$$

As we know from Chapter 1, $\zeta(2m)$ can be expressed in terms of the Bernoulli numbers (see (1.14)):

$$\boxed{\zeta(2m) = \frac{(-1)^{m+1}(2\pi)^{2m}}{2\,(2m)!}B_{2m}, \quad m = 1, 2, \ldots,} \tag{3.19}$$

a relation known to Euler in 1737. It follows that in general we can write

$$\zeta(-m) = -\frac{B_{m+1}}{m+1}, \quad m = 1, 2, \ldots .$$

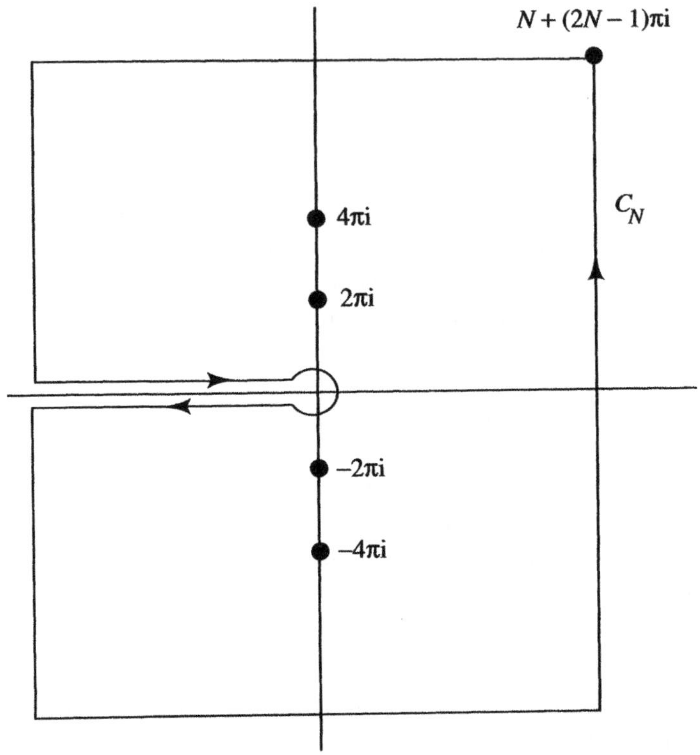

Figure 3.6. Contour for proving the reflection formula (3.18).

Finally, we give an infinite product due to Euler. Assume that $\Re z > 1$. Subtract the series for $2^{-z}\zeta(z)$ from the one in (3.15). Then we obtain

$$\left(1-2^{-z}\right)\zeta(z)=\frac{1}{1^z}+\frac{1}{3^z}+\frac{1}{5^z}+\frac{1}{7^z}+\cdots .$$

Similarly, we obtain

$$\left(1-2^{-z}\right)\left(1-3^{-z}\right)\zeta(z)=\sum\frac{1}{n^z},$$

where now the summation runs over $n\geq 1$, except for multiples of 2 and 3. Now, let ϖ_n denote the n−th prime number, starting with $\varpi_1 = 2$. By repeating the above procedure we obtain

$$\zeta(z)\prod_{n=1}^{m}\left(1-\varpi_n^{-z}\right)=1+\sum\frac{1}{n^z},$$

where in the series no terms are used with $n = 1$ or multiples of the primes

$$\varpi_1, \varpi_2, \ldots, \varpi_m.$$

The sum of this series vanishes as $m \to \infty$ (since $\varpi_m \to \infty$). From this we obtain the required result

$$\frac{1}{\zeta(z)} = \prod_{n=1}^{\infty} \left(1 - \varpi_n^{-z}\right), \quad \Re z > 1.$$

This formula is of fundamental importance for the relation between Riemann's zeta function and the theory of prime numbers.

An immediate consequence is that $\zeta(z)$ does not have zeros in the half-plane $\Re z > 1$. The reflection formula (3.18) makes clear that the only zeros in the half-plane $\Re z < 0$ occur at the points $-2, -4, -6, \ldots$. These are called the *trivial zeros of the zeta functions.*

For the remaining strip $0 \le \Re z \le 1$ we have no information at the moment. Riemann conjectured that in this strip all zeros (it is known that there are infinitely many of them) are located on the line $\Re z = 1/2$. Until now this conjecture has not been proved. An important part of number theory is based on this conjecture. Much time has been spent on attempting to verify the *Riemann hypothesis*, analytically and numerically. For example, at CWI in Amsterdam it has been verified numerically that Riemann's conjecture holds for the first $1\frac{1}{2}$ billion (plus 1) zeros; that is, these zeros are located indeed on the vertical line $\Re z = \frac{1}{2}$.

A generalization of Riemann's zeta function and a few properties of this function are considered in Exercise 3.16.

3.6. Asymptotic Expansions

A well-known result from calculus is *Stirling's formula* (STIRLING (1730, page 135); in fact Stirling obtained a result for the logarithm of the gamma function; see (3.23))

$$\boxed{n! \sim \sqrt{2\pi n}\, n^n e^{-n}, \quad \text{as} \quad n \to \infty.}$$

In many applications this formula proves to be extremely useful. In this section we treat more general forms of this formula, by giving general results for the gamma function and related functions. For instance, Stirling's formula follows from the asymptotic expansion (3.24) below; a proof based on Euler's summation method can be found in Exercise 1.6.

It will appear that the gamma function can be computed very efficiently by using asymptotic expansions. The relation $\Gamma(z) = \Gamma(z+1)/z$ and the reflection formula (3.5) are useful when the argument is not large enough to apply the asymptotic expansions.

Integration of (3.12) gives

$$\ln \Gamma(z+1) = (z+\frac{1}{2})\ln z - z + K + \int_0^\infty \beta(t) e^{-zt}\, dt, \tag{3.20}$$

where K is a constant of integration, which has to be determined. To find K we apply the multiplication formula (3.4), which can be written in the form

$$\ln \frac{2^{2z}\Gamma(z+1)\,\Gamma(z+\frac{1}{2})}{\Gamma(2z+1)} = \frac{1}{2}\ln \pi. \tag{3.21}$$

Since $\beta(t)$ is bounded for $t > 0$, it is easy to verify that the integral in (3.20) vanishes as $z \to \infty$. Substituting (3.20) in (3.21) and letting $z \to \infty$, we obtain $K = \frac{1}{2}\ln 2\pi$. Hence we can write (3.20) in the form

$$\ln \Gamma(z) = \ln\left(\sqrt{2\pi}\, z^{z-\frac{1}{2}} e^{-z}\right) + \int_0^\infty \beta(t) e^{-zt}\, dt. \tag{3.22}$$

Using (3.13) we find, via Watson's lemma (Theorem 2.4) an asymptotic expansion for the logarithm of the gamma function(*Stirling's series*):

$$\boxed{\ln \Gamma(z) \sim \ln\left(\sqrt{2\pi}\, z^{z-\frac{1}{2}} e^{-z}\right) + \sum_{n=1}^{\infty} \frac{B_{2n}}{2n\,(2n-1)}\frac{1}{z^{2n-1}},}$$

as $z \to \infty$. Since the singularities (poles) of $\beta(t)$ are located on the imaginary axis, this expansion holds for $|\arg z| < \pi$.

Because of the importance of this expansion, we explicitly mention an extra number of terms:

$$\begin{aligned}\ln \Gamma(z) \sim \ln\left(\sqrt{2\pi}\, z^{z-\frac{1}{2}} e^{-z}\right) &+ \frac{1}{12z} - \frac{1}{360z^3} + \frac{1}{1260z^5} - \frac{1}{1680z^7}\\ &+ \frac{1}{1188z^9} - \frac{691}{360360z^{11}} + \frac{1}{156z^{13}} - \frac{3617}{122400z^{15}} + \frac{43867}{244188z^{17}} + \cdots,\end{aligned} \tag{3.23}$$

as $z \to \infty$, $|\arg z| < \pi$. Taking the exponential of this result, we get the generalization of Stirling's formula

$$\Gamma(z) \sim \sqrt{2\pi}\, z^{z-\frac{1}{2}} e^{-z} \exp\left(\frac{1}{12z} - \frac{1}{360z^3} + \cdots\right),$$

Table 3.1. Approximating $\Gamma(z)$ via (3.23) and (3.24)

z	(3.23)	(3.24)
1	1.0002878	0.9997110
2	1.0000036	0.9999927
3	2.0000005	1.9999995
4	6.0000002	6.0000009
5	24.0000002	24.0000028

or

$$\Gamma(z) \sim \sqrt{2\pi}\, z^{z-\frac{1}{2}} e^{-z} \left(1 + \frac{1}{12z} + \frac{1}{288z^2} - \frac{139}{51840z^3} - \frac{571}{2488320z^4} + \cdots\right). \tag{3.24}$$

A remarkable feature is that in $\ln\Gamma(z)$ only odd negative powers of z occur, whereas in the expansion of $\Gamma(z)$ both even and odd powers can be seen. That is why (3.23) is much more efficient for numerical calculations than (3.24). Moreover, the remainder of (3.23) (which we did not introduce thus far; see the next subsection) can be estimated more easily than that of (3.24).

It is also useful to have the asymptotic expansion for the reciprocal gamma function. We have

$$\frac{1}{\Gamma(z)} \sim \frac{1}{\sqrt{2\pi}} z^{-z+\frac{1}{2}} e^{z} \left(1 - \frac{1}{12z} + \frac{1}{288z^2} + \frac{139}{51840z^3} - \frac{571}{2488320z^4} + \cdots\right). \tag{3.25}$$

We observe that the series in (3.25) has the same coefficients as (3.24), with different signs of the coefficients with odd index.

To explain this we note that the series in (3.23) has odd powers only. The series for $1/\Gamma(z)$ follows from exponentiation of the series in (3.23) with all signs changed. But changing the signs in the odd series in (3.23) can also be done by formally changing the sign of z. (We do not use (3.23) with z replaced by $-z$; we only perform operations on formal power series to explain the similarity between the series in (3.24) and (3.25).)

In Table 3.1 we show the results of applying (3.23) (with terms up to and including the term $1/1260z^{-5}$) and (3.24) (with terms up to and including $-139/51840z^{-3}$), for $z = 1, 2, 3, 4, 5$. It follows that the accuracy is already quite interesting for these small values of the large asymptotic parameter.

3.6.1. Estimations of the Remainder

Because of the importance of (3.23) for numerical applications we now investigate the remainder, and we construct upper bounds. First we introduce a different representation of the function $\beta(t)$, which shows up in (3.12). We show that this function can be written in the form

$$\beta(t) = P_N(t) + (-1)^N t^{2N} \mu_N(t),$$

where

$$P_N(t) = \sum_{n=1}^{N} \frac{B_{2n}}{(2n)!} t^{2n-2}, \quad \mu_N(t) = \sum_{k=1}^{\infty} \frac{2}{(t^2 + 4k^2\pi^2)(2\pi k)^{2N}}, \quad t \neq 2k\pi i. \tag{3.26}$$

This representation of $\beta(t)$ is in the form of a truncated Taylor series (see (3.13)) where $\mu_N(t)$ is a remainder in a form that suits us quite well in the following analysis. We prove this representation of $\mu_N(t)$ via Taylor's formula for the remainder of a power series:

$$(-1)^N \mu_N(t) = \frac{1}{2\pi i} \int \frac{\beta(\tau)}{(t-\tau)\tau^{2N}}\, d\tau,$$

where, initially, the contour of integration is a closed curve (for instance, an ellipse) that encloses the points $\tau = 0$ and $\tau = t$, and that does not enclose the points $2k\pi i$, $k \in \mathbb{Z}\backslash\{0\}$. The contour of integration is described in the positive sense. We deform the ellipse by bringing the intersection with the positive real axis to $+\infty$; afterwards, the parabola is deformed into a vertical line in the left half planc. This operation is allowed when we take into account the residues from the points $2k\pi i$. On shifting the vertical line to $-\infty$, we see that this makes no contribution. The result is

$$(-1)^N \mu_N(t) = -\sum_{k=1}^{\infty} \left[\frac{1}{(2\pi ik - t)(2\pi ik)^{2N+1}} + \frac{1}{(-2\pi ik - t)(-2\pi ik)^{2N+1}} \right]$$

which easily leads to the desired formula. The proof is valid for $N \geq 0$.

When $N = 0$ we have the well-known result

$$\beta(t) = \sum_{k=1}^{\infty} \frac{2}{t^2 + 4k^2\pi^2}, \quad t \neq 2k\pi i.$$

This is one of the many examples of partial fraction decomposition for a class of trigonometric functions. In this connection, observe that we can write

$$\beta(t) = \frac{1}{t^2} \left(\frac{1}{2} t \coth \frac{1}{2} t - 1 \right).$$

We write (3.22) in the form

$$\ln \Gamma(z) = \ln\left(\sqrt{2\pi}\, z^{z-\frac{1}{2}} e^{-z}\right) + \Phi(z),$$

with

$$\Phi(z) = \int_0^\infty \beta(t) e^{-zt}\, dt, \quad \Re z > 0. \tag{3.27}$$

Using (3.26) we obtain

$$\Phi(z) = \sum_{n=1}^{N} \frac{B_{2n}}{2n\,(2n-1)} \frac{1}{z^{2n-1}} + R_N, \tag{3.28}$$

with

$$R_N = (-1)^N \int_0^\infty e^{-zt} t^{2N} \mu_N(t)\, dt = \frac{2(-1)^N}{z^{2N+1}} \int_0^\infty e^{-u} u^{2N} T_N(u, z)\, du,$$

and

$$T_N(z,u) = \sum_{k=1}^{\infty} \frac{4\pi^2 k^2 z^2}{u^2 + 4\pi^2 k^2 z^2} \frac{1}{(2\pi k)^{2N+2}}.$$

If $z > 0$ we can write $T_N(z,u)$ in the form

$$T_N(z,u) = \theta_N \sum_{k=1}^{\infty} \frac{1}{(2\pi k)^{2N+2}} = \theta_N \frac{(-1)^{N+2} B_{2N+2}}{2\,(2N+2)!},$$

with $0 < \theta_N < 1$. Hence, in this case we have

$$R_N = \theta_N \frac{B_{2N+2}}{2(N+1)(2N+1)} \frac{1}{z^{2N+1}}.$$

In other words, R_N has the sign of the first neglected term in (3.28) and its absolute value is smaller than that term. Moreover, for each n, the value of $\Phi(z)$ lies always between the value of the sum of n terms and that of the sum of $n+1$ terms of the series in (3.28). This follows from the fact that the series is alternating.

This is an ideal situation in asymptotics. In these circumstances one verifies for a real $z-$value which term in (3.23) is smaller than the required precision, and one knows that that term, and all subsequent terms, can be neglected. For example, when $z \geq 10$, all terms in (3.23) after that of z^{-11} can be neglected for obtaining an accuracy of 1.92×10^{-14}, or less.

For complex values of z the situation is somewhat more complicated. To obtain insight in this case we introduce the quantity

$$K_z = \max_{s \geq 0} \left| \frac{z^2}{s^2 + z^2} \right|.$$

Observe that K_z does not change when in $z^2/(s^2+z^2)$ the variables z and/or s are multiplied by arbitrary real numbers ($\neq 0$). Now we obtain

$$|R_n| < \frac{K_z |B_{2n+2}|}{2(n+1)(2n+1)|z|^{2n+1}}.$$

To determine K_z one can use

$$K_z^{-2} = \min_{u\geq 0} \frac{[u^2+(x^2-y^2)]^2+4x^2y^2}{(x^2+y^2)^2}, \qquad z = x+iy,$$

and consequently

$$K_z^{-2} = \begin{cases} 1, & \text{if } x^2 \geq y^2; \\ 4x^2y^2/(x^2+y^2)^2, & \text{if } x^2 < y^2. \end{cases}$$

Hence, if $|\arg z| < \frac{1}{4}\pi$ then (as in the case of real $z = x$) $K_z = 1$. From this it follows that when $|\arg z| < \frac{1}{4}\pi$, the remainder R_n of (3.28) is again smaller than the first neglected term in the series. When $\frac{1}{4}\pi < \arg z < \frac{1}{2}\pi$ the above method gives an increasingly unfavorable estimate of R_N as z approaches the imaginary axis.

The expansion in (3.23) has an asymptotic character in the sector $|\arg z| < \pi$, but this does not follow from the above analysis. See SPIRA (1971), where the following simple result is derived:

$$|R_N| \leq \begin{cases} 2|B_{2N}/(2N-1)|\ |\Im z|^{1-2N}, & \text{if } \Re z < 0,\ \Im z \neq 0; \\ |B_{2N}/(2N-1)|\ |z|^{1-2N}, & \text{if } \Re z \geq 0. \end{cases}$$

Of course, for computations with $\Re z < 0$, the reflection formula (3.5) should be used.

3.6.2. Ratio of Two Gamma Functions

In applications one frequently meets expressions with the *ratio of two gamma functions*. When the arguments of both functions are large it is not always possible to use numerical approximations of both functions, since they may become too large for the computer's number system. Moreover, loss of accuracy may occur when we divide two large numbers obtained via (3.24). This is due to the inaccuracy with which the dominant term (in front of the series in (3.24)) will be computed when z is large.

It is of great help when an algorithm for computing

$$\Gamma^*(z) = \frac{\Gamma(z)}{\sqrt{2\pi}\, z^{z-\frac{1}{2}} e^{-z}}, \qquad \Re z > 0. \tag{3.29}$$

is available. From (3.24) it follows that $\Gamma^*(z) = 1 + \mathcal{O}(1/z)$, as $z \to \infty$.

Suppose that we need to compute $\Gamma(z+a)/\Gamma(z+b)$ for large values of z. Then from (3.29) it follows that

$$\frac{\Gamma(z+a)}{\Gamma(z+b)} = z^{a-b}\frac{\Gamma^*(z+a)}{\Gamma^*(z+b)}Q(z,a,b), \tag{3.30}$$

where

$$Q(z,a,b) = \left(1+\frac{a}{z}\right)^{a-\frac{1}{2}}\left(1+\frac{b}{z}\right)^{-b+\frac{1}{2}} e^{z\left[\ln\left(1+\frac{a}{z}\right)-\frac{a}{z}-\ln\left(1+\frac{b}{z}\right)+\frac{b}{z}\right]}.$$

It is not difficult to verify that $Q(z,a,b) = 1 + \mathcal{O}(1/z)$, as $z \to \infty$; (3.30) shows quite well which contributions play a role in the ratio of the gamma functions.

Although $Q(z,a,b)$ is composed of elementary functions, one should be careful when evaluating the above expression when z is large. The point is that, for small values of z, the function $\ln(1+z) - z$ cannot be accurately computed directly from the log-function (a loss in *relative* accuracy occurs). However, it is rather easy to write a code for the function $\ln(1+z) - z$.

Nevertheless, it is of great importance to have available an asymptotic expansion for the ratio $\Gamma(z+a)/\Gamma(z+b)$. Consider the beta integral in the form

$$\begin{aligned}\frac{\Gamma(z+a)}{\Gamma(z+b)} &= \frac{1}{\Gamma(b-a)}\int_0^1 t^{z+a-1}(1-t)^{b-a-1}\,dt, \quad \Re(b-a) > 0\\ &= \frac{1}{\Gamma(b-a)}\int_0^\infty u^{b-a-1}e^{-zu}f(u)\,du,\end{aligned}$$

where

$$f(u) = e^{-au}\left(\frac{1-e^{-u}}{u}\right)^{b-a-1}.$$

Using the series in (1.3) we obtain

$$f(u) = \sum_{n=0}^\infty c_n u^n, \quad c_n = (-1)^n \frac{B_n^{(a-b+1)}(a)}{n!}.$$

Watson's lemma (in the form of Theorem 2.5) gives the result (under the condition $\Re(b-a) > 0$)

$$\frac{\Gamma(z+a)}{\Gamma(z+b)} \sim z^{a-b}\sum_{n=0}^\infty c_n \frac{\Gamma(b-a+n)}{\Gamma(b-a)}\frac{1}{z^n}, \quad \text{as} \quad z \to \infty \tag{3.31}$$

in the sector $|\arg z| < \pi$.

By representing $f(u)$ in a different way a more efficient expansion is possible. To obtain that expansion we write

$$f(u) = e^{-u(b+a-1)/2}\left[\frac{\sinh(u/2)}{u/2}\right]^{b-a-1}.$$

Again via (1.3) we can write

$$f(u) = e^{-u(b+a-1)/2}\sum_{n=0}^{\infty}(-1)^n C_n u^{2n}, \quad C_n = (-1)^n \frac{B_{2n}^{(2\rho)}(\rho)}{(2n)!},$$

where $\rho = (a-b+1)/2$. Application of Watson's lemma now gives

$$\frac{\Gamma(z+a)}{\Gamma(z+b)} \sim w^{a-b}\sum_{n=0}^{\infty}(-1)^n C_n \frac{\Gamma(b-a+2n)}{\Gamma(b-a)}\frac{1}{w^{2n}}, \quad \text{as} \quad z\to\infty, \tag{3.32}$$

in the sector $|\arg z| < \pi$, with $w = z + (a+b-1)/2$ and $\Re(b-a) > 0$.

Since only even negative powers of the large parameter occur (in this case w), the series in (3.32) gives a more efficient expansion than (3.31). Another favorable feature of (3.32) is that for real a, b and z, with $0 < a-b+1 < 1$ the remainder can be estimated. Let $N = 0, 1, 2, \ldots$ and let R_N be defined by writing

$$\frac{\Gamma(z+a)}{\Gamma(z+b)} = w^{a-b}\left[\sum_{n=0}^{N-1}(-1)^n C_n \frac{\Gamma(b-a+2n)}{\Gamma(b-a)}\frac{1}{w^{2n}} + R_N\right].$$

Then, when $z + \min[a, (a+b-1)/2] > 0$ and $0 < a-b+1 < 1$, we have (FRENZEN (1987))

$$R_N = \theta_N(-1)^N C_N \frac{\Gamma(b-a+2N)}{\Gamma(b-a)}\frac{1}{w^{2N}}, \quad 0 < \theta_N < 1. \tag{3.33}$$

Below we give some coefficients C_n of (3.32). From $0 < \rho < 1/2$ it follows that the series is alternating. This restriction on ρ is not very important, since by recursion the parameters a and b may be changed in order to bring the new ρ in the desired interval. In (3.32) the gamma function ratio $\Gamma(b-a+2n)/\Gamma(b-a)$ can of course be generated by recursion as well.

$$C_0 = 1, \; C_1 = \frac{\rho}{12}, \; C_2 = \frac{\rho}{1440} + \frac{\rho^2}{288}, \; C_3 = \frac{\rho}{90720} + \frac{\rho^2}{17280} + \frac{\rho^3}{10368},$$

$$C_4 = \frac{\rho}{4838400} + \frac{101\rho^2}{87091200} + \frac{\rho^3}{414720} + \frac{\rho^4}{497664},$$

$$C_5 = \frac{\rho}{239500800} + \frac{13\rho^2}{522547200} + \frac{61\rho^3}{1045094400} + \frac{\rho^4}{14929920} + \frac{\rho^5}{29859840},$$

$$C_6 = \frac{691\rho}{7846046208000} + \frac{7999\rho^2}{14485008384000} + \frac{59\rho^3}{41803776000} + \frac{143\rho^4}{75246796800} + \frac{\rho^5}{716636160} + \frac{\rho^6}{2149908480}.$$

3.6.3. Application of the Saddle Point Method

For a short introduction to the saddle point method we refer to §2.2.2. We use Hankel's integral (3.6) for the reciprocal gamma function. We consider positive values of z. A first transformation $s = zt$ gives

$$\frac{1}{\Gamma(z)} = \frac{e^z z^{1-z}}{2\pi i} \int_{-\infty}^{(0^+)} e^{z\phi(t)}\, dt,$$

where

$$\phi(t) = t - 1 - \ln t,$$

and the contour is the same as in Figure 3.3. The saddle point of the integrand is the solution of the equation $\phi'(t) = 0$. The only solution is $t = 1$. We try to define a contour through this saddle point on which $\Im\phi(t)$ is constant. Since $\phi(1) = 0$, this constant must be 0. Writing $t = \rho e^{i\theta}$, we find that the equation $\Im\phi(t) = 0$ is satisfied when the polar coordinates of t satisfy

$$\rho = \frac{\theta}{\sin\theta}, \qquad -\pi < \theta < \pi. \tag{3.34}$$

This defines the path of steepest descent. Next we define a mapping $t \to u(t)$ by writing $\frac{1}{2}u^2 = \phi(t)$. Near the saddle point $t = 1$ we can expand

$$\phi(t) = \frac{1}{2}(t-1)^2 - \frac{1}{3}(t-1)^3 + \dots .$$

Hence, the mapping just defined can be rewritten in terms of

$$u = (t-1)\sqrt{\frac{2(t-1-\ln t)}{(t-1)^2}},$$

where the square root is positive for positive values of t. This definition of the mapping gives a better description of which branch of the function $u(t)$ will be used. For complex values of t the mapping is defined by analytic continuation. Observe that for positive values of t we have $\operatorname{sign} u = \operatorname{sign}(t-1)$ and for complex values $\operatorname{sign}\Im u = \operatorname{sign}\Im t$. The positive $t-$axis is mapped in the $u-$plane onto the whole real axis. The saddle point contour described by (3.34) is mapped onto the whole imaginary axis. So we can write

$$\frac{1}{\Gamma(z)} = \frac{e^z z^{1-z}}{2\pi i} \int_{-i\infty}^{+i\infty} e^{\frac{1}{2}zu^2} f(u)\, du, \tag{3.35}$$

where

$$f(u) = \frac{dt}{du} = \frac{tu}{t-1}. \tag{3.36}$$

We obtain the asymptotic expansion of the reciprocal gamma function by expanding $f(u) = \sum_{n=0}^{\infty} c_n u^n$, and substituting this in the above integral. The result is

$$\frac{1}{\Gamma(z)} \sim \frac{e^z z^{1-z}}{2\pi i} \sum_{n=0}^{\infty} c_{2n}\, i\, (-1)^n\, \Gamma\left(n+\tfrac{1}{2}\right) \left(\tfrac{1}{2}z\right)^{-n-\frac{1}{2}},$$

which can be written in the form

$$\frac{1}{\Gamma(z)} \sim \frac{e^z z^{\frac{1}{2}-z}}{\sqrt{2\pi}} \sum_{n=0}^{\infty} \gamma_n z^{-n}, \quad \gamma_n = (-2)^n \left(\frac{1}{2}\right)_n c_{2n},$$

as $z \to \infty$. The coefficients c_n can be computed by using (3.36). We know that $t = 1 + \sum_{n=0}^{\infty} c_n/(n+1)u^{n+1}$. Then (3.36) gives the relation

$$u\left(1 + \sum_{n=0}^{\infty} \frac{c_n}{n+1} u^{n+1}\right) = \left(\sum_{n=0}^{\infty} \frac{c_n}{n+1} u^{n+1}\right)\left(\sum_{n=0}^{\infty} c_n u^n\right).$$

By equating equal powers of u we find the recursion

$$\frac{c_{n-1}}{n} = \sum_{k=0}^{n} \frac{c_k\, c_{n-k}}{k+1}, \quad n > 0,$$

which can be written in the form

$$c_n = \frac{n+1}{n+2}\left(\frac{1}{n} c_{n-1} - \sum_{k=1}^{n-1} \frac{c_k\, c_{n-k}}{k+1}\right).$$

It follows that

$$\gamma_0 = 1, \quad \gamma_1 = -\frac{1}{12}, \quad \gamma_2 = \frac{1}{288}, \quad \gamma_3 = \frac{139}{51840}, \quad \gamma_4 = -\frac{571}{2488320}. \tag{3.37}$$

Indeed, we find the values as in (3.25).

Using the same transformation $t \mapsto u(t)$ we can derive an expansion for the gamma function itself. Let us write

$$\Gamma(z+1) = z^{z+1} e^{-z} \int_0^{\infty} e^{-z\phi(t)}\, dt,$$

where again, $\phi(t) = t - \ln t - 1$. This representation easily follows from (3.1). The mapping $t \mapsto u(t)$ given by $\frac{1}{2}u^2 = \phi(t)$ now gives

$$\Gamma(z+1) = z^{z+1} e^{-z} \int_{-\infty}^{\infty} e^{-\frac{1}{2}zu^2} f(u)\, du,$$

with f given in (3.36). Further steps are as in the case of the reciprocal gamma function.

From the above analysis it again easily follows that in the asymptotic expansions of $\Gamma(z)$ and $1/\Gamma(z)$ the same coefficients γ_n occur; see also the explanation after (3.25). We have for both functions

$$\boxed{\frac{1}{\Gamma(z)} \sim \frac{e^z z^{\frac{1}{2}-z}}{\sqrt{2\pi}} \sum_{n=0}^{\infty} \gamma_n z^{-n}, \quad \Gamma(z) \sim \sqrt{2\pi}\, e^{-z} z^{z-\frac{1}{2}} \sum_{n=0}^{\infty} (-1)^n \gamma_n z^{-n},}$$

where the first coefficients γ_n are given in (3.37).

The saddle point analysis of the reciprocal gamma function gives a further interesting integral. When we integrate in the t–plane over the saddle point contour with respect to θ, using (3.34), we have

$$dt = \frac{d\left(\rho e^{i\theta}\right)}{d\theta}\, d\theta = e^{i\theta}\left(\frac{d\rho}{d\theta} + i\rho\right) d\theta = [i + h(\theta)]\, d\theta,$$

where $h(\theta)$ is an odd function of θ. It follows that

$$\frac{1}{\Gamma(z)} = \frac{e^z z^{1-z}}{2\pi} \int_{-\pi}^{\pi} e^{-z\Phi(\theta)}\, d\theta, \tag{3.38}$$

where

$$\Phi(\theta) = -\Re\phi(t) = 1 - \theta\cot\theta + \ln\frac{\theta}{\sin\theta}.$$

To evaluate $\Phi(\theta)$ for small values of θ we have

$$\Phi(\theta) = \frac{1}{2}\theta^2 + \frac{1}{36}\theta^4 + \frac{1}{405}\theta^6 + \frac{1}{4200}\theta^8 + \frac{1}{42525}\theta^{10}\ldots,$$

where all coefficients are positive. In general we have

$$\Phi(\theta) = -\sum_{n=0}^{\infty} (-1)^n \frac{2n+1}{2n}\frac{2^{2n}B_{2n}}{(2n)!}\theta^{2n},$$

which follows from well-known expansions for the trigonometric functions. Representation (3.38) is very useful when one wants to evaluate the gamma function by means of a simple quadrature rule. As explained in Exercise 1.7, the trapezoidal rule gives extremely high accuracy in this case.

3.7. Remarks and Comments for Further Reading

3.1. Readers interested in the history of the gamma function should consult the entertaining paper of Davis (1959) or Godefroy (1901). In almost every book on special functions the gamma function receives a lot of attention. The classic WHITTAKER & WATSON (1927) is still an interesting source of information. The proof of the bounds for the remainder in the expansion for the logarithm of the gamma function in §3.6.1 is based on this reference. Our proof, however, is based on (3.26), which is a different starting point.

3.2. The analysis for the ratio of two gamma functions, resulting in (3.31), may also be based on a contour integral. See TRICOMI & ERDÉLYI (1951); FIELDS (1966) has given the more efficient expansion (3.32). FRENZEN (1987) has shown that the remainder in expansion (3.32) can be estimated as in (3.33). In FRENZEN (1992) this is extended to the case of complex parameters.

3.3. Recent papers on asymptotics in connection with the gamma function are BERRY (1991), PARIS & WOOD (1992), and BOYD (1994).

3.4. In §5.9.1 of Chapter 5 we mention so-called q–extensions of the gamma function and the beta integral.

3.8. Exercises and Further Examples

3.1. The *shifted factorial* $(a)_n = \Gamma(a+n)/\Gamma(a)$ is also called *Pochhammer's symbol*. It is defined for $a \in \mathbb{C}$, $n \in \mathbb{N}$. We have $(a)_0 = 1$, and in general

$$(a)_n = a(a+1)(a+2)\cdots(a+n-1) = (a+n-1)(a)_{n-1}, \quad n = 1, 2, \ldots .$$

Verify that for $a \in \mathbb{C}$, $m, n = 0, 1, 2, \ldots$:

$$(-m)_n = \begin{cases} 0, & \text{if } n > m, \\ (-1)^n m!/(m-n)!, & \text{if } n \le m; \end{cases} \tag{1}$$

$$(-a)_n = (-1)^n\,(a-n+1)_n; \tag{2}$$

$$(a)_{2n} = 2^{2n}\left(\frac{1}{2}a\right)_n\left(\frac{1}{2}a+\frac{1}{2}\right)_n; \tag{3}$$

$$(a)_{2n+1} = 2^{2n+1}\left(\frac{1}{2}a\right)_{n+1}\left(\frac{1}{2}a+\frac{1}{2}\right)_n. \tag{4}$$

3.2.

$$\boxed{\int_0^\infty t^{z-1}e^{-\alpha t^x}\,dt = \frac{1}{x}\Gamma\left(\frac{z}{x}\right)\alpha^{-z/x}, \quad \Re\alpha > 0, \quad \Re x > 0, \quad \Re z > 0.}$$

3.3. Let $\Re\alpha > 0, \Re\beta > 0$. Verify the following alternatives for the beta integral:

$$\int_0^\infty t^{\alpha-1}(t+1)^{-\alpha-\beta}\,dt = B(\alpha,\beta); \tag{1}$$

$$\int_y^x (x-t)^{\alpha-1}\,(t-y)^{\beta-1} = B(\alpha,\beta)\,(x-y)^{\alpha+\beta-1},\quad 0 \le y < x; \tag{2}$$

$$2\int_0^{\frac{1}{2}\pi} (\sin t)^{2\alpha-1}(\cos t)^{2\beta-1}\,dt = B(\alpha,\beta). \tag{3}$$

3.4.

$$\boxed{\Gamma\left(-n+\tfrac{1}{2}\right) = (-1)^n\sqrt{\pi}\,2^{2n}\frac{n!}{(2n)!},\qquad n = 0,1,2,\ldots\,.}$$

3.5.

$$\boxed{\Gamma\left(n+\tfrac{1}{2}\right) = \sqrt{\pi}\,2^{-2n}\frac{(2n)!}{(n)!},\qquad n = 0,1,2,\ldots\,.}$$

3.6. The *binomial coefficients* $\binom{\alpha}{n}$ are defined, for $\alpha \in \mathbb{C}$ and $n = 0,1,2,\ldots,$ by

$$\boxed{\binom{\alpha}{n} = \frac{\alpha(\alpha-1)\cdots(\alpha-n+1)}{n!}.}$$

Other useful relations are

$$\begin{aligned}\binom{\alpha}{n} &= (-1)^n\frac{(-\alpha)_n}{n!}\\ &= (-1)^n\frac{\Gamma(n-\alpha)}{n!\,\Gamma(-\alpha)}\\ &= \frac{\Gamma(\alpha+1)}{n!\,\Gamma(\alpha+1-n)}.\end{aligned}$$

Show that the binomial coefficients appear in the expansion:

$$(1+z)^\alpha = \sum_{n=0}^\infty \binom{\alpha}{n} z^n,\qquad |z| < 1.$$

For general complex values of w and z one defines, as for the binomial numbers in Pascal's triangle,

$$\binom{z}{w} = \frac{\Gamma(z+1)}{\Gamma(w+1)\,\Gamma(z-w+1)},\qquad z \ne -1,-2,-3,\ldots\,.$$

3.7. Verify the reflection formula for the gamma function

$$\boxed{\Gamma\left(\frac{1}{2}-z\right)\Gamma\left(\frac{1}{2}+z\right)=\frac{\pi}{\cos\pi z}, \quad z-\frac{1}{2}\notin\mathbb{Z},}$$

with as variant

$$\Gamma(\tfrac{1}{2}-iy)\Gamma(\tfrac{1}{2}+iy)=\frac{\pi}{\cosh\pi y}.$$

3.8. Verify the reflection formula, the more general form of (3.5),

$$\Gamma(z-n)=(-1)^n\frac{\Gamma(z)\,\Gamma(1-z)}{\Gamma(n+1-z)}=\frac{(-1)^n\pi}{\sin\pi z\;\Gamma(n+1-z)}, \qquad n=0,1,2,\ldots\,.$$

3.9. Show, by using the reflection formula (3.5), that

$$|\Gamma(iy)|\sim\sqrt{\frac{2\pi}{|y|}}\;e^{-\pi\,|y|}, \quad \text{as} \quad y\to\pm\infty.$$

3.10. Verify with the help of (3.7) the following infinite products for the sine and cosine functions:

$$\frac{\sin(\pi z)}{\pi z}=\prod_{n=1}^{\infty}\left(1-\frac{z^2}{n^2}\right), \quad \cos(\pi z)=\prod_{n=0}^{\infty}\left[1-\frac{z^2}{(n+\frac{1}{2})^2}\right].$$

3.11. Show that

$$\int_0^{\frac{1}{2}\pi}(\cos t)^x\cos yt\;dt=\frac{\pi}{2^{x+1}}\frac{\Gamma(x+1)}{\Gamma[(x+y)/2+1]\Gamma[(x-y)/2+1]}, \qquad \Re x>-1.$$

The right-hand side can be expressed in terms of the reciprocal of the beta function. To prove this interesting formula integrate $\int(\cos z)^x\exp(iyz)\,dz$ along a contour consisting of the half-lines $\pm\frac{1}{2}\pi+is, s\geq 0$ and the interval $[-\pi/2,\pi/2]$. For the time being, take $\Re y>\Re x>0$. Formula (3.5) is needed also. Analytic continuation gives the result for each complex y, since both left- and right-hand sides are entire functions of y.

3.12.

$$\int_0^{\infty}\frac{\cosh 2yt}{\cosh^{2x}t}\,dt=2^{2x-2}B(x-y,x+y), \quad \Re x>|\Re y|.$$

First write the integral over $(-\infty,\infty)$ and replace $\cosh(2yt)$ with e^{2yt}. The substitution $u=e^t$ and Exercise 3.3 (1) conclude the proof.

3.13. Verify the following contour integral representation:

$$B(p,q)\,\frac{\sin\pi q}{q}=\frac{\Gamma(p)}{\Gamma(p+q)\Gamma(1-q)}=\frac{1}{2\pi i}\int_0^{(1+)}w^{p-1}(w-1)^{q-1}\,dw,$$

where the contour is as in Figure 3.4. At the point where the contour cuts the positive real axis (at the right of 1) the phases of w and $w-1$ are both zero. This representation holds for any complex q and for $\Re p > 0$. By writing in the integral $w = 1/v$, verify that

$$\frac{\Gamma(p)}{\Gamma(p+q)\Gamma(1-q)} = \frac{1}{2\pi i}\int_{c-i\infty}^{c+i\infty} v^{-p-q}(1-v)^{q-1}\,dv, \quad 0 < c < 1.$$

In other words,

$$\frac{1}{qB(p,q)} = \frac{\Gamma(p+q)}{\Gamma(p)\Gamma(q+1)} = \frac{1}{2\pi i}\int_{c-i\infty}^{c+i\infty} v^{-p}(1-v)^{-1-q}\,dv.$$

with $\Re(p+q) > 0$.

3.14. Show that the alternating series

$$\boxed{\eta(z) = \sum_{n=1}^{\infty} \frac{(-1)^{n-1}}{n^z}, \quad \Re z > 0}$$

can be expressed in terms of the Riemann zeta function:

$$\eta(z) = \left(1 - 2^{1-z}\right)\zeta(z).$$

Verify that $\eta(1) = \ln 2$ by using $\lim_{z\to 1}(z-1)\zeta(z) = 1$, an already known result.

3.15. Verify the following integral representation of the Riemann zeta function:

$$\boxed{\zeta(z) = \frac{1}{2} + \frac{1}{z-1} + \frac{1}{\Gamma(z)}\int_0^{\infty} \beta(t)\,t^z e^{-t}\,dt, \quad \Re z > -1,}$$

where $\beta(t)$ is the function used in (3.12). This gives the analytic continuation of the zeta function given in (3.16) and the residue of the pole at $z = 1$.

3.16. The *Hurwitz zeta function* (Hurwitz (1882)), which is also called the *generalized Riemann zeta function*, $\zeta(z,\alpha)$ is defined by

$$\boxed{\zeta(z,\alpha) = \sum_{n=0}^{\infty} (n+\alpha)^{-z}, \quad \Re z > 1, \quad \alpha \neq 0, -1, -2, -3, \ldots .}$$

The function $\zeta(z,\alpha)$ reduces to $\zeta(z)$ for $\alpha = 1$. Show that the following recurrence relation holds

$$\zeta(z, m+\alpha) = \zeta(z,\alpha) - \sum_{n=0}^{m-1} (n+\alpha)^{-z}, \quad m = 1, 2, 3, \ldots .$$

Derive the following integral representation

$$\zeta(z,\alpha) = \frac{1}{\Gamma(z)}\int_0^\infty \frac{t^{z-1}e^{-\alpha t}}{1-e^{-t}}\,dt, \quad \Re z > 1, \quad \Re\alpha > 0,$$

the analog of (3.16). Derive the analogous result of the previous exercise:

$$\zeta(z,\alpha) = \frac{1}{2}\alpha^{-z} + \frac{\alpha^{1-z}}{z-1} + \frac{1}{\Gamma(z)}\int_0^\infty \beta(t)\,t^z e^{-\alpha t}\,dt, \quad \Re z > -1, \quad \Re\alpha > 0.$$

This shows that the residue at the simple pole $z = 1$ again equals 1. Verify the contour integral representation

$$\zeta(z,\alpha) = -\frac{\Gamma(1-z)}{2\pi i}\int_{-\infty}^{(0^+)} \frac{t^{z-1}e^{\alpha t}}{e^t - 1}\,dt,$$

which generalizes (3.17).

3.17. Recall and verify the following properties of the ψ–function:

recurrence relation:

$$\psi(z+1) = \psi(z) + \frac{1}{z}.$$

Multiplication formula:

$$\psi(2z) = \frac{1}{2}\psi(z) + \frac{1}{2}\psi\left(z+\frac{1}{2}\right) + \ln 2.$$

Reflection formula:

$$\psi(z) = \psi(1-z) - \pi\cot(\pi z).$$

Special values:

$$\psi(1) = -\gamma, \quad \psi\left(\frac{1}{2}\right) = -\gamma - 2\ln 2, \quad \psi'(1) = \frac{1}{6}\pi^2, \quad \psi'\left(\frac{1}{2}\right) = \frac{1}{2}\pi^2.$$

Asymptotic expansion:

$$\psi(z) \sim \ln z - \frac{1}{2z} - \sum_{n=1}^{\infty}\frac{B_{2n}}{2nz^{2n}}, \quad z \to \infty, \quad |\arg z| < \pi.$$

3.18. Compute the integrals $\int_0^\infty t^{z-1} f(t)\,dt$ for the cases

$$f(t) = \ln|1-t|, \quad f(t) = \cos t, \quad f(t) = \sin t.$$

For the first case take for a start $-1 < \Re z < 0$, and split up the interval of integration into $(0, 1-\varepsilon), (1+\varepsilon, \infty)$; integrate by parts to remove the logarithm. Substitute on the infinite interval $t \to 1/t$ and combine the results of both intervals; at a certain moment you can let $\varepsilon \to 0$; use (3.11) and the reflection formula from the previous exercise; the answer is: $\pi \cot(\pi z)/z$. For the two remaining cases integrate $\int t^{z-1} \exp(it)\, dt$ over a path composed of $(0, R), (0, iR)$ and a circular arc with radius R in the first quadrant; first take $\Re z > 1$. The answers are:

$$\boxed{\int_0^\infty t^{z-1} \cos t\, dt = \Gamma(z) \cos \frac{1}{2}\pi z, \qquad \int_0^\infty t^{z-1} \sin t\, dt = \Gamma(z) \sin \frac{1}{2}\pi z.}$$

For the sine integral: compute the limit as $z \to 0$. Determine for both integrals the z–domain of validity. Integrals of the kind considered here are *Mellin transforms* (see SNEDDON (1972) for a good introduction, and OBERHETTINGER (1974) for tables of Melllin transforms).

4

Differential Equations

In this chapter we are concerned with the theory of regular and singular points of linear second order differential equations in the complex plane. First we explain how these differential equations arise in mathematical physics.

4.1. Separating the Wave Equation

The functions which will be introduced in later chapters are of importance in physical problems. Many problems in classical mathematical physics are connected with one of the following linear partial differential equations:

- the *Laplace equation* or *potential equation*: $\Delta u = 0$,
- the *diffusion equation* or *equation of conduction of heat*: $\Delta u = u_t$, or
- the *wave equation*: $\Delta u = u_{tt}$.

The symbol Δ is the *Laplace operator*. For instance, in three-dimensional space:

$$\Delta u = \frac{\partial^2 u}{\partial x^2} + \frac{\partial^2 u}{\partial y^2} + \frac{\partial^2 u}{\partial z^2}$$

with a similar form in spaces of other dimensions. The time variable t in the diffusion and wave equations, the so-called *evolution equations*, is often removed by using a Fourier or Laplace transformation, or by introducing special solutions with a time dependent factor e^{ikt}. The result can then be written in terms of the *Helmholtz equation*

$$(\Delta + k^2)v = 0,$$

which is also called the *time-independent wave equation*. Writing, for instance, in the wave equation $u = ve^{ikt}$, with v not depending on t, then v indeed has to satisfy the Helmholtz equation. Similarly, the time-independent function v in the representation $u = ve^{-k^2t}$ of the solution of the diffusion equation should satisfy the Helmholtz equation.

The Helmholtz equation is a special case of the *Schrödinger equation*, a fundamental equation in quantum mechanics, of which the three dimensional time-independent form reads

$$\Delta\psi + \frac{2m}{\hbar}[E - V(x, y, z)]\psi = 0,$$

where ψ is the wave function, m is the mass of the particle in the potential field V, $\hbar$ is Planck's constant and E is the energy. The special functions treated in this book, in particular, the functions from the Chapters 7, 8 and 9, play an important part in the construction of solutions of the Helmholtz and Schrödinger equations.

To obtain a solution by using the *Laplace transformation*, we usually proceed as follows. Let $v(x, y, z, s)$ be the Laplace transform with respect to t of the solution $u(u, y, z, t)$ of the diffusion equation. That is, we write

$$v(x, y, z, s) = \int_0^\infty e^{-st} u(x, y, z, t)\, dt.$$

Then, at least formally,

$$\Delta v(x, y, z, s) = \int_0^\infty e^{-st} \Delta u\, dt = \int_0^\infty e^{-st} u_t(x, y, z, t)\, dt.$$

Integrating by parts in the final integral, we obtain

$$\Delta v(x, y, z, s) = -u_0(x, y, z) + sv(x, y, z, s),$$

where u_0 is the value of u at time $t = 0$ (the *initial value*). When we take $u_0 = 0$, we again arrive at the Helmholtz equation; when $u_0 \neq 0$ we obtain the nonhomogeneous Helmholtz equation. When we can solve the equation for v, we can obtain u by using the formula for the *inverse Laplace transformation*

$$u(x, y, z, t) = \frac{1}{2\pi i} \int_{c-i\infty}^{c+i\infty} e^{st} v(x, y, z, s)\, ds,$$

where the contour of integration is a vertical line at the right of the s–singularities of v.

Example 4.1. The transport of heat in a homogeneous isotropic one dimensional medium (say, an infinite rod) is described by the equation

$$\kappa u_{xx} = u_t, \quad t \geq 0, \quad -\infty < x < \infty$$

where u is the temperature and κ is a coefficient of heat conduction. At time $t = 0$ we prescribe the initial condition $u = u_0 = \text{sign}(x) = \pm 1$, according to the sign of x. We take $\kappa = 1$. The Laplace transform v of u satisfies the equation $v_{xx} = sv - \operatorname{sign}(x)$, which has the solution

$$v(x,s) = Ae^{-x\sqrt{s}} + Be^{x\sqrt{s}} + \frac{1}{s}, \quad x > 0$$

with a similar solution for $x < 0$. Observing that the initial condition implies that u and v are odd functions of x and that the solution should be bounded as $x \to \infty$, we conclude that $A = -1/s, B = 0$, and that, hence,

$$v(x,s) = \frac{1 - e^{-x\sqrt{s}}}{s}, \quad x > 0.$$

The function $u(x,t)$ follows from Laplace inversion, but it is easier to work with the inversion of v_x. From the derived result for $v(x,s)$ it follows that $v_x(x,s) = s^{-1/2}\exp(-x\sqrt{s}\,)$. Hence, by Laplace inversion,

$$u_x(x,t) = \frac{1}{2\pi i}\int_{c-i\infty}^{c+i\infty} e^{st} v_x(x,s)\,ds = \frac{1}{\pi i}\int_{c-i\infty}^{c+i\infty} e^{tw^2 - xw}\,dw, \quad c > 0.$$

The final integral is easily evaluated: $u_x(x,t) = 1/\sqrt{\pi t}\,\exp[-x^2/(4t)]$. Because $u(x,t)$ is an odd function of x, we obtain

$$u(x,t) = \frac{1}{\sqrt{\pi t}}\int_0^x e^{-\xi^2/(4t)}\,d\xi = \operatorname{erf}\frac{x}{2\sqrt{t}},$$

which holds for all $x \in \mathbb{R}$ and $t > 0$. The symbol erf denotes the error function, which will be introduced in §7.3.4.

4.1.1. Separating the Variables

Trial solutions of the Helmholtz or Schrödinger equation can be found by using a method called *separation of variables,* or *Bernoulli's method* or *Fourier's method.* In this method we assume that the solution v can be written as a product of functions of one variable, that is, that the variables can be separated:

$$v(x,y,z) = f_1(x)f_2(y)f_3(z).$$

The Helmholtz equation then reads

$$\frac{1}{f_1}\frac{d^2 f_1}{dx^2} + \frac{1}{f_2}\frac{d^2 f_2}{dy^2} + \frac{1}{f_3}\frac{d^2 f_3}{dz^2} + k^2 = 0,$$

with possible solutions

$$f_1(x) = e^{\pm i\alpha x}, \quad f_2(y) = e^{\pm i\beta y}, \quad f_3(z) = e^{\pm i\gamma z},$$

where the *separation constants* α, β, γ should satisfy $k^2 = \alpha^2 + \beta^2 + \gamma^2$. On account of physical evidence or boundary conditions it may happen that the separation constants cannot be chosen freely. For instance, one condition may be that the solution should be periodic with respect to x with period 2π. Then the range of α may be the set of integers.

A next step is to build a new solution by taking linear combinations of the building blocks f_1, f_2, f_3. By summing or integrating with respect to the separation constants one obtains a solution that may satisfy the imposed boundary conditions.

Example 4.2. Consider the problem of Example 4.1 and write $u(x,t) = f(t)g(x)$. The we obtain $fg_{xx} = f_t g$, or $f/f_t = g/g_{xx}$. Putting both sides equal to a constant $-\lambda^2$, we obtain $f(t) = \exp(-\lambda^2 t)$ and $g(x) = A(\lambda)\sin\lambda x + B(\lambda)\cos\lambda x$. Since u is an odd function of x, we take $B(\lambda) = 0$ and we compose:

$$u(x,t) = \int_0^\infty e^{-\lambda^2 t} \sin\lambda x\, A(\lambda)\, d\lambda.$$

The initial condition $u_o(x) = \text{sign}(x)$ yields the value $A(\lambda) = 2/(\lambda\pi)$, giving

$$u(x,t) = \frac{2}{\pi}\int_0^\infty e^{-\lambda^2 t}\frac{\sin\lambda x}{\lambda}\, d\lambda.$$

It is not difficult to verify that this function again can be written in terms of the error function. For instance, u_x can be evaluated in closed form:

$$u_x(x,t) = \frac{2}{\pi}\int_0^\infty e^{-\lambda^2 t}\cos\lambda x\, d\lambda = \frac{1}{\pi}\int_{-\infty}^\infty e^{-\lambda^2 t + i\lambda x}\, d\lambda = \frac{e^{-x^2/(4t)}}{\sqrt{\pi t}}.$$

In this example the building blocks f, g are elementary transcendental functions. These building blocks are indeed quite natural when the Helmholtz equation is written in terms of a Cartesian system. However, when the Helmholtz equation has to be solved in domains corresponding with interior or exterior parts of configurations such as spheres or cylinders, it often is necessary to write the Helmholtz equation in terms of a different coordinate system (u, v, w), in place of the Cartesian system (x, y, z). When we apply the method of separation of variables to the new form of the Helmholtz equation, usually higher transcendental functions arise as building blocks of solutions.

In Chapter 10, after we have learned more about Legendre and Bessel functions, we treat more boundary and initial value problems. There we also give details on separating the Helmholtz equation for a wide range of coordinate systems, such as cylindrical, spherical, parabolic and spheroidal coordinates.

4.2. Differential Equations in the Complex Plane

By using the above mentioned method of separation of variables in particular system of coordinates, the Helmholtz or Schrödinger equation may split up in several linear ordinary differential equations of the form

$$f'' + p(z)f' + q(z)f = 0. \tag{4.1}$$

A small set of equations of the form (4.1) governs the well-known functions of mathematical physics. We have the following important examples:

the *hypergeometric differential equation*

$$z(1-z)f'' + [c - (a+b+1)z]f' - abf = 0, \tag{4.2}$$

the *Bessel differential equation*

$$z^2 f'' + zf' + (z^2 - \nu^2)f = 0, \tag{4.3}$$

the *Legendre differential equation*

$$\left(1 - z^2\right) f'' - 2zf' + \nu(\nu+1)f = 0, \tag{4.4}$$

the *Kummer differential equation*

$$zf'' + (c-z)f' - af = 0, \quad \text{and} \tag{4.5}$$

the *Hermite differential equation*

$$f'' - 2zf' + 2\nu f = 0. \tag{4.6}$$

The solutions of these equations will be treated in later chapters. Here we are concerned with general properties of the solutions of the equations.

4.2.1. Singular Points

Consider equation (4.1), where $p: G \to \mathbb{C}$ and $q: G \to \mathbb{C}$ are given meromorphic functions in a simply connected domain G of the complex plane. With respect to (4.1) the following questions are relevant:

- Has the differential equation a solution f, that is, can we find a function f defined in G (or in a subdomain of G) which satisfies the differential equation?
- Where is such a solution analytic?
- Is this solution f unique?

As will be shown in this section, the answers to these questions will depend on the nature of the points in G. We define regular and singular points.

Definition 4.1. A point $z_0 \in G$ where p and q are analytic is called a *regular point* of the differential equation. When $z = z_0$ is not a regular point, it is called a *singular point.* When $z = z_0$ is a singular point but both $(z - z_0)p(z)$ and $(z - z_0)^2 q(z)$ are analytic there, then z_0 is called a *regular singular point.* If z_0 is neither a regular point nor a regular singular point, then it is said to be an *irregular singular point.*

Especially important in the theory of special functions are the expansions in terms of power series in the neighborhood of regular singular points. At regular points the solutions can be represented rather straightforwardly in terms of power series. At regular singular points the situation is more complicated. The general set-up for these points is to represent the solution as a formal power series multiplied by an algebraic term. Next the convergence of the series is established.

The proof of existence of solutions of (4.1) will be given in two versions. In §4.2.2 we give a method based on successive approximation; in §4.2.3 we give a power series method, which is also used in §4.2.4. Experience with both methods is of great importance.

It is sufficient to consider the solutions in a neighborhood of the point $z_0 = 0$; a simple transformation $\zeta = z - z_0$ or $\zeta = 1/z$ can map any finite or infinite point to the origin.

4.2.2. Transformation of the Point at Infinity

The transformation $\zeta = 1/z$ yields for (4.1):

$$\frac{d^2 g}{d\zeta^2} + P(\zeta)\frac{dg}{d\zeta} + Q(\zeta) g = 0, \tag{4.7}$$

where

$$g(\zeta) = f(z) = f(1/\zeta), \quad P(\zeta) = \frac{2}{\zeta} - \frac{1}{\zeta^2}\, p(1/\zeta), \quad Q(\zeta) = \frac{1}{\zeta^4}\, q(1/\zeta).$$

A decisive answer about the nature of the point $z = \infty$ of (4.1) is obtained from investigating the functions P and Q in the neighborhood of $\zeta = 0$.

Definition 4.2. The point $z = \infty$ is called a *regular* point, or a regular singular point of differential equation (4.1) when the point $\zeta = 0$ is a regular point, or a regular singular point, respectively, of the differential equation (4.7).

To decide about the nature of the point at infinity, let p, q of (4.1) have the expansions

$$zp(z) = p_0 + p_1/z + p_2/z^2 + \ldots, \qquad z^2 q(z) = q_0 + q_1/z + q_2/z^2 + \ldots,$$

which converge for sufficiently large values of $|z|$. Then P, Q of (4.7) have the expansions

$$\zeta P(\zeta) = 2 - p_0 - p_1\zeta - p_2\zeta^2 + \ldots, \qquad \zeta^2 Q(\zeta) = q_0 + q_1\zeta + q_2\zeta^2 + \ldots,$$

which converge for sufficiently small values of $|\zeta|$. It follows that, when p, q have the above expansions, (4.7) has a regular singular point at zero, and, that, hence, (4.1) has a regular singular point at infinity.

Example 4.3. The hypergeometric differential equation (4.2) has a regular singularity at infinity. The point at infinity is not a regular singularity for Kummer's differential equation (4.5).

4.2.3. The Solution Near a Regular Point

Let $z = 0$ be a regular point of the differential equation (4.1). Then we can find a number $R > 0$ such that the functions p and q are analytic in the disc $|z| < R$. Choose a number r satisfying $0 < r < R$, and let S be the closed disc $|z| \le r$. Instead on f we concentrate on a new function $y = y(z)$ which is defined by

$$f(z) = y(z) \exp\left[-\frac{1}{2}\int_0^z p(\zeta)\, d\zeta\right].$$

Since p is analytic in S, the value of the integral does not depend on the choice of the path of integration, as long this path is contained within S. The differential equation (4.1) is transformed into (verify this!)

$$\frac{d^2y}{dz^2} + J(z)y = 0, \tag{4.8}$$

where

$$J(z) = q(z) - \frac{1}{2}\frac{dp(z)}{dz} - \frac{1}{4}[p(z)]^2.$$

Observe that the new differential equation does not have a term involving the first derivative y'. The transformation of (4.1) into (4.8) is also mentioned in §4.5 where the *Liouville transformation* is given. The function J is defined in G and also is meromorphic there. A regular point of (4.1) is also a regular point of (4.8).

4.2.3.1. Existence of a Solution

We convert the differential equation (4.8) into an integral equation by integrating the differential equation $y'' = -J(z)y$ twice with respect to z. Then we obtain, after integrating by parts, the equation

$$y(z) = b_0 + b_1 z + \int_0^z (\zeta - z)J(\zeta)y(\zeta)\,d\zeta, \tag{4.9}$$

where b_0, b_1 are arbitrary complex numbers. It is also easily verified directly that (4.9) fulfills (4.8). Furthermore we have

$$y(0) = b_0, \quad y'(0) = b_1.$$

The integral equation (4.9) will be solved by the method of *successive approximation*. This method is based on constructing a sequence of functions $\{y_n\}$, $n = 0, 1, 2, \ldots$ defined by

$$\begin{aligned} y_0(z) &= b_0 + b_1 z, \\ y_n(z) &= \int_0^z (\zeta - z)J(\zeta)y_{n-1}(\zeta)\,d\zeta, \quad n \geq 1, \end{aligned}$$

and showing that the sum $Y(z) = \sum_{n=0}^{\infty} y_n(z)$ is a solution of (4.8). This expansion is called the *Liouville-Neumann expansion*.

The quantities b_0 and b_1 are arbitrary complex numbers; the path of integration in the integral defining $y_n(z)$ can be chosen to be the straight line between 0 and z. Since J is known, all functions y_n can be determined in this way. All functions y_n are analytic in S.

Next we introduce constants M and μ by putting

$$M = \max_{z\in S} |J(z)|, \quad \mu = \max_{z\in S} |y_0(z)|.$$

Then we will prove that the functions y_n can be bounded as follows:

$$|y_n(z)| \leq \mu \frac{M^n |z|^{2n}}{n!}, \quad z \in S, \quad n \geq 0. \tag{4.10}$$

For $n = 0$ the bound is trivial. Assume that the bound holds for y_{n-1} with $n \geq 1$. Then it follows, when the path of integration indeed is a straight line,

$$\begin{aligned} |y_n(z)| &= \left| \int_0^z (\zeta - z)J(\zeta)y_{n-1}(\zeta)\,d\zeta \right| \\ &\leq \int_0^{|z|} |\zeta - z|\ |J(\zeta)|\ \frac{\mu M^{n-1}}{(n-1)!}\ |\zeta|^{2n-2}\, |d\zeta| \\ &\leq \frac{\mu M^n}{(n-1)!} |z| \int_0^{|z|} |\zeta|^{2n-2}\ |d\zeta| \\ &= \frac{\mu M^n}{(n-1)!}\ |z| \int_0^{|z|} t^{2n-2}\,dt \leq \frac{\mu M^n}{n!}\ |z|^{2n}. \end{aligned}$$

So, we have verified the induction step, and we conclude that (4.10) holds for $n = 0, 1, 2, \ldots$. Furthermore we have, since $|z| \leq r$ if $z \in S$, the uniform bound

$$|y_n(z)| \leq \mu \frac{M^n r^{2n}}{n!}, \quad z \in S, \quad n \geq 0.$$

Since the series $\sum_{n=0}^{\infty} \mu M^n r^{2n}/n!$ is convergent, it follows that the series $\sum_{n=0}^{\infty} y_n(z)$ is uniformly convergent with respect to $z \in S$. Since all y_n are analytic, the sum $Y(z) = \sum_{n=0}^{\infty} y_n(z)$ constitutes an analytic function in the disc $|z| < r$ and the series may be differentiated term by term. That is,

$$Y^{(k)}(z) = \sum_{n=0}^{\infty} y_n^{(k)}(z), \quad |z| < r, \quad k = 1, 2, 3, \ldots .$$

Since r is an arbitrary number satisfying $0 < r < R$, we infer that Y is analytic in the open disc $|z| < R$.

From the definition of y_n it follows that

$$\begin{aligned} y_0'(z) &= b_1, \quad y_0''(z) = 0, \\ y_n'(z) &= -\int_0^z J(\zeta) y_{n-1}(\zeta)\, d\zeta, \\ y_n''(z) &= -J(z) y_{n-1}(z), \end{aligned}$$

for $n \geq 1$. Hence we obtain

$$Y''(z) = \sum_{n=0}^{\infty} y_n''(z) = -\sum_{n=1}^{\infty} J(z) y_{n-1}(z) = -J(z) Y(z).$$

We conclude that the function Y is a solution of the differential equation (4.8). Also,

$$Y(0) = \sum_{n=0}^{\infty} y_n(0) = b_0, \quad Y'(0) = \sum_{n=0}^{\infty} y_n'(0) = b_1,$$

and, hence, the solution Y fulfills the initial conditions

$$Y(0) = b_0, \quad Y'(0) = b_1. \tag{4.11}$$

With the solution Y there corresponds the function F defined by

$$F(z) = Y(z) \exp\left[-\frac{1}{2}\int_0^z p(\zeta)\, d\zeta\right],$$

a solution of the original differential equation (4.1). The function F is analytic in the disc $|z| < R$ and has initial values

$$\begin{aligned} F(0) &= Y(0) = b_0, \\ F'(0) &= Y'(0) - \tfrac{1}{2} Y(0) p(0) = b_1 - \tfrac{1}{2} b_0 p(0). \end{aligned}$$

We denote these values by a_0, a_1; then F fulfills the initial conditions

$$F(0) = a_0, \quad F'(0) = a_1. \tag{4.12}$$

We now show that the analytic solution F of the differential equation (4.1) with initial conditions (4.12) is ***unique***. Let us assume that the differential equation (4.1) has a second solution F^* with the same initial values a_0 and a_1. Then we know that the function $V(z) = F(z) - F^*(z)$ is analytic in S and satisfies

$$V''(z) + p(z)V'(z) + q(z)V(z) = 0$$

with the initial values

$$V(0) = 0, \quad V'(0) = 0.$$

Substituting $z = 0$ in the differential equation we obtain $V''(0) = 0$. By differentiating the differential equation, and substituting afterwards $z = 0$, we obtain $V'''(0) = 0$. This process can be repeated, and we conclude that $V^{(n)}(0) = 0$, $n = 0, 1, 2, \ldots$. Considering the power series of V we infer that all terms in the series vanish. Hence $V(z) = 0$, that is, $F(z) = F^*(z)$ in the open disc $|z| < R$. This shows that the solution F is determined unambiguously by the initial conditions (4.12).

We summarize the previous results as follows.

Theorem 4.1. *Let the functions p and q be analytic in the open disc $|z| < R$ and let a_0 and a_1 by arbitrary complex numbers. Then, there exists one and only one function f with the properties:*

1) f satisfies the differential equation (4.1);

2) f satisfies the initial conditions $f(0) = a_0$, $f'(0) = a_1$ and

3) f is analytic in $|z| < R$.

Remark 4.1. When applying this theorem we can take R in an optimal way. Then R is the radius of the largest disc around the origin in G, such that no poles of p and q lie inside this disc.

Remark 4.2. When $z_0 \in G$ is an arbitrary regular point, $z_0 \in G$, and p and q are analytic inside a disc with positive radius around z_0, we can use a simple translation and again prove that the solution of (4.1) exists and is analytic inside this disc. Since any two points in the simply connected domain can be joined by a finite sequence of intersecting discs, we can continue the solution of (4.1) throughout G.

4.2.3.2. The Wronskian of Two Solutions

We now apply Theorem 4.1 twice:

1) with initial conditions $a_0 = 1$, $a_1 = 0$; giving a unique solution, which we call f_1;
2) with initial conditions $a_0 = 0$, $a_1 = 1$; giving a unique solution, which we call f_2.

Then the functions f_1 and f_2 are linearly independent solutions of (4.1). To verify this, let $\lambda f_1(z) + \mu f_2(z) = 0$ in $|z| < R$. Then the initial conditions yield necessarily $\lambda = \mu = 0$. Each solution of (4.1) can be written as a linear combination of f_1 and f_2. When, for instance, we try to construct a solution f with initial values $f(0) = A$, $f'(0) = B$ then we have $f = Af_1 + Bf_2$.

Each pair $\{f_1, f_2\}$ such that any solution of (4.1) can be written in the form

$$f(z) = Af_1(z) + Bf_2(z),$$

where A and B are constants is called a *fundamental system* of solutions; that is, the pair $\{f_1, f_2\}$ is a basis for the linear space of solutions.

Let us consider two arbitrary solutions f_1 and f_2 of the second order differential equation (4.1). We construct with this pair the expression

$$W[f_1, f_2](z) = \begin{vmatrix} f_1(z) & f_2(z) \\ f_1'(z) & f_2'(z) \end{vmatrix} = f_1(z)f_2'(z) - f_1'(z)f_2(z), \tag{4.13}$$

which is called the *Wronskian* of the pair $\{f_1, f_2\}$. $W[f_1, f_2]$ plays an important part when investigating the linear independence of a pair of solutions $\{f_1, f_2\}$ of a differential equation. We can consider this expression also for differential equations of higher order (again by using determinants).

It is quite simple to verify, by using (4.1), that the function $W[f_1, f_2]$ is a solution of the equation $w' = -pw$. It follows that, when $z, z_0 \in G$,

$$W[f_1, f_2](z) = C \exp\left[-\int_{z_0}^{z} p(\zeta)\, d\zeta\right], \tag{4.14}$$

which is *Abel's identity* (1827), where C does not depend on z; C equals the value of $W[f_1, f_2](z)$ at the point $z = z_0$. Consequently, the Wronskian vanishes identically (when $C = 0$) or it never vanishes in a domain where p is analytic. Another consequence is that the Wronskian of each pair of solutions $\{f_1, f_2\}$ of (4.1) reduces to a constant when the term with the first derivative in (4.1) is missing ($p = 0$).

Now let $\{f_1, f_2\}$ constitutes a linearly dependent pair in G. This means there exist two complex numbers A and B with $|A| + |B| \neq 0$, such that

$$Af_1(z) + Bf_2(z) = 0, \quad \forall z \in G.$$

Differentiation gives

$$Af_1'(z) + Bf_2'(z) = 0, \quad \forall z \in G.$$

These two equations for A and B can be interpreted as a linear system. We have assumed that a solution $\{A, B\} \neq \{0, 0\}$ exists. Consequently, the determinant of the system should vanish. That is, $W[f_1, f_2](z) = 0$, $z \in G$ when $\{f_1, f_2\}$ constitutes a linearly dependent pair of solutions of (4.1). On the other hand, the Wronskian cannot vanish for a fundamental system $\{f_1, f_2\}$. Summarizing we have:

Theorem 4.2. *The solutions $\{f_1, f_2\}$ of the differential equation (4.1) are linearly independent if and only if the Wronskian (4.13) does not vanish identically in a domain where the solutions are analytic.*

4.2.4. Power Series Expansions Around a Regular Point

The method of this and the following subsection is called *Frobenius method* (FROBENIUS (1873)). We consider a solution of (4.1) with given initial values $f(0) = c_0$, $f'(0) = c_1$. We prove the convergence of the power series $f(z) = \sum_{n=0}^{\infty} c_n z^n$, when $z = 0$ is a regular point. In fact this follows from Theorem 4.1. However, in connection with the treatment of power series around regular singular points it is convenient to have a separate proof.

The series

$$p(z) = \sum_{n=0}^{\infty} p_n z^n, \quad q(z) = \sum_{n=0}^{\infty} q_n z^n$$

are convergent in the disc $|z| < R$. Cauchy's inequality for the coefficients of a Maclaurin series gives the bounds

$$|p_n| \leq Ar^{-n}, \quad |q_n| \leq Br^{-n},$$

where A, B do not depend on n and r is an arbitrary number satisfying $0 < r < R$. Next we pose the induction hypothesis: we can assign positive constants C and m such that for all indices n we have

$$|c_n| \leq C(n+1)^m r^{-n}.$$

The series $\sum (n+1)^m (z/r)^n$ converges in $|z| < r$, and will be used for comparison with the series $\sum_{n=0}^{\infty} c_n z^n$. When the hypothesis is verified we know that the series for f converges.

To verify the bound for $|c_n|$ we use induction with respect to n. The constant C will be adapted such that the bound holds for c_0 and c_1, without regard to the value of $m > 0$. Substitute the series for f, p, q into (4.1) and compare equal powers. When, indeed, f has a convergent power series, the coefficients must satisfy the relation

$$n\,(n+1)\,c_{n+1} + \sum_{k=1}^{n} k\,c_k\,p_{n-k} + \sum_{k=0}^{n-1} c_k\,q_{n-k} = 0, \quad n = 1, 2, \dots .$$

The induction step from n to $n+1$ runs as follows. Using this recursion for c_n we obtain from the above bounds for p_n, q_n and from the hypothesis for c_k, $0 \geq k \geq n$

$$\begin{aligned} |c_{n+1}| &\leq \frac{C}{n\,(n+1)} \left[Ar^{-n} \sum_{k=1}^{n} k(k+1)^m + Br^{-n+1} \sum_{k=1}^{n} k^m \right] \\ &\leq \frac{Cr^{-n}}{n\,(n+1)} \left[A\frac{(n+2)^{m+2}}{m+2} + Br\frac{(n+1)^{m+1}}{m+1} \right] \end{aligned}$$

where we have used the estimate

$$\sum_{k=1}^{n} k^m \leq \int_0^{n+1} x^m\,dx = \frac{(n+1)^{m+1}}{m+1}.$$

The estimate for $|c_{n+1}|$ will be elaborated by writing

$$\begin{aligned} |c_{n+1}| &\leq \frac{Cr^{-n}(n+2)^m}{m+1} \left[A\frac{(n+2)^2}{n\,(n+1)} + Br\frac{1}{n} \right] \\ &\leq Cr^{-n-1}(n+2)^m \,\frac{\frac{9}{2}Ar + Br^2}{m+1}. \end{aligned}$$

So, when we take $m + 1 \geq \frac{9}{2}Ar + Br^2$, we conclude that acceptance of the hypothesis as far as n yields a similar bound for the index value $n+1$. This proves the absolute convergence of $\sum_{n=0}^{\infty} c_n z^n$ in $|z| < r$, and hence in $|z| < R$.

Example 4.4. From the theory of this chapter we conclude that Legendre's differential equation (4.4) has analytic solutions in $|z| < 1$. The solutions hence have convergent power series inside the unit disc. Equation (4.1) is not convenient as a starting point now, since

$$p(z) = -\frac{2z}{1-z^2}, \quad q(z) = \frac{\nu(\nu+1)}{1-z^2}$$

have infinite series expansions. It is more convenient to take (4.4). Substitution of $f(z) = \sum_{n=0}^{\infty} c_n z^n$ into (4.4) yields the simple recursion

$$n(n+1)c_{n+1} = (n - \nu - 1)(n + \nu)c_{n-1}.$$

The starting values $c_0 = 1$, $c_1 = 0$ produce an even function

$$f_1(z) = 1 - \frac{\nu(\nu+1)}{2!}z^2 + \frac{\nu(\nu-2)(\nu+1)(\nu+3)}{4!}z^4 - \dots .$$

The starting values $c_0 = 0$, $c_1 = 1$ give the odd solution

$$f_2(z) = z - \frac{(\nu-1)(\nu+2)}{3!}z^3 + \frac{(\nu-1)(\nu-3)(\nu+2)(\nu+4)}{5!}z^5 - \dots .$$

It is easily verified that this pair $\{f_1, f_2\}$ constitutes a fundamental system with Wronskian

$$W[f_1, f_2](z) = \frac{1}{1-z^2},$$

which indeed never vanishes. For integer values of ν one of these expansions terminates, and we obtain the *Legendre polynomial:*

$$\begin{aligned} f_1(z) &= AP_{2n}(z) \qquad &\text{if} \quad \nu = 2n, \\ f_2(z) &= BP_{2n+1}(z) \qquad &\text{if} \quad \nu = 2n+1, \end{aligned}$$

where A and B do not depend on z. So, when $\nu = n$, one of the above solutions is a polynomial of degree n, and the other solution has an infinite convergent power series in the unit disc.

4.2.5. Power Series Expansions Around a Regular Singular Point

Let $z = 0$ be a regular singular point of (4.1). With a slight change in notation we write the differential equation (4.1) in the form

$$z^2 f'' + zp(z)f' + q(z)f = 0, \tag{4.15}$$

where we assume that p and q are analytic in $|z| < R$ with power series

$$p(z) = \sum_{n=0}^{\infty} p_n z^n, \quad q(z) = \sum_{n=0}^{\infty} q_n z^n.$$

We assume that at least one of the coefficients p_0, q_0 and q_1 is different from 0. We may expect that, for values of z near the regular singular point $z = 0$, the solutions of (4.15) behave as the solutions of the equation

$$z^2 g'' + zp_0 g' + q_0 g = 0.$$

This is Euler's differential equation, with exact solutions $g(z) = z^\mu$, where μ satisfies the quadratic equation

$$\mu(\mu-1) + \mu p_0 + q_0 = 0. \tag{4.16}$$

Actually we try to find a solution of (4.15) of the form

$$f(z) = z^{\mu} \sum_{n=0}^{\infty} c_n z^n, \tag{4.17}$$

in which the series converges in a neighborhood of the origin, and defines an analytic function there. The result is as follows.

Theorem 4.3. *Let the functions p and q be analytic in $|z| < R$. Then (4.15) has a solution of the form (4.17), where μ satisfies equation (4.16), and the series converges for all z in $|z| < R$.*

Equation (4.16) is called the *indicial equation* and the roots μ_1 and μ_2 are called the *exponents of the differential equation* (4.15) at the point $z = 0$.

Proof. Substitution of (4.17) into (4.15) gives for $n \geq 0$

$$(n+\mu)(n+\mu-1)c_n + \sum_{k=0}^{n} (k+\mu)p_{n-k}c_k + \sum_{k=0}^{n} q_{n-k}c_k = 0. \tag{4.18}$$

For $n = 0$ (and $c_0 \neq 0$) this corresponds to the indicial equation (4.16). When μ is a solution of this equation, we can choose c_0 arbitrarily. Usually, c_0 cannot assume the value of f at $z = 0$, since f may be singular at this point. For special functions c_0 will be often chosen so that f has a convenient normalization.

We proceed formally, and we assume that the power series makes sense; that is, we assume that it converges in a neighborhood of $z = 0$. Collecting the coefficients of c_n in (4.18), we obtain $(n+\mu)(n+\mu-1)+(n+\mu)p_0+q_0$. Using (4.16) and the exponents μ_1, μ_2, we can write:

$$\mu(\mu-1) + \mu p_0 + q_0 = (\mu-\mu_1)(\mu-\mu_2).$$

It follows that the coefficients of c_n in (4.18) can be written as

$$(n+\mu)(n+\mu-1) + (n+\mu)p_0 + q_0 = (n+\mu-\mu_1)(n+\mu-\mu_2).$$

Hence, equation (4.18) can be written as

$$(n+\mu-\mu_1)(n+\mu-\mu_2)c_n = -\sum_{k=0}^{n-1} (k+\mu)p_{n-k}c_k - \sum_{k=0}^{n-1} q_{n-k}c_k. \tag{4.19}$$

By choosing $\mu = \mu_1$, (4.19) becomes

$$n(n+\mu_1-\mu_2)c_n = -\sum_{k=0}^{n-1} (k+\mu_1)p_{n-k}c_k - \sum_{k=0}^{n-1} q_{n-k}c_k. \tag{4.20}$$

For $n = 1, 2, 3, \ldots$ the coefficients c_n can be computed successively from this relation. Apart from the free choice of c_0 we obtain just one formal series solution of the differential equation. By choosing, on the other hand, $\mu = \mu_2$ (4.19) becomes

$$n(n + \mu_2 - \mu_1)c_n = -\sum_{k=0}^{n-1}(k + \mu_2)p_{n-k}c_k - \sum_{k=0}^{n-1} q_{n-k}c_k. \tag{4.21}$$

In this way we obtain, in general, a second (independent) solution.

However, the method breaks down when $\mu_1 - \mu_2$ is an element of $\mathbb{Z}$. Let $\mu_1 = \mu_2$. Then both schemes for calculating the coefficients (4.20) and (4.21) are the same. Consequently, in this case at least one of the solutions is not of the form (4.17). Let $m = \mu_1 - \mu_2 = 1, 2, 3, \ldots$; then we find via (4.20) a solution. In (4.21), however, the coefficient of c_n has the form $n(n - m)$. We see that at the value $n = m$ the recurrence relation may break down (it may happen that the right-hand side of (4.21) vanishes as well). When indeed the method breaks down two different solutions of (4.15) cannot have the representation (4.17). A further discussion of this case is given in the next section.

We proceed with (4.20) and we assume that $\mu_1 - \mu_2 \neq -1, -2, -3, \ldots$. We use the estimates for the coefficients p_n, q_n of §4.2.4, that is,

$$|p_n| \leq Ar^{-n}, \quad |q_n| \leq Br^{-n},$$

and we again verify the hypothesis with respect to c_n:

$$|c_n| \leq C(n+1)^m r^{-n}.$$

From (4.20) it follows that

$$\begin{aligned}
|c_n| &\leq \frac{C}{Dn^2}\left[Ar^{-n}\sum_{k=0}^{n-1}|k + \mu_1|(k+1)^m + Br^{-n}\sum_{k=0}^{n-1}(k+1)^m\right] \\
&\leq \frac{Cr^{-n}}{Dn^2}\,\frac{A(n+1)^{m+2} + (A|\mu_1| + B)(n+1)^{m+1}}{m+1} \\
&\leq Cr^{-n}(n+1)^m\frac{4A + 2(A|\mu 1| + B)}{(m+1)D},
\end{aligned}$$

where D is chosen such that $|n + \mu_1 - \mu_2| \geq Dn$. We conclude that, in this case also, it is possible to choose m such that the induction principle can be applied successfully. ∎

We summarize:

1) If the roots μ_1, μ_2 of equation (4.16) satisfy $\mu_1 - \mu_2 \notin \mathbb{Z}$, then we find two solutions

$$f_1(z) = z^{\mu_1} \sum_{n=0}^{\infty} c_n z^n, \quad f_2(z) = z^{\mu_2} \sum_{n=0}^{\infty} d_n z^n.$$

The power series represent analytic functions in $|z| < R$ and may be differentiated term by term. The functions f_1 and f_2 both satisfy the differential equation (4.15).

2) If $\mu_1 - \mu_2 \in \mathbb{Z}$ then we find, using the above method, at least one solution of the form (4.17).

Example 4.5. In Bessel's equation (4.3) $z = 0$ is a regular singular point. The indicial equation (4.16) has the two solutions $\mu_1 = \nu$, $\mu_2 = -\nu$. The method of this section produces for the special choice $c_0 = 2^{-\nu}/\Gamma(\nu+1)$ the *Bessel function*

$$\boxed{J_\nu(z) = \left(\tfrac{1}{2}z\right)^\nu \sum_{n=0}^{\infty} \frac{(-1)^n}{\Gamma(n+\nu+1)\, n!} \left(\tfrac{1}{2}z\right)^{2n}.} \tag{4.22}$$

In general, the pair $\{J_\nu(z), J_{-\nu}(z)\}$ constitutes a fundamental system. However, when $\nu = m \in \mathbb{Z}$ the scheme (4.21) breaks down, since $\mu_2 - \mu_1 = -2m$. From (4.22) it follows however, that when $m = 0, 1, 2, \ldots$, the series actually starts with the term $n = m$; earlier terms vanish owing to the gamma function. In fact we have

$$J_{-m}(z) = (-1)^m J_m(z).$$

Hence, $\{J_m(z), J_{-m}(z)\}$ do not constitute a fundamental system. A second solution of (4.4) has to be obtained in a different way. Also, when $\nu = m + \frac{1}{2}$ we have $\mu_2 - \mu_1 = -2m - 1$, which again is a negative integer. The construction of the second solution does not break down in this case.

When, in the theory of special functions, the construction by power series of the second solution, as described in this section, is not possible, one often uses a special technique to obtain a second solution. First one assumes that $\mu_1 - \mu_2 \notin \mathbb{Z}$. Then a convenient linear combination of two power series solutions is defined. In the present case of the Bessel functions (more on these functions is treated in Chapter 10), the procedure works as follows. Assume that $\nu \notin \mathbb{Z}$ and define as a new solution of Bessel's equation the so-called *Neumann function*:

$$\boxed{Y_\nu(z) = \frac{\cos \nu\pi \, J_\nu(z) - J_{-\nu}(z)}{\sin \nu\pi},} \tag{4.23}$$

a linear combination of two functions defined in (4.22). The cosine function lets the numerator vanishes if ν approaches an integer value. The sine function takes care of an interesting process then. Other periodic functions can be used here, but the choices in (4.23) give the Neumann function a suitable normalization. The limit of the right-hand side of (4.23), when $\nu \to m$, is properly defined, and can be computed by using l'Hôpital's rule. In this way one obtains $Y_m(z)$, which is of the form

$$Y_m(z) = \frac{2}{\pi}\, J_m(z)\, \ln z + z^{-m} \sum_{n=0}^{\infty} c_n z^{2n}.$$

In (9.15) we give a different representation of the Neumann function $Y_m(z)$. We point out, however, that the appearance of the logarithm in the above representation of $Y_m(z)$ is a characteristic feature when the second solution cannot be written in the form (4.17).

4.2.5.1. Further Analysis of the Case $\mu_1 - \mu_2 = m$

In the above example of the Bessel functions we have seen that a logarithmic term occurs via a limiting process when $\mu_1 - \mu_2 = m,\ m = 0, 1, 2, \ldots$. We now discuss more generally what may happen.

Let f_1 be a solution of (4.1) and let it have the form (4.17) with $\mu = \mu_1$. To find a second solution f, we substitute $f = f_1 g$, where g has to be found.

We obtain for g the equation

$$z f_1 g'' + (2z f_1' + p f_1) g' = 0.$$

Integrating this we have

$$\ln g' = -2 \ln f_1 - \int_{z_0}^{z} \zeta^{-1} p(\zeta)\, d\zeta + A,$$

where A is a constant. It follows that g' has the following form:

$$g' = z^{-p_0 - 2\mu_1} h_1(z),$$

where the function $h_1(z)$ is analytic in a neighborhood of $z = 0$. Since $\mu_1 + \mu_2 = 1 - p_0$ we can replace $p_0 + 2\mu_1$ with $m + 1$. The resulting equation is

$$g' = z^{-m-1} h_1(z).$$

In general, in the power series expansion of $h_1(z)$ occurs a term with z^m. It now becomes clear that g may have the representation

$$g(z) = c \ln z + z^{-m} h_2(z)$$

where $h_2(z)$ is analytic at $z = 0$.

In this way the second solution obtains the form

$$f_2(z) = cf_1(z)\ln z + z^{\mu_2}\sum_{n=0}^{\infty} d_n z^n. \tag{4.24}$$

When in the above analysis $m = 0$ we have $d_0 = 0$. This follows from the fact that the term with the logarithm is obtained through $h_1(0)$.

The coefficients d_n may be found using the above construction. However, it is much easier to find the coefficients of (4.24) by substituting (4.24) into the differential equation (4.15) and comparing equal powers of z. In this method one has a free choice for the coefficient d_0, if $m > 0$.

4.2.5.2. Other Singularities

We have seen that the solutions of differential equations with a regular singular point have an algebraic singularity at this point. Special circumstances generate a solution with a logarithmic term. The equation

$$z^3 f'' + zf' - 2f = 0$$

has, at $z = 0$, a singularity of a different type. We have the explicit solution: $f(z) = \exp(1/z)$ and we see that at least one solution has an essential singularity at $z = 0$. This solution cannot be represented by a power series at the origin, whether or not multiplied by algebraic or logarithmic terms.

This chapter deals with *linear* differential equations of the second order. The theory of regular and regular singular points will not work with *nonlinear equations.* Consider, for instance, the equation $y' = 1 + y^2$. No singularities occur in the coefficients of this equation. The solution $y = \tan x$, however, has poles galore.

4.3. Sturm's Comparison Theorem

We discuss *Sturm's comparison theorem* (STURM (1836)) since it is important for investigating the location of zeros of special functions. We apply this theorem in Chapter 9 to obtain information on the zeros of the Bessel functions.

Theorem 4.4. *On the interval (a, b), let y_i be real solutions of the differential equations*

$$y'' + g_i(x)y = 0, \quad i = 1, 2, \tag{4.25}$$

where g_i are continuous and real valued on (a, b). Let $g_1(x) < g_2(x)$ on (a, b). Then between two consecutive zeros of y_1 in (a, b) there is at least one zero of y_2.

Proof. Let y_1 have the consecutive zeros p, q in (a,b). Without loss of generality we assume that $y_1(x) > 0$ on (p,q), from which it follows that $y_1'(p) \geq 0$, $y_1'(q) \leq 0$. Assume now that y_2 has no zero in (p,q); then we may assume that $y_2(x) > 0$ on (p,q). We will show that this assumption leads to a contradiction. Consider (4.25) for both y_i:

$$y_1'' + g_1(x)y_1 = 0, \quad y_2'' + g_2(x)y_2 = 0$$

and multiply the first equation by y_2 and the second one by y_1. Subtracting the results gives

$$y_1''(x)y_2(x) - y_2''(x)y_1(x) = [g_2(x) - g_1(x)]y_1(x)y_2(x).$$

Integrating this on $[p,q]$, we obtain

$$\int_p^q [g_2(x) - g_1(x)]y_1(x)y_2(x)\,dx = y_1'(q)y_2(q) - y_1'(p)y_2(p).$$

The left-hand side of this relation is positive, since

$$g_1(x) < g_2(x), \quad y_1(x) > 0, \quad y_2(x) > 0$$

on (p,q). The right-hand side is non-positive, since

$$y_1'(p) \geq 0, \quad y_1'(q) \leq 0, \quad y_2(p) \geq 0, \quad y_2(q) \geq 0$$

So, we have arrived at a contradiction. It follows that the function y_2 should have at least one zero in (p,q). ∎

4.4. Integrals as Solutions of Differential Equations

Special functions are often introduced as solutions of a linear homogeneous second order differential equation. We discuss a general method for obtaining integral representations of the solutions of such equations. We give an example in which the differential equation considered is that defining the Airy functions. In Chapter 7 this method is used to obtain representations for the confluent hypergeometric functions, and in Chapter 9 to derive integral representations for the Bessel functions (see §9.2 and Remark 9.3). For the Legendre functions (see (8.47) and (8.49)) the method is also used, however without using the general set-up presented in this section.

Let the given differential equation be of the form

$$L_z[y(z)] = \sum_{k=0}^{n} l_k(z)y^{(k)}(z); \tag{4.26}$$

in other words, L_z is a linear differential operator. We introduce two other operators M_t and M_t^* (the adjoint of M_t) by defining

$$M_t[u(t)] = \sum_{k=0}^{n} m_k(t)u^{(k)}(t), \quad M_t^*[v(t)] = \sum_{k=0}^{n}(-1)^k[m_k(t)v(t)]^{(k)}. \tag{4.27}$$

On account of the identity

$$\frac{d}{dt}\sum_{j=0}^{k-1}(-1)^j v^{(j)}(t)u^{(k-j-1)}(t) = v(t)u^{(k)}(t) - (-1)^k u(t)v^{(k)}(t),$$

and using this with $v(t)$ replaced by $m_k(t)v(t)$, we obtain

$$\begin{aligned} &v(t)M_t[u(t)] - u(t)M_t^*[v(t)] \\ &\qquad = \frac{d}{dt}\sum_{k=1}^{n}\sum_{j=0}^{k-1}(-1)^j[m_k(t)v(t)]^{(j)}u^{(k-j-1)}(t) \equiv \frac{d}{dt}P(u,v). \end{aligned} \tag{4.28}$$

This term $P(u,v)$ is called the *bilinear concomitant* related to M_t.

We try to find a solution of (4.26) in the form

$$y(z) = \int_\alpha^\beta K(z,t)v(t)\,dt. \tag{4.29}$$

The kernel $K(z,t)$ is chosen in a clever way, or by trial and error. Anyhow, K should be smooth enough to endure the actions to come. When applying L_z on (4.29), it appears that it is sufficient to find a solution v of the integral equation

$$\int_\alpha^\beta L_z[K(z,t)]v(t)\,dt = 0. \tag{4.30}$$

The crucial step in the method is that we are able to find an operator M_t that satisfies

$$L_z[K(z,t)] = M_t[K(z,t)].$$

If this can be done, we replace (4.30) with

$$\int_\alpha^\beta M_t[K(z,t)]v(t)\,dt = 0 \tag{4.31}$$

and, on account of (4.28), we now can write (4.31) in the form

$$\int_\alpha^\beta \left\{K(z,t)M_t^*[v(t)] + \frac{\partial}{\partial t}P[K(z,t),v(t)]\right\}\,dt = 0. \tag{4.32}$$

Assume next that we can find $v(t)$ such that it solves the equation

$$M_t^*[v(t)] = 0.$$

Then we can integrate the left-hand side of (4.32), giving:

$$P[K(z,t), v(t)]\Big|_{\alpha}^{\beta}.$$

As a final step we choose α and β such that the integrated term vanishes. This may happen when $P[K(z,t), v(t)]$ vanishes at α and β, but also when

$$P[K(z,\alpha), v(\alpha)] = P[K(z,\beta), v(\beta)],$$

because of the special choice $\alpha = \beta$, and since the path of integration is a closed contour on which (and inside which) the function $P[K(z,t), v(t)]$ is analytic.

Example 4.6. Consider the differential equation

$$L_z[y(z)] = y'' - zy = 0 \tag{4.33}$$

and let $K(z,t) = e^{-zt}$, the kernel of the Laplace transformation. We have

$$L_z e^{-zt} = \left(t^2 - z\right) e^{-zt} = \left(t^2 + \frac{d}{dt}\right) e^{-zt} = M_t e^{-zt},$$

which defines our M_t. We see that $M_t^* = -\frac{d}{dt} + t^2$ and that (4.28) with $P(u,v) = uv$ becomes

$$vM_t[u] - uM_t^*[v] = \frac{d}{dt}[uv].$$

We try to find a solution of (4.33) in the form

$$y(z) = \int_{\alpha}^{\beta} e^{-zt} v(t)\, dt.$$

Using the method described in this section we have

$$\begin{aligned} L_z[y] &= \int_{\alpha}^{\beta} \left(t^2 - z\right) e^{-zt} v(t)\, dt \\ &= \int_{\alpha}^{\beta} v(t) M_t[e^{-zt}]\, dt \\ &= \int_{\alpha}^{\beta} \left\{ e^{-zt} M_t^*[v(t)] + \frac{\partial}{\partial t} e^{-zt} v(t) \right\} = 0. \end{aligned}$$

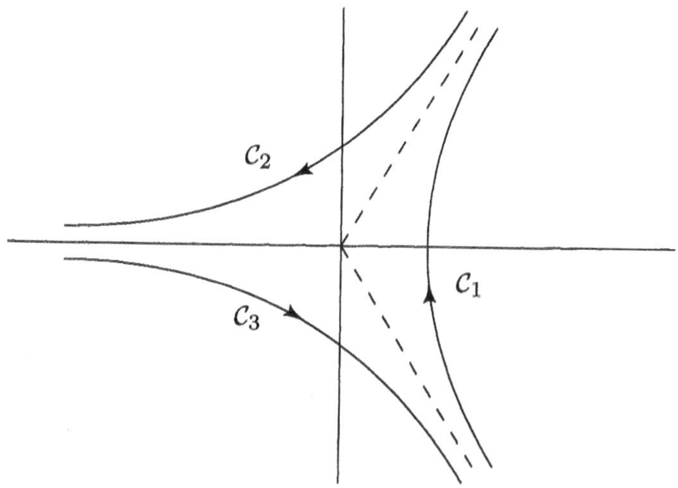

Figure 4.1. Three contours of integration for the Airy integrals in (4.34).

We look for a function v such that

$$M_t^*[v] = -v' + t^2 v = 0.$$

The solution is $v(t) = e^{\frac{1}{3}t^3}$. The values of α and β now follow from

$$e^{-zt+\frac{1}{3}t^3}\Big|_\alpha^\beta = 0.$$

Choosing contours of integration in the complex plane, it appears that we can take contours on which $\Re t^3 \to -\infty$ at the end points. We can select three such contours $\mathcal{C}_i$, which are shown in Figure 4.1. So we arrive at three solutions

$$y_i(z) = \int_{\mathcal{C}_i} e^{-zt+\frac{1}{3}t^3}\, dt. \tag{4.34}$$

These three solutions cannot be linearly independent. Integrating along $\mathcal{C} = \mathcal{C}_1 \cup \mathcal{C}_2 \cup \mathcal{C}_3$ in the shown direction, we easily see that

$$\int_{\mathcal{C}} e^{-zt+\frac{1}{3}t^3}\, dt = 0,$$

which gives $y_1(z) + y_2(z) + y_3(z) = 0$. Equation (4.33) defines the *Airy functions.* A standard notation is

$$\boxed{\mathrm{Ai}(z) = \frac{1}{2\pi i}\int_{\mathcal{C}_1} e^{-zt+\frac{1}{3}t^3}\, dt.} \tag{4.35}$$

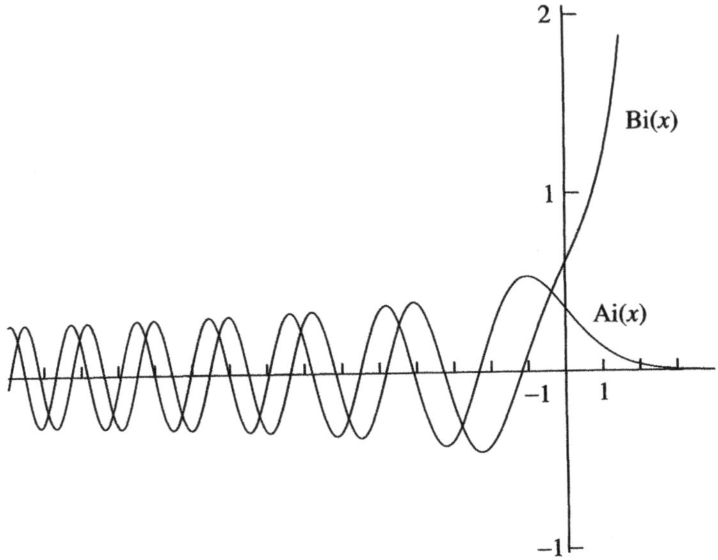

Figure 4.2. Graphs of the Airy functions $\mathrm{Ai}(x)$ and $\mathrm{Bi}(x)$, x real.

The function $\mathrm{Ai}(z)$ is an entire function. This follows from the integral representation (4.35) and also from the theory developed in this chapter: the differential equation in (4.33) does not have finite singular points. When $z = x$ is real $\mathrm{Ai}(z)$ is real. Then we have the real representation

$$\boxed{\mathrm{Ai}(x) = \frac{1}{\pi}\int_0^\infty \cos\left(\frac{1}{3}t^3 + xt\right)\,dt,} \tag{4.36}$$

which can be obtained by deforming contour $\mathcal{C}_1$ into the imaginary axis. A second real solution of (4.33) follows from $\mathrm{Bi}(z) = [(y_3(z) - y_2(z)]/(2\pi)$, which, for real argument, can be written as

$$\boxed{\mathrm{Bi}(x) = \frac{1}{\pi}\int_0^\infty e^{-\frac{1}{3}t^3 + xt}\,dt + \frac{1}{\pi}\int_0^\infty \sin\left(\frac{1}{3}t^3 + xt\right)\,dt.} \tag{4.37}$$

This result follows by deforming contour $\mathcal{C}_2$ into $(-\infty, 0]$ and the positive imaginary axis, and contour $\mathcal{C}_3$ into $(-\infty, 0]$ and the negative imaginary axis. In Figure 4.2 we show the graphs of the functions $\mathrm{Ai}(x), \mathrm{Bi}(x)$. More properties of Airy functions follow from Exercise 9.13.

4.5. The Liouville transformation

Consider the differential equation:

$$Y'' + p(x)Y' + q(x)Y = 0.$$

Substitute

$$Y(x) = W(x)\exp\left[-\frac{1}{2}\int^{x} p(\xi)\,d\xi\right]. \tag{4.38}$$

Then the function $W(x)$ satisfies the equation

$$W'' - Q(x)W = 0, \tag{4.39}$$

where

$$Q(x) = \frac{1}{4}p^2 + \frac{1}{2}p' - q.$$

When x in the final W–equation is transformed by $t(x)$, then, in general, a differential equation is obtained in which the first derivative is present. One has to apply a transformation like (4.38) to the new equation to remove the first derivative. The following transformation (the *Liouville transformation*) of both the dependent and independent variable directly transforms (4.39) into a form in which the first derivative is missing. Assuming that the third derivative of $t(x)$ exists in the considered x–domain, we write:

$$W(x) = \sqrt{\dot{x}}\,w(t), \quad t = t(x),$$

where $\dot{x} = dx/dt$. Then (4.39) becomes

$$\ddot{w} - \psi(t)w = 0, \tag{4.40}$$

where

$$\psi(t) = \dot{x}^2 Q(x) + \sqrt{\dot{x}}\,\frac{d^2}{d\tau^2}\frac{1}{\sqrt{\dot{x}}}. \tag{4.41}$$

The second term in $\psi(x)$ is often expressed in the form

$$\sqrt{\dot{x}}\,\frac{d^2}{d\tau^2}\frac{1}{\sqrt{\dot{x}}} = -\frac{1}{2}\{x, t\},$$

where $\{x, t\}$ is the *Schwarzian derivative*:

$$\{x, t\} = \frac{\dddot{x}}{\dot{x}} - \frac{3}{2}\left(\frac{\ddot{x}}{\dot{x}}\right)^2. \tag{4.42}$$

In OLVER (1974) this transformation is frequently used for obtaining the *Liouville-Green approximation* (also called the *WKB approximation* in relation with connection formulas) for the solutions of the differential equation.

4.6. Remarks and Comments for Further Reading

4.1. This chapter gives the theory of linear second order differential equations, as far as relevant for the theory of special functions. A more extensive introduction to differential equations can be found in, for instance, INCE (1956), BURKILL (1956) or CODDINGTON & LEVINSON (1955). More applications on the theory of regular singular points are given in the next chapter. In §5.5 theoretical aspects of the Riemann-Papperitz equation are considered.

4.2. The approach of §4.4 is taken from HOCHSTADT (1971).

4.7. Exercises and Further Examples

4.1. Show that the equation $\cosh z f'' + f = 0$ has a fundamental system $\{f_1, f_2\}$ given by the power series

$$f_1(z) = 1 - \frac{1}{2}z^2 + \frac{1}{12}z^4 - \frac{13}{720}z^6 + \ldots,$$

$$f_2(z) = z - \frac{1}{6}z^3 + \frac{1}{30}z^5 - \frac{13}{1680}z^7 + \ldots.$$

Compute the Wronskian for this pair and use this result to verify the correctness of the coefficients. Determine the radius of convergence of both power series.

4.2. Show that the *differential equation of Weber*

$$f'' = \left(\frac{1}{4}z^2 + a\right) f$$

has, for all $z \in \mathbb{C}$, two independent solutions

$$f_1(z) = \sum_{n=0}^{\infty} a_{2n} \frac{z^{2n}}{(2n)!}, \quad f_2(z) = \sum_{n=0}^{\infty} a_{2n+1} \frac{z^{2n+1}}{(2n+1)!},$$

with $a_0 = a_1 = 1$, $a_2 = a_3 = a$ and

$$a_{n+2} = a\, a_n + \frac{1}{4}n(n-1)a_{n-2}, \quad n \geq 2.$$

4.3. Determine the solutions of the indicial equation of the equation

$$z^2(z-1)f'' + (\tfrac{3}{2}z - 1)zf' + (z-1)f = 0$$

for the regular singular points $z = 0$ and $z = 1$. Determine in both cases the power series expansions for a fundamental system.

4.4. Show that inside the unit disc the equation

$$z(z-1)f'' + (2z-1)f' + \frac{1}{4}f = 0$$

has a fundamental system $\{f_1, f_2\}$, where

$$f_1(z) = \sum_{n=0}^{\infty} a_n z^n, \quad f_2(z) = f_1(z)\ln z + 4\sum_{n=1}^{\infty}[\psi(2n+1) - \psi(n+1)]a_n z^n,$$

and ψ is the logarithmic derivative of the gamma function (see §3.4) with

$$a_n = \frac{1^2 3^2 \cdots (2n-1)^2}{2^2 4^2 \cdots (2n)^2}.$$

4.5. Consider Hermite's differential equation (4.5) and determine two linearly independent solutions. Show that when $\nu = n \geq 0$ just one element of this pair is a polynomial of degree n. Denote this polynomial by $H_n(z)$, the *Hermite polynomial.* Normalize $H_n(z)$ such that the coefficient of z^n equals 2^n and verify the following special cases:

$$\begin{aligned} &H_0(z) = 1, \quad H_1(z) = 2z, \quad H_2(z) = 4z^2 - 2, \quad H_3(z) = 8z^3 - 12z, \\ &H_4(z) = 16z^4 - 48z^2 + 12, \quad H_5(z) = 32z^5 - 160z^3 + 120z. \end{aligned}$$

4.6. Consider Legendre's differential equation (4.4); $z = \infty$ is a regular singular point. Verify this. Determine two power series solutions in the neighborhood of $z = \infty$, and compare the results with expansions given in Chapter 8 (see, for instance, (8.7) and (8.8)).

4.7. Consider Kummer's differential equation (4.5) and verify that

$$f(z) = \frac{1}{2\pi i}\int_0^{(1+)} e^{zt} t^{a-1}(t-1)^{c-a-1}\,dt, \quad \Re a > 0$$

is a solution. The contour starts and terminates at $t = 0$ and encircles the point $t = 1$ in the positive sense. The many-valued functions of the integrand assume their principal branches; that is, $\arg t$, $\arg(t-1)$ are zero when $t > 0$, $t > 1$, respectively. The function $f(z)$ is connected with Kummer's function $M(a, c, z)$ introduced in Chapter 7. We have

$$M(a, c, z) = \frac{\Gamma(c)\Gamma(1+a-c)}{\Gamma(a)} f(z).$$

Verify this relation by integrating the integral defining $f(z)$ over $(0,1)$ (when $\Re(c-a)>0$) and by comparing the result with (7.7).

4.8. When one solution of a second order linear differential equation (4.1) is known, a second solution can be obtained in terms of this first solution. Consider (4.1) with $p=0$ (this can always be obtained by using a transformation given in (4.38)) and assume that we know a solution $u(z)$. Show that a second solution is given by

$$v(z)=u(z)\int^{z}\frac{d\zeta}{u^2(\zeta)}.$$

Are the solutions u, v linearly independent?

5

Hypergeometric Functions

Many special functions encountered in physics, engineering and probability theory are special cases of hypergeometric functions. In this chapter we give the main properties of the Gauss hypergeometric function. We shortly mention generalizations, also in connection with q-hypergeometric functions. The Legendre functions form a subclass of the Gauss functions, and will be discussed in Chapter 8. Several classical orthogonal polynomials, for instance, Jacobi polynomials, are hypergeometric functions; see Chapter 6.

Familiar examples of applications of hypergeometric functions to physics are given in Chapter 12 on elliptic integrals (where we treat the simple pendulum) and in the chapters on Legendre and Bessel functions (with examples from potential and diffraction theory). These functions naturally arise when separating the Helmholtz or Schrödinger equation. Another class of functions that arise in these problems are the confluent hypergeometric functions; see Chapter 7, where we also mention applications from quantum mechanics.

The central role in all this is played by the hypergeometric differential equation. We will show in §5.5 that any homogeneous linear differential equation of the second order with at most three regular singular points, can be transformed into the hypergeometric differential equation. Moreover, this equation yields in its (confluent) limiting form many other interesting functions of mathematical physics.

In GAUSS (1876), the collected works, Gauss's investigations on the hypergeometric functions can be found. In 1812 he presented the hypergeometric series to the Royal Society of Sciences at Göttingen.

5.1. Definitions and Simple Relations

The *Pochhammer symbol* or *shifted factorial* is defined by

$$(a)_n = a(a+1)(a+2)\dots(a+n-1), \quad n \geq 0, \tag{5.1}$$

with $(a)_0 = 1$. Hence

$$(a)_n = \Gamma(a+n)/\Gamma(a) \quad n \geq 0.$$

This has already been introduced in Exercise 3.1, and is very useful in the notation for hypergeometric functions.

The first steps are

$$(1-z)^{-1} = \sum_{n=0}^{\infty} z^n, \quad (1-z)^{-a} = \sum_{n=0}^{\infty} \binom{-a}{n} (-z)^n = \sum_{n=0}^{\infty} \frac{(a)_n}{n!} z^n, \tag{5.2}$$

which lead to the following generalization of the geometric series:

$$\boxed{\begin{aligned} F(a,b;c;z) &= \sum_{n=0}^{\infty} \frac{(a)_n (b)_n}{(c)_n \, n!} z^n \\ &= 1 + \frac{ab}{c} z + \frac{a(a+1)\, b(b+1)}{c(c+1)\, 2!} z^2 + \ldots \end{aligned} \quad , \quad |z| < 1.} \tag{5.3}$$

This is the *hypergeometric function*, which is named after Gauss.

Verify from (3.31) that the radius of convergence of (5.3) indeed equals unity.

In definition (5.3) a, b and c may assume all complex values with the exception $c = 0, -1, -2, \ldots$. However, it is easily shown that the function

$$\frac{1}{\Gamma(c)} F(a,b;c;z)$$

is an entire function in all three parameters a, b and c. For instance, we have

$$\begin{aligned} \lim_{c \to 0} \frac{1}{\Gamma(c)} F(a,b;c;z) &= abz \sum_{n=0}^{\infty} \frac{(a+1)_n (b+1)_n}{(n+1)!\, n!} z^n \\ &= abzF(a+1, b+1; 2; z). \end{aligned}$$

When a or b are non-positive integers the series in (5.3) terminates, and F reduces to a polynomial. We have (see Exercise 3.1 (1) for the interpretation of $(-m)_n$),

$$F(-m,b;c;z) = \sum_{n=0}^{m} \frac{(-m)_n \, (b)_n}{(c)_n \; n!} z^n = \sum_{n=0}^{m} (-1)^n \binom{m}{n} \frac{(b)_n}{(c)_n} z^n.$$

Several orthogonal polynomials can be expressed in terms of the Gauss hypergeometric function; see (6.35) for a relation with the Jacobi polynomials.

Example 5.1. It will not be difficult to verify the following special cases (always for $|z| < 1$):

$$
\begin{aligned}
F(a, b; b; z) &= (1-z)^{-a}, \\
F(1, 1; 2; z) &= -\frac{\ln(1-z)}{z}, \\
F\left(\frac{1}{2}, 1; \frac{3}{2}; z^2\right) &= \frac{1}{2z} \ln\left(\frac{1+z}{1-z}\right), \\
F\left(\frac{1}{2}, 1; \frac{3}{2}; -z^2\right) &= \frac{\arctan z}{z}, \\
F\left(\frac{1}{2}, \frac{1}{2}; \frac{3}{2}; z^2\right) &= \frac{\arcsin z}{z}, \\
F\left(\frac{1}{2}, \frac{1}{2}; \frac{3}{2}; -z^2\right) &= \frac{\ln\left(z + \sqrt{1+z^2}\right)}{z}.
\end{aligned}
$$

We observe that in all these examples the F–functions become singular when their arguments assume the value 1. In general, the point $z = 1$ is an algebraic or logarithmic singularity of $F(a, b; c; z)$, and the F–function is many-valued due to this singularity. Inside the unit disc, where (5.3) defines an analytic function, no problems arise in this connection. Outside the unit disc we need a cut in the complex plane to define the principal branch of the F–function. We take the cut from $+1$ to $+\infty$.

Remark 5.1. As mentioned above, the definition of F breaks down when $c = 0, -1, -2, \ldots$. However, when a or b are also equal to a non-positive integer the definition may have a meaning. Let $a = -m$ and $c = -m - k$, with k, m non-negative integers. When $k = 0$ (that is, $a = c$), F reduces to $(1-z)^{-b}$; see (5.2). When $k > 0$, F reduces to a polynomial:

$$
F(-m, b; -m-k; z) = \sum_{n=0}^{m} \binom{m}{n} \frac{(b)_n\,(m+k-n)!}{(m+k)!} z^n.
$$

When we take $k = 0$ in this result, we obtain $\sum_{n=0}^{m} (b)_n z^n/n!$, the first part of the power series of $(1-z)^{-b}$ (again, see (5.2)). Hence, one should be careful in the interpretation of the F–function when the parameters c and a (or b) assume negative integer values. The answer may depend on whether c and a (or b) are independent or not; see also Remark 5.2.

5.2. Analytic Continuation

We are concerned with the analytic continuation of F outside the unit disc, and we want to know the nature of the singularities of F. In the above examples we have seen algebraic and logarithmic singularities. The following *integral representation of the hypergeometric function*, due to Euler, is an important tool for deriving numerous properties of F. We have (EULER (1748))

$$F(a,b;c;z) = \frac{\Gamma(c)}{\Gamma(b)\Gamma(c-b)} \int_0^1 t^{b-1}(1-t)^{c-b-1}(1-tz)^{-a}\,dt, \tag{5.4}$$

where

$$\Re c > \Re b > 0, \quad |\arg(1-z)| < \pi.$$

To prove this we use the second series in (5.2) and the beta integral given in (3.2) and (3.3):

$$\begin{aligned} F(a,b;c;z) &= \frac{\Gamma(c)}{\Gamma(b)} \sum_{n=0}^{\infty} z^n \frac{(a)_n}{n!} \frac{\Gamma(b+n)}{\Gamma(c+n)} \\ &= \frac{\Gamma(c)}{\Gamma(b)\Gamma(c-b)} \sum_{n=0}^{\infty} z^n \frac{(a)_n}{n!} \int_0^1 t^{b+n-1}(1-t)^{c-b-1}\,dt. \end{aligned}$$

Since for $\Re b > 1$, $\Re(c-b) > 1$ and $|z| < 1$ the series

$$\sum_{n=0}^{\infty} U_n(t), \quad U_n(t) = z^n \frac{(a)_n}{n!} t^{b+n-1}(1-t)^{c-b-1}$$

converges uniformly with respect to $t \in [0,1]$, we are able to interchange the order of integration and summation for these values of b, c and z. Furthermore, observe that the right-hand side of (5.4) is defined for all complex values of z, with the exception of the interval $[1,\infty)$. According to Theorem 2.3, the integral is an analytic function of z in $\mathbb{C} \setminus [1,\infty)$. Next we apply analytic continuation with respect to b, c and z in order to arrive at the conditions announced after (5.4). Hence we have obtained the analytic continuation of F, *qua* function of z, outside the unit disc. It appears that the point $z = 1$ is the only finite regular singular point of F.

5.2.1. Three Functional Relations

The hypergeometric function satisfies a great number of relations, of which a few simple examples will now be given. First we observe that (5.3) is symmetric in a and b, giving $F(a,b;c;z) = F(b,a;c;z)$. Furthermore we have

$$\begin{aligned} F(a,b;c;z) &= (1-z)^{-a} F\left(a, c-b; c; \frac{z}{z-1}\right) \\ &= (1-z)^{-b} F\left(c-a, b; c; \frac{z}{z-1}\right) \\ &= (1-z)^{c-a-b} F(c-a, c-b; c; z). \end{aligned} \tag{5.5}$$

The proof follows by substituting $t = 1 - s$ in (5.4). Then we obtain

$$\int_0^1 s^{c-b-1}(1-s)^{b-1}[(1-z)+sz]^{-a}\,ds$$
$$= (1-z)^{-a}\int_0^1 s^{c-b-1}(1-s)^{b-1}\left(1-\frac{sz}{z-1}\right)^{-a}\,ds$$

giving the first relation. The second one follows from the first one and from the symmetry in $F(a,b;c;z)$ with respect to a and b. The third relation in (5.5) follows from using the first or second relation twice.

Although, in general, $z = 1$ is a regular singular point of F, under certain conditions the limit for $z = x \to 1$, $x > 0$ may exist. From (5.4) it follows

$$\lim_{x\uparrow 1} F(a,b;c;x) = \frac{\Gamma(c)\Gamma(c-a-b)}{\Gamma(c-a)\Gamma(c-b)}, \tag{5.6}$$

which holds when $\Re(c-a-b) > 0$. This condition is also sufficient for the convergence at $z = 1$ of the series in (5.3). The important result (5.6) is due to Gauss.

Remark 5.2. The relations in (5.5) are derived from (5.4), and are valid under the condition given after (5.4). For instance, it is not difficult to verify that the second of (5.5) does not hold when $a = -1, c = -2$ (see Remark 5.1). We have $F(-1,b;-2;z) = 1 + \frac{1}{2}bz$, whereas the second line of (5.5) gives a wrong answer.

5.2.2. A Contour Integral Representation

A more general integral is the loop integral defined by

$$F(a,b;c;z) = \frac{\Gamma(c)\Gamma(1+b-c)}{2\pi i\,\Gamma(b)}\int_0^{(1+)} t^{b-1}(t-1)^{c-b-1}(1-tz)^{-a}\,dt,\ \Re b > 0,$$

where the contour starts and terminates at $t = 0$ and encircles the point $t = 1$ in the positive direction. The point $1/z$ should be outside the contour. The many-valued functions of the integrand assume their principal branches: $\arg(1-tz)$ tends to zero when $z \to 0$, and $\arg t$, $\arg(t-1)$ are zero at the point where the contour cuts the real positive axis (at the right of 1). Observe that no condition on c is needed, whereas in (5.4) we need $\Re(c-b) > 0$. The proof of the above representations runs as for (5.4), with the help of the loop integral for the beta function; see Exercise 3.13.

5.3. The Hypergeometric Differential Equation

To derive the differential equation for the hypergeometric function it is convenient to introduce the differential operator $\vartheta = z\,d/dz$. We have $\vartheta z^\mu = \mu z^\mu$. We observe that $\vartheta(\vartheta + c - 1)z^n = n(n + c - 1)z^n$. Hence

$$\begin{aligned}\vartheta(\vartheta + c - 1)F(a, b; c; z) &= \sum_{n=1}^{\infty} n(n + c - 1)\frac{(a)_n\,(b)_n}{(c)_n\,n!}z^n \\ &= \sum_{n=0}^{\infty} \frac{(a)_n\,(b)_n}{(c)_n\,n!}(a + n)(b + n)z^{n+1} \\ &= z(\vartheta + a)(\vartheta + b)F(a, b; c; z).\end{aligned}$$

It follows that F satisfies the differential equation

$$\vartheta(\vartheta + c - 1)F = z(\vartheta + a)(\vartheta + b)F. \tag{5.7}$$

In explicit form (5.7) reads

$$\boxed{z(1 - z)F'' + [(c - (a + b + 1)z]F' - abF = 0,} \tag{5.8}$$

the *hypergeometric differential equation*, which was given by Gauss.

With equal ease we can show that a second solution of (5.7) or (5.8) is of the form $z^{1-c}G$, where again G is a hypergeometric function. Indeed, the substitution $F = z^{1-c}G$ in (5.7) on the one hand gives the result

$$\vartheta(\vartheta + c - 1)z^{1-c}G = z^{1-c}(\vartheta + 1 - c)\vartheta G,$$

and on the other

$$(\vartheta + a)(\vartheta + b)z^{1-c}G = z^{1-c}(\vartheta + a - c + 1)(\vartheta + b - c + 1)G.$$

Hence

$$\vartheta(\vartheta + 1 - c)G = z(\vartheta + a - c + 1)(\vartheta + b - c + 1)G.$$

But this is nothing other than a reparameterization of the hypergeometric differential equation, of which $F(a - c + 1, b - c + 1; 2 - c; z)$ is a solution. It follows that, in addition to $F(a, b; c; z)$, a second solution of (5.7) or (5.8) is given by $z^{1-c}F(a - c + 1, b - c + 1; 2 - c; z)$. When $c = 1$ this does not yield a new solution, but, in general, the second solution of (5.8) appears to be of the form

$$PF(a, b; c; z) + Qz^{1-c}F(a - c + 1, b - c + 1; 2 - c; z), \tag{5.9}$$

where P and Q are independent of z.

Next we observe that with the help of (5.8) and (5.9) we can express a hypergeometric function with argument $1-z$ or $1/z$ in terms of functions with argument z. For example, when in (5.8) we introduce the new variable $z' = 1 - z$ we obtain a hypergeometric differential equation, but now with parameters a, b and $a+b-c+1$. Hence, besides the solutions in (5.9) we have $F(a,b;a+b-c+1;1-z)$ as a solution as well. It follows that we can find numbers P and Q, which do not depend on z, such that

$$F(a,b;a+b-c+1;1-z) = PF(a,b;c;z) \\ + Qz^{1-c}F(a-c+1,b-c+1;2-c;z).$$

To find P and Q we substitute $z=0$ and $z=1$ and use (5.6), under the conditions $\Re(c-a-b)>0, \Re c<1$, which relations can be relaxed by using analytic continuation with respect to the parameters. Instead of giving the values of P and Q, we observe that it is more convenient to write the above relation in a form that has the function $F(a,b;c;z)$ at the left-hand side. In addition, we observe that the relations in (5.5) can be used to obtain more relations. The following list is the result of such manipulations. Let

$$A = \frac{\Gamma(c)\Gamma(c-a-b)}{\Gamma(c-a)\Gamma(c-b)}, \qquad B = \frac{\Gamma(c)\Gamma(a+b-c)}{\Gamma(a)\Gamma(b)},$$
$$C = \frac{\Gamma(c)\Gamma(b-a)}{\Gamma(b)\Gamma(c-a)}, \qquad D = \frac{\Gamma(c)\Gamma(a-b)}{\Gamma(a)\Gamma(c-b)}.$$

Then

$$\begin{aligned} F(a,b;c;z) &= A\,F(a,b;a+b-c+1;1-z) \\ &\quad + B\,(1-z)^{c-a-b}\,F(c-a,c-b;c-a-b+1;1-z) \qquad (5.10) \\ &= C\,(-z)^{-a}\,F(a,1-c+a;1-b+a;1/z) \\ &\quad + D\,(-z)^{-b}\,F(b,1-c+b;1-a+b;1/z) \qquad (5.11) \\ &= C\,(1-z)^{-a}\,F[a,c-b;a-b+1;1/(1-z)] \\ &\quad + D\,(1-z)^{-b}\,F[b,c-a;b-a+1;1/(1-z)] \qquad (5.12) \\ &= A\,z^{-a}\,F(a,a-c+1;a+b-c+1;1-1/z) \\ &\quad + B\,z^{a-c}(1-z)^{c-a-b}\,F(c-a,1-a;c-a-b+1;1-1/z). \qquad (5.13) \end{aligned}$$

There are restrictions on the phases of z or $1-z$.

(5.10) holds when $|\arg(1-z)|<\pi$,
(5.11) holds when $|\arg(-z)|<\pi$,
(5.12) holds when $|\arg(1-z)|<\pi$,
(5.13) holds when $|\arg(1-z)|<\pi$ and $|\arg z|<\pi$.

The relation (5.11) yields for $F(a,b;c;z)$ a representation with convergent series expansion when $|z| > 1$; that is, (5.11) can be viewed as an asymptotic representation for large values of $|z|$. On the other hand, the relations in (5.5) and (5.10) – (5.13) also supply us with interesting formulas for the numerical evaluation of the F–functions. For instance, when $\frac{1}{2} < z < 1$, then the convergence of (5.3) is not so good. In that case (5.10) gives a way out. Also for complex values of z we can often find a representation for $F(a,b;c;z)$ in terms of hypergeometric functions with argument w, such that $|w| \le \frac{1}{2}$; see also Remark 5.6 in §5.10.

Something goes wrong in the above formulas when $-c = 0, 1, 2, \ldots$. As explained in §5.1, this is inherent in the definition of the F–function. A different source of trouble is more interesting and arises, for instance, when in (5.10), $c = a + b + m$, $m \in \mathbb{Z}$. Then A or B become undefined, whereas the left-hand side remains defined. In fact, one or more terms in the series of the right-hand side have to control this behavior of A and B, and have to remove the singularities. Assume, for instance, that $m = 1$; then B of (5.10) (the term multiplying the second series) is not defined. In the first F–function on the right-hand side a zero value occurs at the c–location. Introducing a limit process $c = a + b + 1 + \varepsilon$, $\varepsilon \to 0$ we can still define (5.10). We give the result of this limit for (5.10), when $c = a + b$:

$$
\begin{aligned}
F(a,b;a+b;z) &= \frac{\Gamma(a+b)}{\Gamma(a)\Gamma(b)} \sum_{n=0}^{\infty} \frac{(a)_n\,(b)_n}{n!\;n!} C_n (1-z)^n, \\
C_n &= 2\psi(n+1) - \psi(a+n) - \psi(b+n) - \ln(1-z),
\end{aligned}
\tag{5.14}
$$

with the conditions $|1-z| < 1$, $|\arg(1-z)| < \pi$. This result can be used to compute the *elliptic integral* $K(k)$ presented in Exercise 5.2 and in §12.1.

In the remaining relations similar removable singularities occur. As in (5.14) logarithmic terms always arise. For example, in (5.11) a logarithmic term appears when $b - a = 0, 1, 2, \ldots$. More formulas of this form can be found in the literature, for instance, in Bateman Project, (1953, Vol. I) or Abramowitz & Stegun (1964, p. 559 – 560).

As remarked in §5.1 the F–function reduces to a polynomial when a or b assume non-positive integer values. When $c - a$ or $c - b$ equal non-positive integer values it follows from (5.5) that $F(a,b;c;z)$ reduces to a polynomial multiplied by an algebraic factor of the form $(1-z)^p$.

5.4. The Singular Points of the Differential Equation

The appearance of logarithmic terms as in (5.14) can be explained further with the help of the theory of Chapter 4. The fact is that the points 0, 1 and ∞

are three regular singular points for the hypergeometric differential equation (5.8). If we consider the indicial equation (4.16) for these three points, we arrive at the scheme

$$\begin{array}{lll} z=0 & \mu_1=0 & \mu_2=1-c \\ z=1 & \mu_1=0 & \mu_2=c-a-b \\ z=\infty & \mu_1=a & \mu_2=b \end{array}$$

It follows that when none of the numbers c, $c-a-b$, $a-b$ assume integer values, the difference of the indices $\mu_1-\mu_2$ can never assume integer values. Under these circumstances the approach of §4.4 yields two power series expansions for a fundamental system, which, obviously, for the present case can be expressed in terms of F–functions. We have the following sets of fundamental pairs (still under the condition that c, $c-a-b$, $a-b$ do not assume integer values):

$$\begin{array}{lll} z=0 & f_1(z)= & F(a,b;c;z) \\ & f_2(z)= & z^{1-c}F(a-c+1,b-c+1;2-c;z) \end{array}$$

$$\begin{array}{lll} z=1 & f_1(z)= & F(a,b;a+b+1-c;1-z) \\ & f_2(z)= & (1-z)^{c-a-b}F(c-b,c-a;c-a-b+1;1-z) \end{array}$$

$$\begin{array}{lll} z=\infty & f_1(z)= & z^{-a}F\left(a,a-c+1;a-b+1;z^{-1}\right) \\ & f_2(z)= & z^{-b}F\left(b,b-c+1;b-a+1;z^{-1}\right) \end{array}$$

These six solutions of the hypergeometric differential equation can be transformed through the three relations in (5.5). This gives a total number of 24 solutions, the basic forms, already given by Kummer in 1836. The formulas (5.10) – (5.13) can be used for analytic continuation of these solutions.

When one of the numbers a, b, $c-a$, $c-b$ equals a negative integer, then at least one of these 24 solutions is of the form $z^{\alpha}(1-z)^{\beta}p(z)$, where $p(z)$ is a polynomial.

When, for one of the regular singular points, the difference of the two indices $\mu_1-\mu_2$ equals an integer, then the corresponding fundamental system has one member in which logarithmic terms occur. This is in full agreement with the theory of the previous chapter. We give a fundamental system for $c=1$:

$$f_1(z)=F(a,b;1;z),$$

$$f_2(z)=F(a,b;1;z)\ln z+\sum_{n=1}^{\infty}\frac{(a)_n(b)_n}{n!\;n!}C_n z^n,$$

$$C_n=\psi(a+n)-\psi(a)+\psi(b+n)-\psi(b)-2\psi(n+1)+2\psi(1).$$

A complete list of other examples will not be given here. See BATEMAN PROJECT, Vol. I (1953) or ABRAMOWITZ & STEGUN (1964, p. 564), for more information.

5.5. The Riemann-Papperitz Equation

The importance of the differential equation (5.8) comes, among other reasons, from the following theorem.

Theorem 5.1. *Any homogeneous linear differential equation of the second order with at most three singularities (inclusive perhaps of the point at infinity), which are regular singular points, can be transformed into the hypergeometric differential equation (5.8).*

Proof. First we consider the equation

$$\frac{d^2 f}{dz^2} + p(z)\frac{df}{dz} + q(z)f = 0, \tag{5.15}$$

of which we assume that the three finite points ξ, η and ζ are regular singular points. The indices corresponding with these points are denoted by the pairs (α_1, α_2), (β_1, β_2) and (γ_1, γ_2). We have implicitly assumed in the formulation of the theorem that the only singularities of p and q (perhaps the point at infinity) are poles. On account of a theorem from the theory of functions (see, for instance, COPSON (1935, §5.56)) we conclude that p and q are rational functions. It follows that

$$p(z) = \frac{P(z)}{(z-\xi)(z-\eta)(z-\zeta)}, \quad q(z) = \frac{Q(z)}{(z-\xi)^2(z-\eta)^2(z-\zeta)^2},$$

where P and Q are polynomials. Since we have assumed that the point at infinity is a regular point, the functions $2z - z^2 p(z)$ and $z^4 q(z)$ should be analytic at ∞ (this follows from the fact that P and Q of (4.7) have to be analytic at the regular point 0). From this we infer that P and Q are polynomials of degree 2 or lower, and that the coefficient of z^2 in P equals 2. Hence

$$p(z) = \frac{A}{z-\xi} + \frac{B}{z-\eta} + \frac{C}{z-\zeta}$$

and

$$(z-\xi)(z-\eta)(z-\zeta)q(z) = \frac{D}{z-\xi} + \frac{E}{z-\eta} + \frac{F}{z-\zeta},$$

with

$$A + B + C = 2. \tag{5.16}$$

The numbers A, B, C, D, E and F, of course, depend on the indices of the singular points. From the indicial equation of the point ξ (see (4.16))

$$\mu(\mu-1)+A\mu+\frac{D}{(\xi-\eta)(\xi-\zeta)}=0$$

it follows that

$$A=1-\alpha_1-\alpha_2,\quad D=(\xi-\eta)(\xi-\zeta)\alpha_1\alpha_2.$$

For the remaining points $z=\eta$ and $z=\zeta$ we obtain in the same way

$$B=1-\beta_1-\beta_2,\quad E=(\eta-\zeta)(\eta-\xi)\beta_1\beta_2,$$
$$C=1-\gamma_1-\gamma_2,\quad F=(\zeta-\xi)(\zeta-\eta)\gamma_1\gamma_2.$$

From (5.16) it follows that the indices cannot be chosen arbitrarily. They have to satisfy

$$\alpha_1+\alpha_2+\beta_1+\beta_2+\gamma_1+\gamma_2=1. \tag{5.17}$$

With all these relations, we find for equation (5.15) the form

$$\begin{aligned} f''&+\left(\frac{1-\alpha_1-\alpha_2}{z-\xi}+\frac{1-\beta_1-\beta_2}{z-\eta}+\frac{1-\gamma_1-\gamma_2}{z-\zeta}\right)f' \\ &-\left[\frac{\alpha_1\alpha_2}{(z-\xi)(\eta-\zeta)}+\frac{\beta_1\beta_2}{(z-\eta)(\zeta-\xi)}+\frac{\gamma_1\gamma_2}{(z-\zeta)(\xi-\eta)}\right] \\ &\times\frac{(\xi-\eta)(\eta-\zeta)(\zeta-\xi)}{(z-\xi)(z-\eta)(z-\zeta)}f=0. \end{aligned} \tag{5.18}$$

Next we introduce the following transformations:

$$t=\frac{(\zeta-\eta)(z-\xi)}{(\zeta-\xi)(z-\eta)},\quad F=t^{-\alpha_1}(1-t)^{-\gamma_1}f. \tag{5.19}$$

The first transformation is a *fractional linear transformation* (sometimes called a *bilinear transformation*), which maps the extended z–plane one-to-one to the extended t–plane. The differential equation in the new variables F and t again is of the second order and linear. The only singularities are the points which correspond with $z=\xi,\eta$ and ζ. This means, the points $t=0,\infty$ and 1, respectively. These singularities are regular, as can be easily verified. From the second relation in (5.19) it follows that the indices in these points are

$$(0,\alpha_2-\alpha_1),\quad(\alpha_1+\beta_1+\gamma_1,\alpha_1+\beta_2+\gamma_1),\quad(0,\gamma_2-\gamma_1),$$

respectively. The transformations (5.19) lead to the differential equation

$$F''+\left(\frac{1-\alpha_2+\alpha_1}{t}+\frac{1-\gamma_2+\gamma_1}{t-1}\right)F'+\frac{(\alpha_1+\beta_1+c_1)(\alpha_1+\beta_2+\gamma_1)}{t(t-1)}F=0. \tag{5.20}$$

Finally, from (5.17) it follows that (5.20) has the form (5.8) with

$$a=\alpha_1+\beta_1+\gamma_1,\quad b=\alpha_1+\beta_2+\gamma_1,\quad c=1+\alpha_1-\alpha_2.$$

This proves the theorem. ■

Equation (5.18) was first given by PAPPERITZ (1885) and is called the *Riemann-Papperitz equation.* In a notation due to RIEMANN (1857) we write

$$f = \mathcal{P}\left\{\begin{matrix} \xi & \eta & \zeta & \\ \alpha_1 & \beta_1 & \gamma_1 & z \\ \alpha_2 & \beta_2 & \gamma_2 & \end{matrix}\right\}. \tag{5.21}$$

The singularities occur in the first row; their ordering is not significant. The corresponding indices are in the second and third row.

The theorem does not cover the case in which $z = \infty$ is one of the three regular singular points. However, it can easily be verified that, when we have two finite regular singular points at ξ and η and one at ∞, the differential equation (5.18) takes the form

$$\begin{aligned} f'' &+ \left(\frac{1-\alpha_1-\alpha_2}{z-\xi} + \frac{1-\beta_1-\beta_2}{z-\eta}\right) f' \\ &+ \left[\frac{\alpha_1\alpha_2(\xi-\eta)}{z-\xi} - \frac{\beta_1\beta_2(\xi-\eta)}{z-\eta} + \gamma_1\gamma_2\right] \frac{f}{(z-\xi)(z-\eta)} = 0. \end{aligned} \tag{5.22}$$

This equation is the limit of (5.18) as $\zeta \to \infty$ and it has the scheme (5.21) with ζ replaced by ∞. Equation (5.22) can also be transformed into (5.8).

In Riemann's notation the hypergeometric differential equation (5.8) can be represented by the scheme

$$f = \mathcal{P}\left\{\begin{matrix} 0 & 1 & \infty & \\ 0 & 0 & a & z \\ 1-c & c-a-b & b & \end{matrix}\right\}. \tag{5.23}$$

We give another example of how to compute with this scheme. The transformation $f = (1-z)^\rho g$, with f satisfying (5.23), gives for g a differential equation of which the regular singular points are the same as those for f, but with different indices. It is obvious that the indices at the point $z = 0$ remain the same. At the point $z = 1$ they are lowered by the quantity ρ and at the point at infinity they are raised by the same quantity. The function g is not necessarily of hypergeometric type, since in general both the indices at $z = 1$ are different from zero. Choosing, however, $\rho = c - a - b$ then we have for g the scheme

$$g = \mathcal{P}\left\{\begin{matrix} 0 & 1 & \infty & \\ 0 & a+b-c & c-b & z \\ 1-c & 0 & c-a & \end{matrix}\right\}.$$

Since the indices for $z = 1$ can be interchanged, it follows that this scheme indeed corresponds to that for a hypergeometric function. The solution that

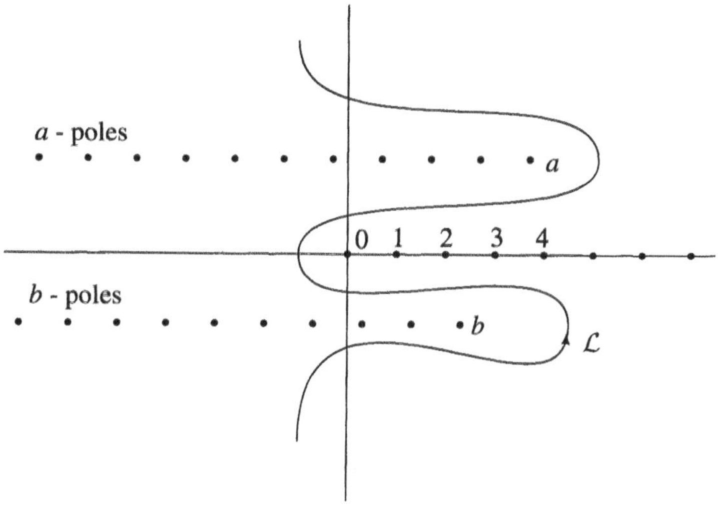

Figure 5.1. Possible contour of integration for (5.24) when $a = 3.7 + 2i$, $b = 2.3 - 1.5i$.

is regular at the origin is $g = F(c-a, c-b; c; z)$. And again we arrive at the third line of (5.5).

5.6. Barnes' Contour Integral for *F(a,b;c;z)*

We consider the integral (Barnes (1908))

$$\Phi(a,b;c;z) = \frac{1}{2\pi i}\frac{\Gamma(c)}{\Gamma(a)\Gamma(b)}\int_{\mathcal{L}} \frac{\Gamma(a+s)\Gamma(b+s)\Gamma(-s)}{\Gamma(c+s)} z^s\, ds, \tag{5.24}$$

where $\mathcal{L}$ runs from $-i\infty$ to $+i\infty$, such that $\mathcal{L}$ separates the poles of $\Gamma(a+s)$, $\Gamma(b+s)$ (at $s = -a-n$, $s = -b-m$, with $m, n = 0, 1, 2, \ldots$) from those of $\Gamma(-s)$ at $s = 0, 1, 2, \ldots$. We assume that a, b, c are different from $0, -1, -2, \ldots$ and that $|\arg z| < \pi$. In general, the contour cannot be a vertical line, but a contour that meanders around the poles of the gamma functions according to the description above. For an orientation one may consider $a = 3.7 + 2i$, $b = 2.3 - 1.5i$. In that case one can take $\mathcal{L}$ to be the contour as shown in Figure 5.1.

From (3.31) and (3.5) it follows that the integrand has the estimate

$$\mathcal{O}\left[|s|^{a+b-c-1} e^{-\arg(z)\Im s - \pi|\Im s|}\right], \quad s \to \infty, \quad s \in \mathcal{L}$$

Hence, according to Theorem 2.3, we know that $\Phi(a,b;c;z)$ is an analytic function of z in the domain described by $|\arg z| < \pi$.

Now use (3.5) and consider

$$\Phi_N(a, b; c; z) = \frac{1}{2\pi i} \frac{\Gamma(c)}{\Gamma(a)\Gamma(b)} \int_{C_N} \frac{\Gamma(a+s)\Gamma(b+s)}{\Gamma(c+s)\Gamma(1+s)} \frac{\pi z^s}{\sin s\pi} \, ds,$$

where C_N is the semi-circle with radius $N + \frac{1}{2}$ at the right of the imaginary axis. From (3.31) we know that

$$\frac{\Gamma(a+s)\Gamma(b+s)}{\Gamma(c+s)\Gamma(1+s)} \frac{\pi z^s}{\sin s\pi} = \mathcal{O}\left(N^{a+b-c-1}\right) \frac{z^s}{\sin s\pi},$$

as $N \to \infty$, for all values of $\arg s \in [-\pi/2, \pi/2]$. Writing $s = (N + \frac{1}{2})e^{i\theta}$ and taking $|z| < 1$, with $|\arg z| \le \pi - \delta$, we then have

$$\frac{z^s}{\sin s\pi} = \begin{cases} \mathcal{O}\left[e^{(N+\frac{1}{2}) \ln |z|/\sqrt{2}}\right], & \text{if } 0 \le |\theta| \le \frac{1}{4}\pi; \\ \mathcal{O}\left[e^{-\delta(N+\frac{1}{2})/\sqrt{2}}\right], & \text{if } \frac{1}{4}\pi \le |\theta| \le \frac{1}{2}\pi. \end{cases}$$

Hence if $\ln |z|$ is negative ($|z| < 1$) then the integrand tends to zero sufficiently rapidly to ensure that $\int_{C_N} \to 0$, as $N \to \infty$. Applying the method of residues inside a contour consisting of $\mathcal{L}$ and C_N (integrating in negative, that is, clockwise direction), where for $s = n = 0, 1, 2, \ldots$ the poles of $1/\sin s\pi$ are located with residues $(-1)^n$, yields

$$\lim_{N\to\infty} \Phi_N(a, b; c; z) = \Phi(a, b; c; z) = F(a, b; c; -z).$$

The condition $|z| < 1$, $|\arg z| < \pi$ can be replaced by $|\arg z| < \pi$ by invoking the principle of analytic continuation.

Using a similar technique we can take a semi-circle on the left of the imaginary axis. Then we can take into account the residues of the poles of the functions $\Gamma(a+s)$ and $\Gamma(b+s)$. When the poles of the gamma functions do not coincide, two series of residues arise. These series have negative powers of z and can be written as $F-$ functions:

$$\begin{aligned} F(a, b; c; -z) &= z^{-a} \frac{\Gamma(c)\Gamma(b-a)}{\Gamma(b)\Gamma(c-a)} F\left(a, 1-c+a; 1-b+a; -\frac{1}{z}\right) \\ &+ z^{-b} \frac{\Gamma(c)\Gamma(a-b)}{\Gamma(a)\Gamma(c-b)} F\left(b, 1-c+b; 1-a+b; -\frac{1}{z}\right). \end{aligned}$$

This corresponds to (5.11). When the poles of the gamma functions coincide (this happens when $b - a$ equals an integer), then poles of the second order occur, of which the residues contain logarithmic terms in z. In this way we become acquainted for the third time with the logarithmic terms in the representations of the F-functions.

Contour integrals, as in (5.24), play an important part in the theory of hypergeometric functions, and also in the theory of generalized hypergeometric functions. Integrals of the type (5.24) are called *Mellin-Barnes integrals.* One can interpret (5.24) in terms of the inversion formula of the *Mellin transform.* We recall the Mellin transformation pair (SNEDDON (1972))

$$g(z) = \int_0^\infty t^{z-1} f(t)\, dt, \quad f(t) = \frac{1}{2\pi i}\int_{c-i\infty}^{c+i\infty} t^{-z} g(z)\, dz,$$

and obtain

$$\int_0^\infty F\left(a, b; c; -\frac{1}{x}\right) x^{s-1}\, dx = \frac{\Gamma(a+s)\Gamma(b+s)\Gamma(-s)\Gamma(c)}{\Gamma(a)\Gamma(b)\Gamma(c+s)}.$$

This can also be written in the form

$$\int_0^\infty F\left(a, b; c; -z\right) z^{s-1} dz = \frac{\Gamma(a-s)\,\Gamma(b-s)\,\Gamma(s)\,\Gamma(c)}{\Gamma(a)\,\Gamma(b)\,\Gamma(c-s)}. \tag{5.25}$$

This result holds for $\Re s > 0, \Re(a-s) > 0, \Re(b-s) > 0$. This follows from the behavior of F at $z = 0$ and at $z = -\infty$; see (5.11).

5.7. Recurrence Relations

The six functions

$$F\left(a \pm 1, b; c; z\right), \quad F\left(a, b \pm 1; c; z\right), \quad F\left(a, b; c \pm 1; z\right)$$

are called neighbors of $F(a, b; c; z)$. We use the notation F for $F(a, b; c; z)$ and $F(a+), F(a-)$ to denote the F-function with a replaced by $a+1, a-1$, respectively, and so on. Gauss proved that there exists a linear relation between F and two of its neighbors. The coefficients are linear functions of z. There are fifteen of such relations, also called *contiguous relations.* Only four of them are independent, since the other ones can be obtained from these four by elimination and since F is symmetric with respect to a and b. Thanks to this symmetry it is sufficient to have available only nine of the fifteen contiguous relations. These nine are as follows.

$$
\begin{aligned}
&(c-a)F(a-)+(2a-c-az+bz)F+a(z-1)F(a+)=0,\\
&c(c-1)(z-1)F(c-)+c[c-1-(2c-a-b-1)z]F\\
&\qquad\qquad\qquad +(c-a)(c-b)zF(c+)=0,\\
&c[a+(b-c)z]F-ac(1-z)F(a+)+(c-a)(c-b)zF(c+)=0,\\
&c(1-z)F-cF(a-)+(c-b)zF(c+)=0,\\
&(b-a)F+aF(a+)-bF(b+)=0,\\
&(c-a-b)F+a(1-z)F(a+)-(c-b)F(b-)=0,\\
&(c-a-1)F+aF(a+)-(c-1)F(c-)=0,\\
&(b-a)(1-z)F-(c-a)F(a-)+(c-b)F(b-)=0,\\
&[a-1+(b+1-c)z]F+(c-a)F(a-)-(c-1)(1-z)F(c-)=0.
\end{aligned}
$$

As always, one needs to be careful when $c = 0, -1, -2, \ldots$; see §5.1. For instance, from the second recurrence relation it does not follow that $F(a, b; 1; z) = 0$. The above relations can be systematically verified by expanding the hypergeometric functions in power series and showing that the coefficients of all z−powers vanish identically. Computer algebra is, as always, a very useful tool in this method. Other proofs can be based on integrating by parts in the integral (5.4).

5.8. Quadratic Transformations

In §5.4 we mentioned Kummer's 24 solutions of the hypergeometric differential equation. They form the complete set of bilinear transformations (that is, of the form $(\alpha+\beta z)/(\gamma+\delta z)$), with which the hypergeometric equation can be transformed into another equation of hypergeometric type. GOURSAT (1881) made a thorough study of another kind of transformation, although the basic ideas are due to Gauss and Kummer. We will discuss a few examples.

We introduce the following transformations:

$$z = \frac{4\zeta}{(1+\zeta)^2}, \qquad f = (1+\zeta)^{-2a} g.$$

Then (5.8) can be cast in the form

$$
\begin{aligned}
\zeta(1-\zeta)^2\frac{d^2g}{d\zeta^2} &+ \left[c-(4b-2c)\zeta+(c-4a-2)\zeta^2\right]\frac{dg}{d\zeta}\\
&- 2a[2b-c+(2a-c+1)\zeta]g = 0.
\end{aligned}
\tag{5.26}
$$

Now, let $b = a + \frac{1}{2}$. Then this equation reads

$$\zeta(1-\zeta)\frac{d^2g}{d\zeta^2} + [c-(4a-c+2)\zeta]\frac{dg}{d\zeta} - 2a(2a+1-c)g = 0. \tag{5.27}$$

A solution is $F(2a, 2a+1-c; c; \zeta)$. When c does not assume a non-positive integer value, the equations (5.8) and (5.27) have one and only one solution that is analytic at the origin. Hence we obtain the *quadratic transformation*

$$F(2a, 2a+1-c; c; z) = (1+z)^{-2a} F\left[a, a+\frac{1}{2}; c; \frac{4z}{(1+z)^2}\right]. \tag{5.28}$$

Verify that another (inverse) version of this transformation reads

$$F\left(a, a+\frac{1}{2}; c; z\right) = \left(\frac{1}{2}+\frac{1}{2}\sqrt{1-z}\right)^{-2a} F\left(2a, 2a-c+1; c; \frac{1-\sqrt{1-z}}{1+\sqrt{1-z}}\right).$$

By using the relations in (5.5) we can obtain many more examples.

Example 5.2. Applying the second transformation of (5.5) to the left-hand side of the above relation we obtain

$$F\left(a, a+\frac{1}{2}; c; z\right) = (1-z)^{-a-\frac{1}{2}} F\left(a+\frac{1}{2}, c-a; c; \zeta\right), \quad \zeta = \frac{z}{z-1}.$$

Combining the right-hand sides of the above formula and the above quadratic transformation, and introducing new parameters $\alpha = a + \frac{1}{2}$, $\beta = c - a$, we obtain

$$\begin{aligned} F\left(\alpha, \beta; \alpha+\beta-\frac{1}{2}; z\right) &= (1-z)^{-\frac{1}{2}} \left(\frac{1}{2}+\frac{1}{2}\sqrt{1-z}\right)^{1-2\alpha} \\ &\quad \times F\left(2\alpha-1, \alpha-\beta+\frac{1}{2}; \alpha+\beta-\frac{1}{2}; \frac{\sqrt{1-z}-1}{\sqrt{1-z}+1}\right). \end{aligned}$$

See Exercise 5.7 (1) for another example.

One can show that a quadratic transformation exist for the F–functions if and only if the numbers

$$1-c, \quad a-b, \quad a+b-c$$

satisfy one of the following properties:

- one of them is equal to $\pm\frac{1}{2}$,
- one of them is equal to another one or equal to the opposite of another one.

For formula (5.27) we took $b = a + \frac{1}{2}$. We can also take $b = c/2$. To see this, consider (5.26) and substitute $z = \zeta^2$. The result is given in Exercise 5.7 (2).

Summarizing, we conclude that for each of the following hypergeometric functions a quadratic transformation exists:

$F\left(a, b; \frac{1}{2}; z\right)$	$F\left(a, a+\frac{1}{2}; c; z\right)$	$F\left(a, b; a+b-\frac{1}{2}; z\right)$
$F\left(a, b; \frac{3}{2}; z\right)$	$F\left(a, a-\frac{1}{2}; c; z\right)$	$F\left(a, b; a+b+\frac{1}{2}; z\right)$
$F\left(a, b; 2a; z\right)$	$F\left(a, b; b-a+1; z\right)$	$F\left(a, 1-a; c; z\right)$
$F\left(a, b; 2b; z\right)$	$F\left(a, b; a-b+1; z\right)$	$F\left(a, b; \frac{1}{2}(a+b+1); z\right)$

The first two rows correspond with the first criterion; in the second column the first two cases are not different, due to symmetry $F(a, b; c; z) = F(b, a; c; z)$; the same for the final two cases in the first column and the second column. Observe that (5.28) covers two cases: the first and fourth case of the second column. Examples of corresponding transformations of all cases in the table follow from (5.28) (and the examples given there), and Exercises 5.7–5.9. The results hold always in a (unspecified) neighborhood of the origin. The domain of validity can be extended by using the principle of analytic continuation. See also Exercise 5.8 for a warning and more information on this point.

5.9. Generalized Hypergeometric Functions

In the theory of special functions the following generalization of the Gauss hypergeometric function is used. Let $p, q = 0, 1, 2, \ldots$ with $p \leq q+1$. Then we have

$$
{}_pF_q(a_1, \ldots, a_p; b_1, \ldots, b_q; z) = \sum_{n=0}^{\infty} \frac{(a_1)_n \cdots (a_p)_n}{(b_1)_n \cdots (b_q)_n} \frac{z^n}{n!}. \tag{5.29}
$$

It is clear that the Gauss function corresponds to $p = 2$ and $q = 1$. If $p = q+1$ then the radius of convergence of the ${}_pF_q$–series is again unity. If $p < q+1$ then the radius of convergence is equal to ∞.

The above generalization contains many elementary and special functions. For instance,

$$
e^z = {}_0F_0(\cdot; \cdot; z).
$$

The Bessel functions, Whittaker functions and many orthogonal polynomials can be written as generalized hypergeometric functions. The ${}_pF_q$– function is also used if $p > q+1$. On the one hand as a formal series, on the other as a terminating series. When one of the parameters $a_1, \ldots, a_p$ is equal to a non-positive integer number the series in (5.29) terminates. Also, when $p > q+1$, the notation is useful in asymptotic expansions, in which convergence is not relevant. LUKE (1969) is a good reference for generalized hypergeometric functions.

The differential equation of the ${}_pF_q$–function is a generalization of (5.8). Using the notation of (5.7) we have

$$\vartheta(\vartheta+b_1-1)(\vartheta+b_2-1)\cdots(\vartheta+b_q-1)F = z(\vartheta+a_1)(\vartheta+a_2)\cdots(\vartheta+a_p)F. \quad (5.30)$$

The origin is a singular point; when $p > q + 1$ this point is not regular.

5.9.1. A First Introduction to q–functions

A completely different generalization is due to HEINE (1846), (1847). Nowadays it is called the q–hypergeometric function. We start with Heine's generalization of the ordinary Gauss function:

$$1+\frac{(1-q^a)(1-q^b)}{(1-q)(1-q^c)}z+\frac{(1-q^a)(1-q^{a+1})(1-q^b)(1-q^{b+1})}{(1-q)(1-q^2)(1-q^c)(1-q^{c+1})}z^2+\ldots. \quad (5.31)$$

We assume that the following conditions on convergence hold: $|q| < 1$, $|z| < 1$. Note that

$$\lim_{q\to 1}\frac{1-q^a}{1-q} = a.$$

On account of this limit we conclude that, when $q \to 1$, each term in (5.31) converges to a corresponding term in the well-known series of the Gauss function.

Remark 5.3. Generalization (5.29) is characterized by the fact that the series is of the form $\sum_{n=0}^{\infty} a_n z^n$, in which the ratio a_{n+1}/a_n is a rational function of n. In (5.31) the ratio of successive terms is a rational function of q^n.

The present-day theory of q–functions is based on the following starting points. First we consider the q–variant of the shifted factorial (5.1). Let $(a;q)_0 = 1$ and, in general,

$$(a;q)_n = (1-a)(1-aq)\cdots(1-aq^{n-1}), \quad n = 1, 2, \ldots. \quad (5.32)$$

When we replace a with q^a we obtain a product that turns up in the series (5.31). When $n = \infty$ we have

$$(a;q)_\infty = \prod_{n=0}^{\infty}(1-aq^n),$$

which converges if $|q| < 1$.

The q–analog of (5.29) is ${}_r\phi_s$, the *generalized basic hypergeometric function*. It is the generalization of Heine's series (5.31) and defined by

$$\begin{aligned} &{}_r\phi_s(a_1,\ldots,a_r;b_1,\ldots,b_s;q,z) \\ &= \sum_{n=0}^{\infty}\frac{(a_1;q)_n\cdots(a_r;q)_n}{(q;q)_n(b_1;q)_n\cdots(b_s;q)_n}\, z^n\left[(-1)^n q^{\binom{n}{2}}\right]^{1+s-r}. \end{aligned} \quad (5.33)$$

When $0 < |q| < 1$, the ${}_r\phi_s$ converges absolutely for all z if $r \le s$ and for $|z| < 1$ if $r = s + 1$. When $|q| > 1$, the series converges absolutely if

$$|z| < \frac{|b_1 \cdots b_s|}{|a_1 \cdots a_r|}.$$

The series in (5.33) may terminate, as in the case of ordinary hypergeometric functions, for certain values of the parameters.

In the q–theory the number q is called the *base*, and the function in (5.33) is called a *basic hypergeometric function.* Using the relation

$$(a; q^{-1})_n = (-a)^n \, q^{-\binom{n}{2}} \, (a^{-1}; q)_n$$

the series in (5.33) with base q can be transformed into a series with base q^{-1}. Hence, it is sufficient to consider q–functions with q satisfying $|q| < 1$.

Many special functions can be generalized in terms of (5.33). For instance, the q–binomial series generalization of the second series in (5.2) reads

$${}_1\phi_0(a; -; q, z) = \sum_{n=0}^{\infty} \frac{(a; q)_n}{(q; q)_n} z^n, \quad |z| < 1.$$

This function reduces to the one in (5.2) when we replace a with q^a and take the limit $q \to 1^-$. In fact we have the relation

$${}_1\phi_0(a; -; q, z) = \frac{(az; q)_\infty}{(z; q)_\infty}.$$

In the theory of q–functions often a relation exists between infinite products and infinite series. Euler and Gauss have already discovered such relations. An important class of special functions sharing this property is the class of the (Jacobi) theta functions (see Chapter 12).

A great deal of current research in special function theory takes place in this q–setting. For instance, one is interested in more q–analogs of the well-known special functions. For $0 < q < 1$ one defines the q–gamma function as follows:

$$\Gamma_q(x) = \frac{(q; q)_\infty}{(q^x; q)_\infty} (1 - q)^{1-x}.$$

This definition is not quite obvious without further preparations. The infinite product representation (3.9) of the gamma function is an important source of inspiration here. It can be verified that $\lim_{q \to 1^-} \Gamma_q(x) = \Gamma(x)$ (see KOORNWINDER (1990a)). The q–beta integral now follows more easily (see (3.3)):

$$B_q(x, y) = \frac{\Gamma_q(x)\Gamma_q(y)}{\Gamma_q(x + y)}, \quad 0 < q < 1, \quad \Re x, y > 0.$$

5.10. Remarks and Comments for Further Reading

5.1. The Gauss hypergeometric functions arise in physical problems when Legendre functions are used (see Chapter 9 and §7.3.2). Other occurrences arise in the form of elliptic integrals; see Chapter 12, where a simple pendulum is considered. See also the introductory texts of SEABORN (1991) and LAWDEN (1989).

5.2. Asymptotic representations of $F(a, b; c; z)$ for large values of z follow from (5.11). These representations are convergent when $\Re(c - a - b) > 0$. However, convergence is not needed for an asymptotic expansion. When c is large (with respect to a, b, z, the definition in (5.3) gives an asymptotic representation. When the parameters a, b are large, the asymptotic problem is much more complicated. In WAGNER (1990) a special case is considered, with further references to the literature. When the hypergeometric functions reduce to other well-known functions, such as Legendre functions or Jacobi polynomials, much more information about the asymptotic behavior is available.

5.3. The approach of §5.5 is based on OLVER (1974).

5.4. A nice treatment of the $q-$ analogs of the gamma and beta functions can be found in ASKEY (1980), where more references are given. See GASPER & RAHMAN (1990) for a good introduction to $q-$functions.

5.5. KOORNWINDER (1990b) demonstrates how hypergeometric series can be manipulated in *Maple*, a software package for computer algebra.

5.6. The *numerical evaluation of hypergeometric functions* can be based on the series expansions (5.3) and the transformation formulas given in (5.5), (5.10) – (5.13). For real values of x one can always use one or two power series with argument w such that $|w| < \frac{1}{2}$. For instance, we can use the scheme

if	$x < -2$	then use (5.11),	$b - a \notin \mathbb{Z}$,
else if	$x < -1$	then use (5.12),	$b - a \notin \mathbb{Z}$,
else if	$x < \frac{1}{3}$	then use (5.5),	
else if	$x < \frac{1}{2}$	then use (5.3),	
else if	$x < 1.5$	then use (5.10),	$c - a - b \notin \mathbb{Z}$,
else if	$x < 2$	then use (5.13),	$c - a - b \notin \mathbb{Z}$,
else		use (5.11),	$b - a \notin \mathbb{Z}$.

When $x > 1$ the function $F(a, b; c; x)$ is complex, unless it reduces to a polynomial. When $c - a - b \in \mathbb{Z}$ (5.10) and (5.13) reduce to expansions involving logarithms, as shown in (5.14) for (5.10) and the case $c = a + b$. A similar situation happens for the expansions (5.11) and (5.12) when $b - a \in \mathbb{Z}$.

5.11. Exercises and Further Examples

5.1. Verify the following relations:

$$\frac{d}{dz}F(a,b;c;z) = \frac{ab}{c}F(a+1,b+1;c+1;z),$$
$$\frac{d^n}{dz^n}F(a,b;c;z) = \frac{(a)_n\,(b)_n}{(c)_n}F(a+n,b+n;c+n;z),$$
$$F(a,b+1;c;z) = F(a,b;c;z) + \frac{az}{c}F(a+1,b+1;c+1;z).$$

Prove the third formula by using the relations of §5.5 or directly, by verifying that

$$\frac{(a)_n\,(b+1)_n}{(c)_n\,n!} - \frac{(a)_n\,(b)_n}{(c)_n\,n!} = \frac{a}{c}\frac{(a+1)_{n-1}(b+1)_{n-1}}{(c+1)_{n-1}(n-1)!}.$$

5.2. Show that for $|k| < 1$ the *complete elliptic integrals* (more information is given in §12.1)

$$\boxed{K(k) = \int_0^1 \frac{dt}{\sqrt{(1-t^2)(1-k^2t^2)}} = \int_0^{\pi/2} \frac{d\phi}{\sqrt{1-k^2\sin^2\phi}},}$$

$$\boxed{E(k) = \int_0^1 \sqrt{\frac{1-k^2t^2}{1-t^2}}\,dt = \int_0^{\pi/2} \sqrt{1-k^2\sin^2\phi}\,d\phi}$$

can be written in the form

$$K(k) = \tfrac{1}{2}\pi F\left(\tfrac{1}{2},\tfrac{1}{2};1;k^2\right), \quad E(k) = \tfrac{1}{2}\pi F\left(-\tfrac{1}{2},\tfrac{1}{2};1;k^2\right).$$

5.3. Show that the *incomplete beta function*

$$\boxed{B_x(p,q) = \int_0^x t^{p-1}(1-t)^{q-1}\,dt, \quad 0 < x < 1, \quad \Re p > 0}$$

can be expressed as follows:

$$\begin{aligned}
B_x(p,q) &= \frac{x^p}{p}F(p,1-q;p+1;x)\\
&= \frac{1}{p}x^p(1-x)^{q-1}F\left(1,1-q;p+1;\frac{x}{x-1}\right).\\
&= \frac{1}{p}x^p(1-x)^q F(p+q,1;p+1;x).
\end{aligned}$$

5.4. Show that

$$\cos 2at = F(a, -a; \frac{1}{2}; \sin^2 t). \tag{1}$$

Observe, however, that both left-hand and right-hand sides are periodic functions with respect to t, with different periods. Indeed, the results hold in a limited region of the complex domain. Verify, with the help of (5.6) and Exercise 3.7, that the relation holds if $t = \pi/2$. Put $z = \sin^2 t$ and derive by using (5.8) a differential equation with respect to t. Verify that

$$\frac{dF}{dz} = \frac{1}{\sin 2t}\dot{F}, \quad \frac{d^2F}{dz^2} = \frac{1}{\sin^2 2t}\ddot{F} - \frac{2\cos 2t}{\sin^3 2t}\dot{F},$$

where $\dot{F}, \ddot{F}$ denote derivatives with respect to t. The differential equation

$$\ddot{F} + 4a^2 F = 0$$

is the result. A solution is the left-hand side of (1). As a further exercise, show that

$$F\left(a, 1-a; \frac{1}{2}; -\sinh^2 t\right) = \frac{\cosh(2a-1)t}{\cosh t}. \tag{2}$$

5.5. Verify with the help of (5.4)

$$F(b, a; a-b+1; -1) = \sqrt{\pi}\,\frac{2^{-a}\Gamma(1+a-b)}{\Gamma\left(1+\frac{1}{2}a-b\right)\Gamma\left(\frac{1}{2}+\frac{1}{2}a\right)}.$$

5.6. Give a direct proof of (5.25) by substituting an infinite series with the help of the first line in (5.5). In the proof you will need (5.6) and Exercise 3.3. Show that

$$(1+t)^a = \frac{1}{2\pi i}\frac{1}{\Gamma(-a)}\int_{\mathcal{L}} \Gamma(s-a)\Gamma(-s)t^s\,ds$$

and specify the the path of integration in terms of the complex parameter a.

5.7. Apply the first formula of (5.5) on the right-hand side of (5.28). Derive from the result the quadratic transformation:

$$F\left(a,b;a-b+1;z\right)=(1-z)^{-a}F\left[\frac{1}{2}a,\frac{1}{2}a-b+\frac{1}{2};a-b+1;-\frac{4z}{(1-z)^2}\right]. \tag{1}$$

Put in (5.26) $c=2b$ and $z=\zeta^2$. Derive from this result the following quadratic transformation:

$$F\left(a,a-b+\frac{1}{2};b+\frac{1}{2};z^2\right)=(1+z)^{-2a}F\left[a,b;2b;\frac{4z}{(1+z)^2}\right], \tag{2}$$

with inverse

$$\begin{aligned}F\left(a,b;2b;z\right)&=\left(\frac{1}{2}+\frac{1}{2}\sqrt{1-z}\right)^{-2a}\\&\quad\times F\left[a,a-b+\frac{1}{2};b+\frac{1}{2};\left(\frac{1-\sqrt{1-z}}{1+\sqrt{1-z}}\right)^2\right].\end{aligned} \tag{3}$$

5.8. Derive the following quadratic transformations:

$$F\left(2a,2b;a+b+\frac{1}{2};z\right)=F\left(a,b;a+b+\frac{1}{2};4z(1-z)\right), \tag{1}$$

$$F\left(a,1-a;c;z\right)=(1-z)^{c-1}F\left(\frac{1}{2}c-\frac{1}{2}a,\frac{1}{2}c+\frac{1}{2}a-\frac{1}{2};c;4z(1-z)\right), \tag{2}$$

$$F\left(a,b;a+b+\frac{1}{2};z\right)=F\left(2a,2b;a+b+\frac{1}{2};\frac{1}{2}-\frac{1}{2}\sqrt{1-z}\right). \tag{3}$$

Hints. Consider Exercise 5.7 (1); put $c=a+b+\frac{1}{2}$ and replace z with $z/(z-1)$. Next apply the first transformation of (5.5) on the left-hand side of (1). (3) is the inverse of (1). To prove (2), apply the third line of (5.5) on the left-hand side of (1), and take $c=a+b+\frac{1}{2}$. Verify using (5.6) that for $z=1$ the left-hand side of (1) reduces to $\cos[\pi(a-b)]/\cos[\pi(a+b)]$, whereas the right-hand side is equal to 1. This paradox is explained by the fact that the quadratic transformation holds only in a limited domain D around the origin, this domain being defined by the connected subset of $D=\{z\in\mathbb{C}\mid |z|<1\wedge|4z(1-z)|<1\}$ containing the point $z=0$. Obviously, $z=1$ does not belong to this domain. Verify that the interval $(\frac{1}{2},1]$ does not belong to this domain.

5.9. Verify the quadratic transformations for the cases $c=\frac{1}{2},\frac{3}{2}$:

$$\begin{aligned}\frac{2\Gamma(\frac{1}{2})\Gamma(a+b+\frac{1}{2})}{\Gamma(a+\frac{1}{2})\Gamma(b+\frac{1}{2})}F\left(a,b;\frac{1}{2};z\right)&=F\left(2a,2b;a+b+\frac{1}{2};\frac{1}{2}(1+\sqrt{z})\right)\\&\quad+F\left(2a,2b;a+b+\frac{1}{2};\frac{1}{2}(1-\sqrt{z})\right)\end{aligned}$$

$$\frac{2\Gamma(-\frac{1}{2})\Gamma(a+b-\frac{1}{2})}{\Gamma(a-\frac{1}{2})\Gamma(b-\frac{1}{2})}\sqrt{z}\,F\left(a,b;\frac{3}{2};z\right)$$
$$=F\left(2a-1,2b-1;a+b-\frac{1}{2};\frac{1}{2}(1-\sqrt{z})\right)$$
$$-F\left(2a-1,2b-1;a+b-\frac{1}{2};\frac{1}{2}(1+\sqrt{z})\right)$$

Observe that these transformations involve two F–functions in the right-hand sides. To verify these forms apply to Exercise 5.8 (1) the transformation $z \to \frac{1}{2}(1+\sqrt{z}\,)$ and obtain the form

$$F\left(a,b;a+b+\frac{1}{2};1-z\right)=F\left[2a,2b;a+b+\frac{1}{2};\frac{1}{2}(1+\sqrt{z}\,)\right].$$

Apply (5.10) to the left-hand side to obtain two F–functions, one with $c=\frac{1}{2}$ and one with $c=\frac{3}{2}$. Repeat this with $\sqrt{z}$ replaced by $-\sqrt{z}$ to obtain a similar relation, from which the F–functions with $c=\frac{1}{2}$ and $c=\frac{3}{2}$ can be solved.

5.10. Put in (5.28) $c=2a$; then the left-hand side becomes $1/(1-z)$. Show that

$$F\left(a,a+\frac{1}{2};2a;\zeta\right)=\frac{2^{2a-1}}{\sqrt{1-\zeta}}\left(1+\sqrt{1-\zeta}\right)^{1-2a}.$$

Show that

$$F\left(a,a+\frac{1}{2};2a+1;\zeta\right)=2^{2a}\left(1+\sqrt{1-\zeta}\right)^{-2a}.$$

5.11. Verify with the help of (5.6) that

$$F\left(-n,b;c;1\right)=\frac{(c-b)_n}{(c)_n},\quad n=1,2,3,\ldots.$$

From this we obtain the remarkable relation

$$(a+b)_n=\sum_{k=0}^{n}\binom{n}{k}(a)_k(b)_{n-k}.$$

It is remarkable, since the notation has a striking similarity to Newton's binomial formula.

5.12. Show that for $\Re c>\Re\lambda>0$, $|z|<1$

$$F\left(a,b;c;z\right)=\frac{\Gamma(c)}{\Gamma(\lambda)\Gamma(c-\lambda)}\int_0^1 x^{\lambda-1}(1-x)^{c-\lambda-1}F\left(a,b;\lambda;xz\right)dx,$$

by using (3.2) and (5.3). Extend the z–domain of validity to $|\arg(1-z)|<\pi$, $z\neq 1$. Take $\lambda=b$ and compare your result with Euler's integral (5.4).

5.13. Prove Barnes' lemma (BARNES (1908); see also WHITTAKER & WATSON (1927), page 289):

$$\frac{1}{2\pi i}\int_{-i\infty}^{i\infty}\Gamma(a+s)\Gamma(b+s)\Gamma(c-s)\Gamma(d-s)\,ds$$
$$=\frac{\Gamma(a+c)\Gamma(a+d)\Gamma(b+c)\Gamma(b+d)}{\Gamma(a+b+c+d)},$$

where the path of integration is curved so that the poles of $\Gamma(c-s)\Gamma(d-s)$ lie on the right of the path and the poles of $\Gamma(a+s)\Gamma(b+s)$ lie on the left. It is supposed that a, b, c, d are such that no pole of the first set coincides with any pole of the second set.

6

Orthogonal Polynomials

Orthogonal polynomials are of great importance in mathematical physics, approximation theory, the theory of numerical quadrature, etc., and are the subject of an enormous literature. An important application occurs in physics when we consider the *Schrödinger equation* for a linear harmonic oscillator of mass m, angular frequency ω_0 and total energy E, that is,

$$\frac{d^2\psi}{dx^2} + \left(\frac{2mE}{\hbar^2} - \frac{m^2\omega_0^2}{\hbar^2}x^2\right)\psi = 0,$$

where ψ is the wave function and $\hbar$ is Planck's constant. In quantum mechanics it is required to find the values of E for which this equation has bounded solutions in the interval $-\infty < x < \infty$. The eigenvalues that make this possible are $E = E_n = (n+\frac{1}{2})\hbar\omega_0$ and the corresponding eigenfunctions are related with Hermite polynomials: $\psi_n(x) = \exp(-t^2/2)H_n(t), t = \sqrt{(\hbar/(m\omega_0)}\,x$; see Exercise 6.9.

The $\{\psi_n\}$ constitute an orthogonal set on $\mathbb{R}$; the Hermite polynomials $\{H_n\}$ constitute an orthogonal set of polynomials on $\mathbb{R}$ with respect to the Gaussian weight function $\exp(-t^2)$. In this chapter we give general properties of orthogonal polynomials and we give further details on the classical orthogonal polynomials.

6.1. General Orthogonal Polynomials

Before discussing the well-known classical orthogonal polynomials we treat the basic concepts which are of vital importance for all orthogonal polynomials to be considered.

We take a real interval (a, b) - where $a = -\infty$ and/or $b = +\infty$ are accepted - and a function w: $(a, b) \to [0, \infty)$, the *weight function*, with the property that the integral

$$\int_a^b w(x)x^k\,dx$$

exists for all $k = 0, 1, 2, \ldots$. The integration can be formulated in the sense of Lebesgue-Stieltjes (with a measure $d\mu(x)$ in place of $w(x)\,dx$), but here we assume that the function w is Riemann integrable.

We introduce the linear space $\mathcal{P}$ of polynomials of the real parameter x with real coefficients. Let $f, g \in \mathcal{P}$. Then we call

$$\langle f, g\rangle = \int_a^b w(x)f(x)g(x)\,dx \tag{6.1}$$

the *inner product* of f and g (always with respect to w). The following properties can be easily verified:

$$\langle f, g\rangle = \langle g, f\rangle; \quad \langle \alpha f + \beta g, h\rangle = \alpha\langle f, h\rangle + \beta\langle g, h\rangle$$

when $\alpha, \beta \in \mathbb{R}$, $f, g, h \in \mathcal{P}$. Furthermore we have

$$\langle f, f\rangle = \int_a^b w(x)f^2(x)\,dx \geq 0, \quad \langle f, f\rangle = 0 \Rightarrow f = 0.$$

We say that $\|f\| = \sqrt{\langle f, f\rangle}$ is the *norm* of f and that $f, g \in \mathcal{P}$ are *orthogonal polynomials* if $\langle f, g\rangle = 0$.

We now construct, starting from the linear independent set of polynomials

$$f_0(x) = 1,\ f_1(x) = x,\ f_2(x) = x^2,\ f_3(x) = x^3, \ldots,\ f_n(x) = x^n \tag{6.2}$$

a new set

$$p_0, p_1, p_2, \ldots, p_n$$

of orthogonal polynomials corresponding to the inner product (6.1). This process can be executed for any $n = 0, 1, 2, \ldots$ and it is known as the *Gram-Schmidt orthogonalization method.*

Put

$$p_0 = \frac{f_0}{\|f_0\|}, \quad p_1 = \frac{f_1 - \langle f_1, p_0\rangle p_0}{\|f_1 - \langle f_1, p_0\rangle p_0\|},$$

and, in general,

$$p_k = \frac{f_k - \sum_{i=0}^{k-1}\langle f_k, p_i\rangle p_i}{\left\|f_k - \sum_{i=0}^{k-1}\langle f_k, p_i\rangle p_i\right\|}.$$

Obviously, p_k is a polynomial of exact degree k, with $\|p_k\| = 1$ and $\langle p_1, p_0\rangle = 0$. Using the induction principle it is not difficult to prove that

$$\langle p_j, p_k\rangle = 0, \quad j \neq k.$$

The set $\{p_k\}$ constructed in this manner is *orthonormal*: all p_k have norms equal to 1.

Let $f \in \mathcal{P}$ have degree n. Then we can express f uniquely in terms of the polynomials p_k:

$$f(x) = \sum_{k=0}^{n} \alpha_k p_k(x), \quad \alpha_k = \langle f, p_k \rangle, \quad 0 \le k \le n. \tag{6.3}$$

We further remark that, for $k = 1, 2, 3, \ldots$,

$$\langle p_k, f_j \rangle = \int_a^b w(x) p_k(x) x^j \, dx = 0, \quad j < k, \tag{6.4}$$

where the f_j are introduced in (6.2).

Let k_n be the coefficient of x^n in p_n. That is,

$$p_n(x) = k_n x^n + \cdots . \tag{6.5}$$

Theorem 6.1. *The orthonormal polynomials $\{p_k\}$ satisfy the following recurrence relation*

$$\boxed{p_{n+1} - (a_n x + b_n) p_n + c_n p_{n-1} = 0, \quad n = 1, 2, \ldots,} \tag{6.6}$$

where

$$a_n = \frac{k_{n+1}}{k_n}, \quad b_n = -a_n \langle x p_n, p_n \rangle, \quad c_n = \frac{a_n}{a_{n-1}}, \quad c_0 = 0.$$

Proof. It is clear that

$$p_{n+1}(x) - a_n x p_n(x) = \sum_{k=0}^{n} \alpha_k x^k = \sum_{k=0}^{n} \beta_k p_k(x)$$

for certain α_k, β_k, since the term with x^{n+1} in the left-hand side cancels. Using the orthonormal relations we can write

$$-a_n \langle x p_n, p_j \rangle = \beta_j, \quad j \le n.$$

Referring to (6.4) we have, however,

$$\langle x p_n, p_j \rangle = \int_a^b w(x)\, x\, p_n(x)\, p_j(x)\, dx = \langle p_n, x p_j \rangle = 0, \quad j \le n-2,$$

since $xp_j(x)$ has degree $\le n-1$. This implies

$$\beta_0 = \beta_1 = \cdots = \beta_{n-2} = 0.$$

We observe that $b_n = \beta_n$ and we put $c_n = -\beta_{n-1}$. This shows the validity of (6.6). The expression for c_n is found by using

$$c_n = a_n \langle xp_n, p_{n-1}\rangle = a_n(p_n, xp_{n-1}),$$

with

$$xp_{n-1}(x) = k_{n-1}x^n + \ldots = \frac{k_{n-1}}{k_n}\left[k_n x^n + \ldots\right]$$
$$= \frac{1}{a_{n-1}}\left[p_n(x) + \sum_{j=0}^{n-1}\gamma_j p_j(x)\right]$$

for certain numbers γ_j. Hence

$$c_n = \frac{a_n}{a_{n-1}}\left\langle p_n, p_n + \sum_{j=0}^{n-1}\gamma_j p_j\right\rangle = \frac{a_n}{a_{n-1}}.$$

■

Remark 6.1. Conversely, when $a_n > 0$, $c_n > 0$ and the family of polynomials $\{p_n\}$ satisfies (6.6), then there exists a weight function with which $\{p_n\}$ becomes a family of orthogonal polynomials. This weight function need not be a regular function but may be defined as a measure, for instance in the sense of Stieltjes-Lebesgue. See FAVARD (1935).

The next result is the *Christoffel-Darboux formula*. First we introduce a function which plays a significant part in the theory of orthogonal polynomials:

$$\boxed{K_n(x,y) = \sum_{k=0}^{n} p_k(x)p_k(y).} \tag{6.7}$$

It is possible to find a closed expression for this function. We have

Theorem 6.2. *(Christoffel-Darboux)*

$$\boxed{K_n(x,y) = \frac{k_n}{k_{n+1}}\,\frac{p_n(y)p_{n+1}(x) - p_n(x)p_{n+1}(y)}{x-y}.} \tag{6.8}$$

Proof. The formula is correct for $n = 0$. This is easily verified by using $p_0(x) = k_0$, $p_1(x) = k_1x + c$, for certain number c. Next we use induction with respect to n. For the induction step of $n-1$ to n we use the recurrence relation (6.6) in

$$\frac{p_n(y)p_{n+1}(x) - p_n(x)p_{n+1}(y)}{x-y} = a_n p_n(x)p_n(y) + c_n \frac{p_{n-1}(y)p_n(x) - p_{n-1}(x)p_n(y)}{x-y}.$$

This proves the theorem. ∎

It is difficult to use (6.8) when $x = y$. However, a simple application of l'Hôpital's rule gives the result

$$K_n(x,x) = \frac{k_n}{k_{n+1}}[p_n(x)p'_{n+1}(x) - p'_n(x)p_{n+1}(x)] = \sum_{k=0}^{n} p_k^2(x) > 0, \tag{6.9}$$

which confirms (6.7) when $x = y$.

The function $K_n(x,y)$ is called the *reproducing kernel* and a similar function plays an important role in the theory of linear operators. Here we mention the following property. For each $f \in \mathcal{P}$ of degree n we have using (6.3)

$$\langle f, K_n(\,\cdot\,,y)\rangle = \sum_{k=0}^{n} \langle f, p_k\rangle p_k(y) = f(y). \tag{6.10}$$

This explains the name *reproducing* kernel.

6.1.1. Zeros of Orthogonal Polynomials

An interesting aspect in the theory of orthogonal polynomials is connected with the zeros of the polynomials. In this respect we have the following fundamental property.

Theorem 6.3. *p_n has n real simple zeros x_k satisfying*

$$a < x_k < b, \quad 1 \le k \le n.$$

Proof. We assume that k zeros of p_n are located inside the interval (a,b), $0 \le k < n$, and that the function p_n changes sign at these zeros. That is, we assume that the zeros have odd multiplicity. When we succeed in proving that $k = n$, we are done, since then p_n has exactly n zeros in (a,b) at which p_n changes sign. But the polynomial p_n has just n zeros. Thus it follows that the zeros are simple and inside (a, b). Now, assume that $k < n$ and consider the polynomial

$$q_k(x) = (x - x_1)(x - x_2)\cdots(x - x_k) = \sum_{j=0}^{k} \alpha_j p_j(x).$$

This is a polynomial of degree k with the property $\langle p_n, q_k \rangle = 0$. Observe that $p_n(x)q_k(x)$ cannot change sign on (a, b), since the zeros of this function have even multiplicity. Assume that (without loss of generality) $p_n(x)q_k(x) \geq 0$. Then we see that

$$\langle p_n, q_k \rangle = \int_a^b w(x)p_n(x)q_k(x)\, dx > 0.$$

This leads to a contradiction, and it follows that $k = n$. ■

We have another result, that describes the relation between the zeros of p_n and p_{n+1}.

Theorem 6.4. *The zeros of p_n and p_{n+1} alternate on the interval (a, b), and p_n and p_{n+1} do not have common zeros.*

Proof. Let x_r and x_{r+1} denote two successive zeros of p_n. Then, by virtue of (6.9),

$$\sum_{k=0}^{n} p_k^2(x_r) = \frac{k_n}{k_{n+1}}[-p_n'(x_r)p_{n+1}(x_r)] > 0,$$
$$\sum_{k=0}^{n} p_k^2(x_{r+1}) = \frac{k_n}{k_{n+1}}[-p_n'(x_{r+1})p_{n+1}(x_{r+1})] > 0.$$

Since x_r and x_{r+1} are *successive* zeros of p_n, it follows that $p_n'(x_r)$ and $p_n'(x_{r+1})$ are of opposite sign. Therefore $p_{n+1}(x_r)$ and $p_{n+1}(x_{r+1})$ also have opposite sign. Since p_{n+1} is continuous, this polynomial must vanish at least once between x_r and x_{r+1}. Verify further that p_{n+1} has one zero at the left of x_1 and one at the right of x_n. The remaining part, to verify that p_n and p_{n+1} cannot have a common zero x_r, is left to the reader. ■

6.1.2. Gauss Quadrature

Here we give the main ingredients of the *Gauss quadrature* formulas, in which the zeros of the orthogonal polynomials are of decisive importance.

Let $x_1, x_2, \ldots, x_n$ be the zeros of p_n and and take a polynomial $f \in \mathcal{P}$ of degree $\leq 2n - 1$. We construct a new polynomial F by means of the *Lagrange interpolation formula:*

$$F(x) = \sum_{k=1}^{n} f(x_k) \frac{p_n(x)}{(x - x_k)p_n'(x_k)}.$$

It is clear that F is a polynomial of degree $n - 1$ and that

$$F(x_k) = \lim_{x \to x_k} F(x) = f(x_k), \quad k = 1, 2, \ldots, n.$$

Hence $F - f$ is a polynomial of degree $\leq 2n - 1$ and $x_1, x_2, \ldots, x_n$ are the zeros of $F - f$. It follows that

$$r(x) = \frac{F(x) - f(x)}{p_n(x)}$$

is a polynomial of degree $n - 1$. Hence we can write

$$f(x) = \sum_{k=1}^{n} f(x_k) \frac{p_n(x)}{(x - x_k)p_n'(x_k)} - r(x)p_n(x).$$

We integrate this formula and introduce the quantities

$$\lambda_{k,n} = \int_a^b w(x) \frac{p_n(x)}{(x - x_k)p_n'(x_k)} \, dx, \tag{6.11}$$

the so-called *Christoffel numbers.* Since

$$\int_a^b w(x)r(x)p_n(x) \, dx = 0$$

we obtain the required rule of Gauss quadrature:

$$\int_a^b w(x)f(x) \, dx = \sum_{k=1}^{n} \lambda_{k,n} f(x_k). \tag{6.12}$$

This rule tells us that for a polynomial of degree $\leq 2n-1$ the integral is exactly equal to the expression in the right-hand side of (6.12). In other words: the n-points Gauss quadrature formula gives an exact result for polynomials having degree $\leq 2n - 1$.

When f is not a polynomial the rule is also very useful for approximating integrals. When f has at least $2n$ continuous derivatives on (a, b), one has

$$\int_a^b w(x)f(x) \, dx = \sum_{k=1}^{n} \lambda_{k,n} f(x_k) + R_n. \tag{6.13}$$

It can be shown that a number $\xi \in (a, b)$ exists such that $R_n = C_n f^{(2n)}(\xi)$, where C_n does not depend on x. For this result, and for much more information on the numerical aspects of Gauss quadrature, we refer to STROUD & SECREST(1966).

Example 6.1. Consider the computation of the integral

$$\int_{-1}^{1} \frac{dx}{x+3} = \ln 2.$$

Take $w(x) = 1$ and $n = 2$. Verify that the first few orthogonal polynomials are given by

$$p_0(x) = \sqrt{\frac{1}{2}}, \quad p_1(x) = \sqrt{\frac{3}{2}}\,x, \quad p_2(x) = \sqrt{\frac{5}{8}}\,(3x^2 - 1).$$

The zeros of p_2 are $x_1 = -\frac{1}{3}\sqrt{3}$, $x_2 = \frac{1}{3}\sqrt{3}$, and we have

$$\lambda_{1,2} = \int_{-1}^{1} \frac{3x^2 - 1}{(x + \frac{1}{3}\sqrt{3})(-2\sqrt{3})}\,dx = 1,$$

$$\lambda_{2,2} = \int_{-1}^{1} \frac{3x^2 - 1}{(x - \frac{1}{3}\sqrt{3})(+2\sqrt{3})}\,dx = 1.$$

Then by (6.13)

$$\int_{-1}^{1} \frac{dx}{x+3} \approx \frac{1}{3 - \frac{1}{3}\sqrt{3}} + \frac{1}{3 + \frac{1}{3}\sqrt{3}} = \frac{9}{13} = 0.6923\ldots\,.$$

Comparing this with $\ln 2 = 0.6930\ldots$, we conclude that an accuracy of 0.15% is obtained.

Representation (6.11) is not the ideal form for computing the Christoffel numbers. Much more attractive forms are available. From the third and fourth formula of the following theorem it follows that the numbers are positive. Especially (*iii*) gives, combined with the recurrence relation of the polynomials, a rather stable representation for numerical calculations.

Theorem 6.5.

$$\lambda_{k,n} = \int_a^b w(x) \frac{p_n(x)}{(x - x_k)p_n'(x_k)}\,dx \qquad (i)$$

$$= -\frac{k_{n+1}}{k_n} \frac{1}{p_n'(x_k)p_{n+1}(x_k)} \qquad (ii)$$

$$= \Big[\sum_{j=0}^{n-1} p_j^2(x_k)\Big]^{-1} \qquad (iii)$$

$$= \int_a^b w(x) \Big[\frac{p_n(x)}{(x - x_k)p_n'(x_k)}\Big]^2 dx. \qquad (iv)$$

Proof. Since $p_n(x_k) = 0$, by Theorem 6.2

$$K_n(x, x_k) = \sum_{k=0}^{n-1} p_k(x)p_k(x_k) = \frac{k_n}{k_{n+1}} \frac{-p_n(x)p_{n+1}(x_k)}{x - x_k}, \tag{6.14}$$

which gives

$$\frac{p_n(x)}{x - x_k} = -\frac{K_n(x, x_k)k_{n+1}}{k_n p_{n+1}(x_k)}.$$

From (i) it then follows that

$$\lambda_{k,n} = -\frac{k_{n+1}}{k_n p_n'(x_k)p_{n+1}(x_k)} \int_a^b w(x)K_n(x, x_k)\, dx.$$

Applying now (6.10) with $f = 1$ we obtain (ii). Taking in (6.14) the limit $x \to x_k$ we obtain

$$K_n(x_k, x_k) = \sum_{j=0}^{n-1} p_j^2(x_k) = -\frac{k_n}{k_{n+1}} p_n'(x_k)p_{n+1}(x_k),$$

which easily yields (iii). The fourth relation follows from applying Gauss quadrature to (6.13) (verify that $R_n = 0$ in this case):

$$\int_a^b w(x) \left[\frac{p_n(x)}{(x - x_k)p_n'(x_k)}\right]^2 dx = \sum_{j=1}^{n} \lambda_{j,n} \lim_{x \to x_j} \left[\frac{p_n(x)}{(x - x_k)p_n'(x_k)}\right]^2 = \lambda_{k,n}.$$

■

Remark 6.2. When applying (iii) one should realize that the underlying orthogonal polynomials are *orthonormal.* The well-known classical orthogonal polynomials are usually considered in a standard form, which does not show the orthonormal versions. Of course, (iii) can be adapted to the circumstances by means of the norm of the non-orthonormal polynomials.

6.2. Classical Orthogonal Polynomials

The orthogonal polynomials associated with the names of Jacobi, Gegenbauer, Chebyshev, Legendre, Laguerre and Hermite are called the *classical orthogonal polynomials.* They will be discussed in the remaining part of this chapter. The separate families share many features.

The following points are characteristic of the classical orthogonal polynomials:

(*i*) the family $\{p_n'\}$ is also an orthogonal system;

(*ii*) p_n satisfies a second order linear differential equation

$$A(x)y'' + B(x)y' + \lambda_n y = 0,$$

where A and B not depend on n and λ_n does not depend on x;

(*iii*) there is a *Rodrigues formula* of the form

$$p_n(x) = \frac{1}{K_n w(x)} \left(\frac{d}{dx}\right)^n [w(x)X^n(x)], \tag{6.15}$$

where X is a polynomial in x with coefficients not depending on n, and K_n does not depend on x.

These three properties are so characteristic that any system of orthogonal polynomials having these three properties, can be reduced to a system of classical orthogonal polynomials. For a recent reference, see Al-Salam's contribution to NEVAI (1990).

6.3. Definitions by the Rodrigues Formula

First we will consider a finite interval (a, b); it is convenient to take $(-1, 1)$. For the function X in the Rodrigues formula we take $X(x) = 1 - x^2$. It is easily verified that with this choice of X the function p_n of (6.15) satisfies

$$\int_{-1}^{1} w(x)x^k p_n(x)\, dx = 0, \quad 0 \le k < n. \tag{6.16}$$

The proof follows from integrating by parts repeatedly:

$$\begin{aligned} &\int_{-1}^{1} x^k \left(\frac{d}{dx}\right)^n \left[w(x)\left(1-x^2\right)^n\right] dx \\ &= \sum_{l=0}^{k} (-1)^l k(k-1)\dots(k-l+1) \left(\frac{d}{dx}\right)^{n-l-1} \left[w(x)\left(1-x^2\right)^n\right]\Bigg|_{-1}^{1} = 0. \end{aligned} \tag{6.17}$$

We assume, of course, that w is sufficiently regular at ± 1. It follows that p_n is orthogonal with respect to any polynomial of degree less than n.

Observe that, at the moment, no weight function is specified. In fact, the present choice of X gives limited possibilities for w. An essential condition is that the right-hand side of (6.15) must produce polynomials. For p_1 we find

$$K_1 p_1(x) = \frac{w'(x)}{w(x)}\left(1 - x^2\right) - 2x.$$

When this has to be a linear function, the only possibility is

$$w(x) = (1-x)^\alpha (1+x)^\beta,$$

where α and β are constants. Indeed, this w produces in (6.15) a polynomial for any value of n. (To prove this, apply Leibniz' rule for repeatedly differentiating products.) Moreover, when $\alpha > -1$, $\beta > -1$, all integrated terms in (6.17) vanish. This weight function leads to the family of orthogonal polynomials $\{p_n\}$ defined on the interval $(-1, 1)$: the *Jacobi polynomials.* The numbers K_n in (6.15) are usually taken equal to $K_n = (-1)^n 2^n n!$. This gives the definition based on the Rodrigues formula

$$\boxed{P_n^{(\alpha,\beta)}(x) = \frac{(-1)^n}{2^n\, n!}\, \frac{1}{(1-x)^\alpha (1+x)^\beta} \left(\frac{d}{dx}\right)^n \left[(1-x)^{\alpha+n}(1+x)^{\beta+n}\right].} \tag{6.18}$$

The choice of K_n is, here and in later cases, not always obvious. It does not influence the orthogonality, of course, but the orthonormality. The norm of the Jacobi polynomial follows from Exercise 6.11. Usually one does not use orthonormal systems for the classical polynomials. Simple coefficients of the polynomials and attractive formulas are of more importance. It is not difficult to verify that in the present case we have the pleasant normalization and symmetry

$$\boxed{P_n^{(\alpha,\beta)}(1) = \binom{n+\alpha}{n}, \quad P_n^{(\alpha,\beta)}(-x) = (-1)^n P_n^{(\beta,\alpha)}(x).} \tag{6.19}$$

When $\alpha = \beta = 0$ we have the *Legendre polynomials*

$$\boxed{P_n(x) = \frac{(-1)^n}{2^n n!} \left(\frac{d}{dx}\right)^n \left(1-x^2\right)^n.} \tag{6.20}$$

The graphs of the first ten Legendre polynomials on the interval $[-1, 1]$ are given in Figure 6.1.

When $\alpha = \beta = -\frac{1}{2}$ and a different K_n we obtain the *Chebyshev polynomials of the first kind*

$$\boxed{T_n(x) = \frac{(-1)^n\, n!\, 2^n}{(2n)!} \sqrt{1-x^2} \left(\frac{d}{dx}\right)^n \left(1-x^2\right)^{n-\frac{1}{2}}.} \tag{6.21}$$

The graphs of the first ten Chebyshev polynomials on the interval $[-1, 1]$ are given in Figure 6.2.

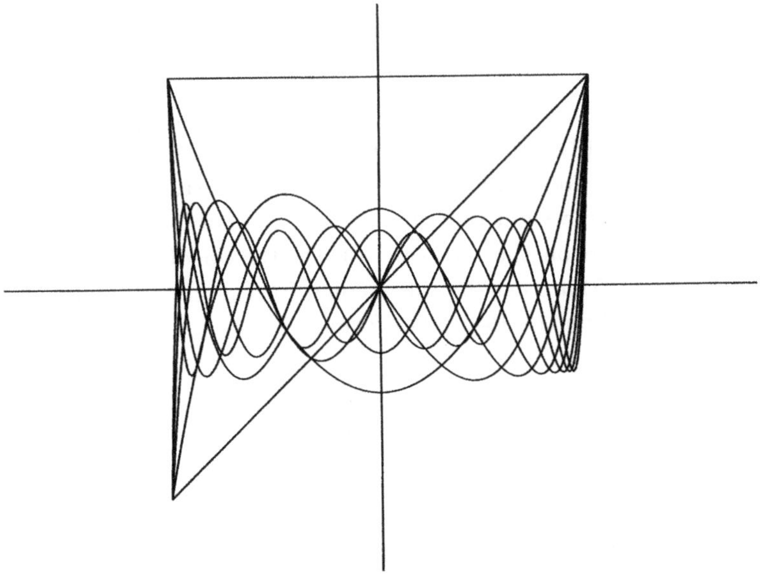

Figure 6.1. Graphs of the Legendre polynomials $P_n(x), n = 0, 1, 2, \ldots, 10$ on $[-1, 1]$.

The Legendre and Chebyshev polynomials are special cases of the *Gegenbauer polynomials* or *ultraspherical polynomials*, which follow from (6.18) by taking $\alpha = \beta = \gamma - \frac{1}{2}$ and with adapted K_n:

$$\boxed{C_n^\gamma(x) = \frac{(-1)^n\,(2\gamma)_n}{2^n n!\left(\gamma+\frac{1}{2}\right)_n}\left(1-x^2\right)^{\frac{1}{2}-\gamma}\left(\frac{d}{dx}\right)^n\left(1-x^2\right)^{n+\gamma-\frac{1}{2}}.} \tag{6.22}$$

When $\gamma = 0$ these polynomials are equal to 0 (if $n > 0$). The following limit holds:

$$\lim_{\gamma\to 0}\frac{1}{\gamma}C_n^\gamma(x) = \frac{2}{n}T_n(x), \quad n > 0.$$

The choice $X(x) = 1 - x^2$, that gave us the Jacobi polynomials, has zeros at ± 1. For the semi-infinite interval $(0, \infty)$ we take $X(x) = x$, giving for p_1

$$K_1 p_1(x) = \frac{w'(x)x}{w(x)} + 1.$$

This becomes a linear function if $w(x) = x^\alpha e^{\beta x}$. For convergence of the integrals (for instance (6.1)), β should be negative, say $\beta = -1$. This gives

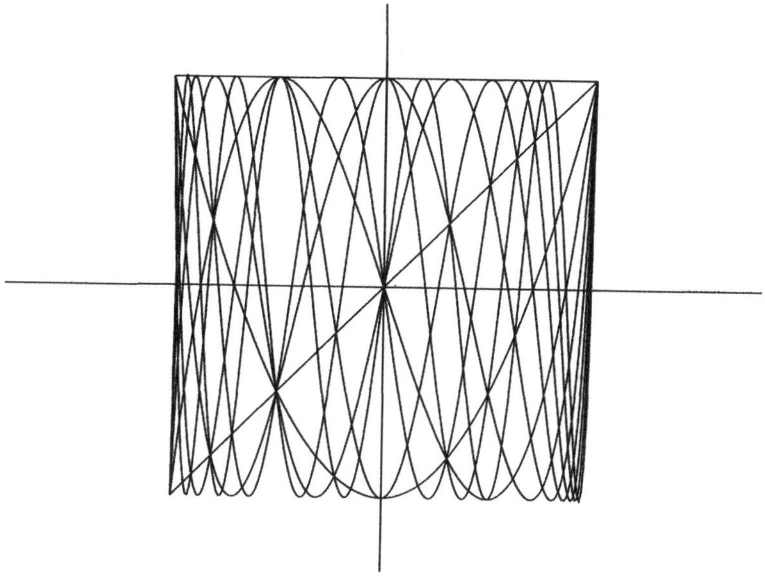

Figure 6.2. Graphs of the Chebyshev polynomials $T_n(x), n = 0, 1, 2, \ldots, 10$ on $[-1, 1]$.

the definition of the *Laguerre polynomials*

$$\boxed{L_n^\alpha(x) = \frac{1}{n!} x^{-\alpha} e^x \left(\frac{d}{dx}\right)^n \left[e^{-x} x^{n+\alpha}\right].} \tag{6.23}$$

The norm of the Laguerre polynomial follows from Exercise 6.5.

Choosing $X = \textit{constant}$ we finally arrive for the interval $(-\infty, \infty)$ at

$$K_1 p_1(x) = \frac{w'(x)}{w(x)}.$$

This yields a linear function when w is an exponential function with a quadratic function as argument. On account of symmetry and normalization one takes $w(x) = \exp(-x^2)$, although $w(x) = \exp(-\frac{1}{2}x^2)$ also is convenient, for instance, in physics. The first variant gives the *Hermite polynomials*

$$\boxed{H_n(x) = (-1)^n e^{x^2} \left(\frac{d}{dx}\right)^n e^{-x^2}.} \tag{6.24}$$

The second variant leads to

$$He_n(x) = (-1)^n e^{\frac{1}{2}x^2} \left(\frac{d}{dx}\right)^n e^{-\frac{1}{2}x^2}. \tag{6.25}$$

The relation between the two families is:

$$He_n(x) = 2^{-n/2} H_n(x/\sqrt{2}), \quad H_n(x) = 2^{n/2} He_n(x\sqrt{2}).$$

We summarize the results in Table 6.1. For each polynomial we give the weight function $w(x)$ and (a, b), the interval of orthogonality. We also give a few extra polynomials that are useful in applications, and which can be obtained from the classical ones by simple transformations. In some important cases we refer to explicit representations, which will be given in later sections.

6.4. Recurrence Relations

On account of the general theory we know that the classical orthogonal polynomials introduced in the previous section satisfy a recurrence relation. We recall (6.6):

$$p_{n+1} - (a_n x + b_n) p_n + c_n p_{n-1} = 0, \quad n = 1, 2, \ldots,.$$

We give the first two values of p_0 and p_1 for the classical orthogonal polynomials. The coefficients of the recurrence relations are given without proof, since the Rodrigues formula is not very suitable for obtaining the coefficients. We will point out in Exercise 6.1 how to obtain the coefficients a_n, b_n, c_n in the recurrence relation for the case of the Jacobi polynomials.

Jacobi:

$$a_n = \frac{(2n+\alpha+\beta+1)\,(2n+\alpha+\beta+2)}{(2n+2)\,(n+\alpha+\beta+1)}$$

$$b_n = \frac{(2n+\alpha+\beta+1)\,(\alpha^2-\beta^2)}{(2n+2)\,(n+\alpha+\beta+1)\,(2n+\alpha+\beta)}$$

$$c_n = \frac{2(n+\alpha)\,(n+\beta)\,(2n+\alpha+\beta+2)}{(2n+2)\,(n+\alpha+\beta+1)\,(2n+\alpha+\beta)}$$

$$P_0^{(\alpha,\beta)}(x) = 1, \quad P_1^{(\alpha,\beta)}(x) = \frac{1}{2}(\alpha-\beta) + \left[1 + \frac{1}{2}(\alpha+\beta)\right] x.$$

Gegenbauer:

$$a_n = \frac{2(n+\gamma)}{n+1}, \quad b_n = 0, \quad c_n = \frac{n+2\gamma-1}{n+1}$$

$$C_0^\gamma(x) = 1, \quad C_1^\gamma(x) = 2\gamma x.$$

Table 6.1. The Classical Orthogonal Polynomials and Some Variants

Name	$w(x)$	(a,b)
Jacobi: $P_n^{(\alpha,\beta)}(x)$ (see (6.35))	$(1-x)^\alpha(1+x)^\beta$	$(-1,1)$
Jacobi: (shifted) $R_n^{(\alpha,\beta)}(x) = P_n^{(\alpha,\beta)}(2x-1)$	$(1-x)^\alpha x^\beta$	$(0,1)$
Gegenbauer: $C_n^\gamma(x) = \frac{(2\gamma)_n}{(\gamma+\frac{1}{2})_n} P_n^{(\gamma-\frac{1}{2},\gamma-\frac{1}{2})}(x)$	$(1-x^2)^{\gamma-\frac{1}{2}}$	$(-1,1)$
Legendre: $P_n(x) = P_n^{(0,0)}(x)$	1	$(-1,1)$
Chebyshev: (first kind) $T_n(x) = \frac{n!}{(\frac{1}{2})_n} P_n^{(-\frac{1}{2},-\frac{1}{2})}(x)$	$1/\sqrt{(1-x^2)}$	$(-1,1)$
Chebyshev: (first kind, shifted) $T_n^*(x) = T_n(2x-1)$	$1/\sqrt{x(1-x)}$	$(-1,1)$
Chebyshev: (second kind) $U_n(x) = \frac{(n+1)!}{(\frac{3}{2})_n} P_n^{(\frac{1}{2},\frac{1}{2})}(x)$	$\sqrt{(1-x^2)}$	$(-1,1)$
Chebyshev: (second kind, shifted) $U_n^*(x) = U_n(2x-1)$	$\sqrt{x(1-x)}$	$(0,1)$
Laguerre: $L_n^\alpha(x)$ (see (6.40) or Remark 6.3)	$x^\alpha e^{-x}$	$(0,\infty)$
Hermite: $H_n(x)$ (see (6.41) or Remark 6.3)	e^{-x^2}	$(-\infty,\infty)$
Hermite: (variant) $He_n(x) = 2^{-n/2} H_n(x/\sqrt{2})$	$e^{-\frac{1}{2}x^2}$	$(-\infty,\infty)$

Chebyshev:

$$a_n = 2, \quad b_n = 0, \quad c_n = 1$$

$$T_0(x) = 1, \quad T_1(x) = x.$$

Legendre:

$$a_n = \frac{2n+1}{n+1}, \quad b_n = 0, \quad c_n = \frac{n}{n+1}$$

$$P_0(x) = 1, \quad P_1(x) = x.$$

Laguerre:

$$a_n = -\frac{1}{n+1}, \quad b_n = \frac{2n+\alpha+1}{n+1}, \quad c_n = \frac{n+\alpha}{n+1}$$

$$L_0^\alpha(x) = 1, \quad L_1^\alpha(x) = 1 + \alpha - x.$$

Hermite:

$$a_n = 2, \quad b_n = 0, \quad c_n = 2n$$

$$H_0(x) = 1, \quad H_1(x) = 2x.$$

The results for the polynomials on the finite interval all follow from the result for the Jacobi polynomials. Many results for the Laguerre and Hermite polynomials follow also from Jacobi polynomials by taking special limits. The Jacobi polynomials constitute a very rich class, for which many beautiful results are available.

The Christoffel-Darboux formula (6.8) is given for general orthonormal polynomials. To obtain the formula for Jacobi polynomials we first observe that the polynomial $p_n(x)$ given by

$$p_n(x) = \frac{P_n^{(\alpha,\beta)}(x)}{\sqrt{h_n^{(\alpha,\beta)}}}, \quad h_n^{(\alpha,\beta)} = \frac{2^{\alpha+\beta+1}}{2n+\alpha+\beta+1} \frac{\Gamma(n+\alpha+1)\,\Gamma(n+\beta+1)}{\Gamma(n+1)\,\Gamma(n+\alpha+\beta+1)}$$

is an orthonormal polynomial; see Exercise 6.11. The value of k_n in (6.5) follows from (6.36):

$$k_n = \lim_{x\to\infty} \frac{p_n(x)}{x^n} = \frac{2^{-n}}{\sqrt{h_n^{(\alpha,\beta)}}} \binom{2n+\alpha+\beta}{n}.$$

It follows that the *Christoffel-Darboux formula for the Jacobi polynomials* reads:

$$\sum_{k=0}^{n} \frac{P_k^{(\alpha,\beta)}(x)\,P_k^{(\alpha,\beta)}(y)}{h_k^{(\alpha,\beta)}} = A_n^{(\alpha,\beta)} \frac{P_n^{(\alpha,\beta)}(y)\,P_{n+1}^{(\alpha,\beta)}(x) - P_n^{(\alpha,\beta)}(x)\,P_{n+1}^{(\alpha,\beta)}(y)}{x-y}$$

where

$$A_n^{(\alpha,\beta)} = \frac{k_n}{k_{n+1}} \frac{1}{\sqrt{h_n^{(\alpha,\beta)} h_{n+1}^{(\alpha,\beta)}}}$$
$$= \frac{2^{-\alpha-\beta}}{2n+\alpha+\beta+2} \frac{\Gamma(n+2)\Gamma(n+\alpha+\beta+2)}{\Gamma(n+\alpha+1)\Gamma(n+\beta+1)}.$$

Similar formulas hold for the other polynomials.

6.5. Differential Equations

The next step is the derivation of the differential equation and this leads to a connection with the hypergeometric functions.

Consider the expression

$$\frac{d}{dx}[(1-x^2)w(x)p_n'(x)] = w(x)[(1-x^2)p_n''(x) + \{\beta - \alpha - (\alpha+\beta+2)x\}p_n'(x)], \tag{6.26}$$

where $w(x) = (1-x)^\alpha(1+x)^\beta$, $p_n(x) = P_n^{(\alpha,\beta)}(x)$. The expression between square brackets on the right-hand side is a polynomial of degree n. Hence we can write

$$\frac{d}{dx}\left[\left(1-x^2\right)w(x)p_n'(x)\right] = w(x)\sum_{j=0}^{n} \alpha_j p_j(x) \tag{6.27}$$

for certain α_j. By (6.3),

$$\alpha_k \|p_k\|^2 = \int_{-1}^{1} p_k(x)\frac{d}{dx}\left[\left(1-x^2\right)w(x)p_n'(x)\right] dx.$$

Integration by parts twice (observe that the integrated terms vanish) yields

$$\alpha_k \|p_k\|^2 = \int_{-1}^{1} p_n(x)\frac{d}{dx}\left[\left(1-x^2\right)w(x)p_k'(x)\right] dx.$$

But

$$\frac{d}{dx}[\left(1-x^2\right)w(x)p_k'(x)] = w(x)p(x),$$

where p is a polynomial of degree k. Hence $\alpha_k = 0$, for $k < n$. It follows that (6.27) can be written as

$$\frac{1}{w(x)}\frac{d}{dx}\left[\left(1-x^2\right)w(x)p_n'(x)\right] = \alpha_n p_n(x). \tag{6.28}$$

This equation can be interpreted as an *eigenvalue equation* for the operator defined by the left-hand side of (6.28). To compute α_n we substitute (6.5) and we take care of the coefficient of x^n in the left-hand side and right-hand side of (6.28). This gives $\alpha_n = -n(n+\alpha+\beta+1)$. With this result we have found the differential equation. For the Gegenbauer, Legendre and Chebyshev polynomials the differential equation follows easily by selecting the parameters. For the Hermite and Laguerre polynomials we can proceed as above. The differential equations below are defined for any complex value of n. When $n = 0, 1, 2, \ldots$ the equations have one and only one polynomial solution. When $n \notin \mathbb{N}$ the equations define more general special functions.

Jacobi: $y = P_n^{(\alpha,\beta)}(x)$ is a solution of

$$\left(1-x^2\right)y'' + [(\beta-\alpha) - (\alpha+\beta+2)x]y' + n(n+\alpha+\beta+1)y = 0. \tag{6.29}$$

Gegenbauer: $y = C_n^\gamma(x)$ is a solution of

$$\left(1-x^2\right)y'' - (2\gamma+1)xy' + n(n+2\gamma)y = 0. \tag{6.30}$$

Chebyshev: $y = T_n(x)$ is a solution of

$$\left(1-x^2\right)y'' - xy' + n^2 y = 0. \tag{6.31}$$

Legendre: $y = P_n(x)$ is a solution of

$$\left(1-x^2\right)y'' - 2xy' + n(n+1)y = 0. \tag{6.32}$$

Laguerre: $y = L_n^\alpha(x)$ is a solution of

$$xy'' + (\alpha+1-x)y' + ny = 0. \tag{6.33}$$

Hermite: $y = H_n(x)$ is a solution of

$$y'' - 2xy' + 2ny = 0. \tag{6.34}$$

When in (6.29) we substitute $z = (1-x)/2$, $w(z) = y(x)$ then we obtain:

$$z(1-z)w'' + [\alpha+1-(\alpha+\beta+2)z]w' + n(n+\alpha+\beta+1)w = 0.$$

This is the hypergeometric differential equation (see (5.8)) with

$$a = -n, \quad b = \alpha+\beta+n+1, \quad c = \alpha+1.$$

Taking into account the normalization (6.19) and the fact that the Jacobi polynomials are regular at $z = 0\,(x = 1)$ we arrive at the important result

$$\boxed{P_n^{(\alpha,\beta)}(x) = \binom{n+\alpha}{n} F\left(-n, \alpha+\beta+n+1; \alpha+1; \frac{1-x}{2}\right).} \tag{6.35}$$

6.6. Explicit Representations

With (6.35) we immediately have the explicit representation:

$$\boxed{P_n^{(\alpha,\beta)}(x) = \binom{n+\alpha}{n} \sum_{k=0}^{n} \frac{(-n)_k (n+\alpha+\beta+1)_k}{(\alpha+1)_k\; k!} \left(\frac{1-x}{2}\right)^k.}$$

This can also be written (see Exercises 3.1 and 3.6) as:

$$P_n^{(\alpha,\beta)}(x) = \frac{\Gamma(n+\alpha+1)}{n!\,\Gamma(n+\alpha+\beta+1)} \sum_{k=0}^{n} \binom{n}{k} \frac{\Gamma(n+k+\alpha+\beta+1)}{\Gamma(k+\alpha+1)} \left(\frac{x-1}{2}\right)^k. \tag{6.36}$$

Now (6.35) is available, many transformations are possible by invoking the formulas in (5.5). In this way we will not find a polynomial representation in terms of powers of x. Until now an attractive representation of the coefficients has not been found.

From (6.35) and (5.4) it is not possible to give a simple integral representation of the Jacobi polynomials, because the condition with respect to the b and c parameters in ((5.4) cannot be satisfied. However, by using the loop integral for the F–function in §5.2, we find

$$P_n^{(\alpha,\beta)}(x) = \frac{\Gamma(n+\alpha+1)\Gamma(n+\beta+1)}{n!\,\Gamma(n+\alpha+\beta+1)} \int_0^{(1+)} \frac{t^{n+\alpha+\beta}(1-tz)^n}{(t-1)^{n+\beta+1}}\, dt,$$

where $z = (1-x)/2$. This integral holds when $\Re(n+\alpha+\beta) > -1$. The contour starts and terminates at $t = 0$ and encircles the point $t = 1$ in positive direction.

Of course, we now also have explicit representations for the remaining polynomials that are orthogonal on $(-1, 1)$. We need not give the representations in powers of $1-x$, since for other polynomials the representations in powers of x are known.

Remark 6.3. The representations for the Hermite and Laguerre polynomials do not follow at once from that of the Jacobi polynomials. With the help of (6.30)–(6.34) it is not difficult, however, to obtain explicit polynomial solutions of these equations. When we look at Kummer's equation (7.4) we see that the Laguerre polynomial is a ${}_1F_1$–function:

$$\boxed{L_n^\alpha(x) = \binom{n+\alpha}{n} {}_1F_1(-n;\alpha+1;x),}$$

where the binomial coefficient is chosen for normalization and on account of convention; see also Exercise 7.10. The Laguerre polynomial can also be obtained as a certain limit of the Jacobi polynomial; see Exercise 6.10. The Hermite polynomial can be obtained from the Laguerre polynomial $L_n^\alpha(x)$ when we take $\alpha = \pm\frac{1}{2}$; see Exercise 6.7.

Gegenbauer:

$$\begin{aligned} C_n^\gamma(x) &= \frac{(2\gamma)_n}{\left(\gamma+\frac{1}{2}\right)_n} P_n^{(\gamma-\frac{1}{2},\gamma-\frac{1}{2})}(x) \\ &= \frac{(2\gamma)_n}{n!} F\left(-n, n+2\gamma; \gamma+\frac{1}{2}; \frac{1}{2}-\frac{1}{2}x\right) \\ &= \frac{1}{\Gamma(\gamma)} \sum_{k=0}^{\lfloor n/2 \rfloor} \frac{(-1)^k \Gamma(n-k+\gamma)}{k!\,(n-2k)!}(2x)^{n-2k}, \end{aligned} \tag{6.37}$$

where $\lfloor x \rfloor$ is the integer satisfying $\lfloor x \rfloor \le x < \lfloor x \rfloor + 1$ with $x \in \mathbb{R}$.

Chebyshev:

$$\begin{aligned} T_n(x) &= \frac{n!}{\left(\frac{1}{2}\right)_n} P_n^{(-\frac{1}{2},-\frac{1}{2})}(x) \\ &= F\left(-n, n; \frac{1}{2}; \frac{1}{2}-\frac{1}{2}x\right) \\ &= \frac{n}{2} \sum_{k=0}^{\lfloor n/2 \rfloor} \frac{(-1)^k (n-k-1)!}{k!\,(n-2k)!}(2x)^{n-2k} \\ &= \sum_{k=0}^{\lfloor n/2 \rfloor} \binom{n}{2k} x^{n-2k}\left(x^2-1\right)^k. \end{aligned} \tag{6.38}$$

Legendre:

$$\begin{aligned} P_n(x) &= P_n^{(0,0)}(x) \\ &= F\left(-n, n+1; 1; \frac{1}{2}-\frac{1}{2}x\right) \\ &= \sum_{k=0}^{\lfloor n/2 \rfloor} \frac{(-1)^k (2n-2k)!}{2^k\, k!\,(n-k)!\,(n-2k)!} x^{n-2k} \\ &= n! \sum_{k=0}^{\lfloor n/2 \rfloor} \frac{x^{n-2k}\left(x^2-1\right)^k}{2^{2k}(k!)^2(n-2k)!}. \end{aligned} \tag{6.39}$$

Laguerre:

$$\begin{aligned} L_n^\alpha(x) &= \binom{n+\alpha}{n}\, {}_1F_1(-n;\alpha+1;x) \\ &= \sum_{k=0}^{n} (-1)^k \binom{n+\alpha}{n-k} \frac{x^k}{k!}. \end{aligned} \tag{6.40}$$

Hermite:

$$\begin{aligned} H_{2n}(x) &= \frac{(-1)^n (2n)!}{n!}\, {}_1F_1\left(-n;\frac{1}{2};x^2\right) \\ H_{2n+1}(x) &= \frac{(-1)^n (2n+1)!}{n!}\, 2x\, {}_1F_1\left(-n;\frac{3}{2};x^2\right) \\ H_n(x) &= n! \sum_{k=0}^{\lfloor n/2 \rfloor} \frac{(-1)^k}{k!\,(n-2k)!} (2x)^{n-2k}. \end{aligned} \tag{6.41}$$

The third representation in (6.38) is only defined for $n \ge 1$. The final representations in (6.38) and (6.39) follow from transformation formulas of hypergeometric functions. For instance, we can apply the quadratic transformation (1) of Exercise 5.8 with $a = -n/2, b = (n+1)/2$ on the second line of (6.39). Applying next the first transformation of (5.5) we obtain the final line in (6.39).

We give another representation of the Chebyshev polynomials, which is unique. Put in (6.31) $x = \cos\theta$, $0 \le \theta \le \pi$, $w(\theta) = y(x)$. Then the equation for w becomes

$$\frac{d^2 w}{d\theta^2} + n^2 w = 0$$

with solutions $w = \cos n\theta, w = \sin n\theta$. Taking into account the values of T_0 and T_1 we find

$$\boxed{T_n(\cos\theta) = \cos n\theta, \quad \text{that is,} \quad T_n(x) = \cos(n \arccos x).} \tag{6.42}$$

Many nice properties of the Chebyshev polynomials follow directly from those of the elementary trigonometric functions. The solution $w = \sin n\theta$ does not yield a polynomial solution of the above differential equation. However, the function

$$\boxed{U_n(\cos\theta) = \frac{\sin(n+1)\theta}{\sin\theta}, \qquad n = 0, 1, 2, \ldots}$$

is a polynomial in $x = \cos\theta$, and is called the *Chebyshev polynomial of the second kind.* It is associated with the weight function $\sqrt{1-x^2}$ on $(-1,1)$. In fact, we have

$$U_n(x) = \frac{(n+1)!}{\left(\frac{3}{2}\right)_n} P_n^{\left(\frac{1}{2},\frac{1}{2}\right)}(x).$$

6.7. Generating Functions

A generating function is a function $F(x,t)$ having a convergent power series in t of the form

$$F(t,x) = \sum_{n=0}^{\infty} a_n p_n(x) t^n$$

where in the present case $\{p_n\}$ is a family of classical orthogonal polynomials. The radius of convergence may be finite or infinite. Many generating functions can be obtained by using the Rodrigues formulas. Often a convenient starting point is the Cauchy integral for a function and its derivatives:

$$f(z) = \frac{1}{2\pi i} \int \frac{f(\zeta)\, d\zeta}{(\zeta - z)}, \quad f^{(n)}(z) = \frac{n!}{2\pi i} \int \frac{f(\zeta)\, d\zeta}{(\zeta - z)^{n+1}}$$

integrated along a suitable contour in the complex plane. We demonstrate the method for the Laguerre polynomials. The result also gives a verification of the recurrence relation for the Laguerre polynomials.

From (6.23) it follows that

$$L_n^{\alpha}(x) = \frac{x^{-\alpha} e^{x}}{2\pi i} \int \frac{e^{-\zeta} \zeta^{n+\alpha}}{(\zeta - x)^{n+1}}\, d\zeta$$

where we integrate along a circle around the (complex) point $x \neq 0$ with radius less than $|x|$ (to keep the origin outside the circle). Next we set up the summation:

$$\sum_{n=0}^{\infty} L_n^{\alpha}(x) t^n = \frac{x^{-\alpha} e^{x}}{2\pi i} \int \frac{e^{-\zeta} \zeta^{\alpha}}{\zeta - x} \sum_{n=0}^{\infty} \frac{t^n \zeta^n}{(\zeta - x)^n}\, d\zeta.$$

With ζ on the contour of integration we can write

$$\zeta = x + r \exp(i\theta), \quad 0 < r < |x|, \quad 0 \le \theta < 2\pi.$$

Then we can choose t so small that the geometric series uniformly converges with respect to ζ. That is, we can make

$$\left| \frac{t\zeta}{\zeta - x} \right| = \left| \frac{t[x + r \exp(i\theta)]}{r \exp(i\theta)} \right|$$

smaller than unity, uniformly with respect to ζ. Performing the summation we obtain

$$\sum_{n=0}^{\infty} L_n^{\alpha}(x) t^n = \frac{x^{-\alpha} e^{x}}{2\pi i (1-t)} \int \frac{e^{-\zeta} \zeta^{\alpha}}{\zeta - x/(1-t)}\, d\zeta.$$

Calculating the integral by the residue method, we obtain the generating function

$$(1-t)^{-\alpha-1}e^{-tx/(1-t)} = \sum_{n=0}^{\infty} L_n^{\alpha}(x)t^n, \quad \alpha, x \in \mathbb{C}, \quad |t| < 1. \tag{6.43}$$

Analytic continuation gives the domain of validity of the parameters. When $x = 0$ a direct proof follows from

$$L_n^{\alpha}(0) = \binom{n+\alpha}{n} = \frac{(\alpha+1)_n}{n!}.$$

Let us denote the left-hand side of (6.43) by $F(t)$. Then F satisfies the differential equation

$$(1-t)^2F' = [(\alpha+1)(1-t) - x]F$$

and we know that F has a convergent power series $\sum_{n=0}^{\infty} c_n t^n$ for $x, \alpha \in \mathbb{C}$ and $|t| < 1$. Substitution of this power series gives for the coefficients c_n the same recurrence relation as that given for the Laguerre polynomials in §6.4. This method for finding the coefficients in the recurrence relation is usually more efficient than a method based on the Rodrigues formula.

For the Hermite polynomials we find in a similar way

$$e^{2xz-z^2} = \sum_{n=0}^{\infty} \frac{H_n(x)}{n!} z^n, \quad x, z \in \mathbb{C}. \tag{6.44}$$

For the Gegenbauer polynomials we have for $\gamma \neq 0$ (see Exercise 6.2)

$$(1-2xz+z^2)^{-\gamma} = \sum_{n=0}^{\infty} C_n^{\gamma}(x)z^n, \quad -1 \le x \le 1, \quad |z| < 1, \tag{6.45}$$

a formula that is often used as the definition for the Gegenbauer polynomials. For the Legendre polynomials we thus obtain

$$\frac{1}{\sqrt{1-2xz+z^2}} = \sum_{n=0}^{\infty} P_n(x)z^n, \quad -1 \le x \le 1, \quad |z| < 1. \tag{6.46}$$

A direct proof based on the Rodrigues formula is not difficult for the Legendre polynomials (Exercise 6.3).

6.8. Legendre Polynomials

We derive a few extra results for Legendre polynomials: integral representations, bounds and an asymptotic expansion.

6.8.1. The Norm of the Legendre Polynomials

The generating functions can sometimes be used to compute the norms of the polynomials. We give an example based on (6.46). By squaring the series, a new power series arises:

$$\frac{1}{1-2xz+z^2} = \sum_{n=0}^{\infty} c_n z^n, \quad c_n = \sum_{k=0}^{n} P_k(x)P_{n-k}(x).$$

Integration with respect to x over $[-1,1]$ yields only a result for the even powers of z (use the orthogonality). In c_{2n} only the product $P_n^2(x)$ makes a contribution:

$$\sum_{n=0}^{\infty} \|P_n(x)\|^2 z^{2n} = \int_{-1}^{1} \frac{dx}{1-2xz+z^2} = \frac{1}{z}\ln\frac{1+z}{1-z} = \sum_{n=0}^{\infty} \frac{2}{2n+1} z^{2n}.$$

Hence

$$\int_{-1}^{1} P_n(x)P_m(x)\,dx = \frac{2}{2n+1}\delta_{n,m}. \tag{6.47}$$

See Exercise 6.5 for obtaining the norm of the Laguerre polynomials in this way.

6.8.2. Integral Expressions for the Legendre Polynomials

We observe that the generating function in (6.46) has two forms:

$$\begin{aligned} \frac{1}{\sqrt{1-2xz+z^2}} &= \sum_{n=0}^{\infty} P_n(x)z^n, \qquad -1\le x\le 1, \quad |z|<1, \\ &= \sum_{n=0}^{\infty} P_n(x)z^{-n-1}, \quad -1\le x\le 1, \quad |z|>1. \end{aligned} \tag{6.48}$$

We see that, writing $x=\cos\theta, 0<\theta<\pi$, and applying Cauchy's theorem on the coefficients in the second power series,

$$P_n(\cos\theta) = \frac{1}{2\pi i}\oint \frac{z^n}{\sqrt{1-2z\cos\theta+z^2}}\,dz, \tag{6.49}$$

where the integral is taken (in positive sense) along a circle $|z|=R$, $R>1$. By deforming the contour of integration around the segment joining the points

$\exp \pm i\theta$, the singular points of the integrand, we can obtain *Laplace's formula for the Legendre polynomial*

$$\boxed{P_n(\cos\theta) = \frac{1}{\pi}\int_0^{\pi} (\cos\theta + i\sin\theta\,\cos\psi)^n\,d\psi.} \tag{6.50}$$

This formula is derived in Chapter 8 by using a different method; see (8.42). A direct verification follows from expanding the integrand by using the binomial theorem and comparing the result with the final line in (6.39).

Another interesting pair of integrals is

$$\boxed{P_n(\cos\theta) = \frac{2}{\pi}\int_0^{\theta} \frac{\cos\left(n+\frac{1}{2}\right)t}{\sqrt{2\cos t - 2\cos\theta}}\,dt = \frac{2}{\pi}\int_\theta^{\pi} \frac{\sin\left(n+\frac{1}{2}\right)t}{\sqrt{2\cos\theta - 2\cos t}}\,dt,} \tag{6.51}$$

which are known as the *Dirichlet-Mehler formula for the Legendre polynomial.* A proof can be based on (6.49), by deforming the contour around the circular arc $-\theta \le \arg z \le \theta$ of the unit circle. This gives the first formula. The second formula follows from the first one by a simple change of variables: $t \to \pi - t$ and $\theta \to \pi - \theta$ and using $P_n(\cos\theta) = (-1)^n P_n(-\cos\theta)$.

A direct proof of the first formula in (6.51) follows from Laplace's formula (6.50) by using the change of variables $\psi \to t$ given by $\exp(it) = \cos\theta + i\sin\theta\,\cos\psi$. Initially, the ψ–interval $[0,\pi]$ is mapped to a path in the complex plane from θ to $-\theta$. This path can be deformed into a real interval.

6.8.3. Some Bounds on Legendre Polynomials

Laplace's formula yields a few simple properties of the Legendre polynomials. First we observe that

$$\boxed{|P_n(x)| \le 1, \quad -1 \le x \le 1, \quad n = 0,1,2,\ldots,} \tag{6.52}$$

which easily follows from the fact that the maximum modulus of the integrand in Laplace's integral equals unity. A sharper bound, which is not sharp near the end points $x = \pm 1$, follows from observing

$$|\cos\theta + i\sin\theta\,\cos\psi|^n = |1 - \sin^2\theta\sin^2\psi|^{n/2}.$$

Hence

$$|P_n(\cos\theta)| \le \frac{2}{\pi}\int_0^{\frac{1}{2}\pi} |1 - \sin^2\theta\sin^2\psi|^{n/2}\,d\psi.$$

But

$$1 - \sin^2\theta\sin^2\psi < 1 - \frac{4\psi^2\sin^2\theta}{\pi^2} < \exp\left(-\frac{4\psi^2\sin^2\theta}{\pi^2}\right),$$

since $\sin\psi > 2\psi/\pi$, $0 < \psi < \frac{1}{2}\pi$ and $1 - t < \exp(-t)$, $t > 0$. Thus we obtain

$$|P_n(\cos\theta)| < \frac{2}{\pi}\int_0^{\frac{1}{2}\pi} e^{\left(-\frac{2n\psi^2\sin^2\theta}{\pi^2}\right)}\,d\psi < \frac{2}{\pi}\int_0^{\infty} e^{\left(-\frac{2n\psi^2\sin^2\theta}{\pi^2}\right)}\,d\psi.$$

This gives

$$\boxed{|P_n(x)| < \sqrt{\frac{\pi}{2n\,(1-x^2)}}, \quad -1 < x < 1, \quad n = 1, 2, 3, \ldots .} \tag{6.53}$$

6.8.4. An Asymptotic Expansion as n is Large

We use the first Dirichlet-Mehler formula in (6.51) to derive an asymptotic expansion of $P_n(\cos\theta)$ as $n \to \infty$. First we write

$$P_n(\cos\theta) = \frac{1}{\pi}\int_{-\theta}^{\theta} \frac{e^{i(n+\frac{1}{2})t}}{\sqrt{2\cos t - 2\cos\theta}}\,dt. \tag{6.54}$$

In fact the above integral can be expanded by using the method of stationary phase, but we prefer a method based on Watson's lemma. First we choose a different path of integration on which the integral is free of strong oscillations. Consider the integral $\oint f(t)\,dt$, where f is the integrand in (6.54), and the closed contour lies in the upper half plane $\Im t > 0$. Since f is analytic in $\Im t > 0$ the integral is zero. We choose the contour along the two half lines $\Re t = \pm\theta, \Im t > 0$ and the interval $(-\theta, \theta)$. It is not difficult to verify that

$$\pi\sqrt{2}\,P_n(\cos\theta) = \left[\int_{-\theta}^{-\theta+i\infty} - \int_{\theta}^{\theta+i\infty}\right] \frac{e^{i(n+\frac{1}{2})\tau}}{\sqrt{\cos\tau - \cos\theta}}\,d\tau.$$

Substituting $\tau = -\theta + it, \tau = \theta + it$, and observing that we can take twice the real part of the first integral, we obtain

$$P_n(\cos\theta) = \frac{2}{\pi\sqrt{2}}\Re\left[\int_0^{\infty} \frac{e^{-i(n+\frac{1}{2})\theta+\frac{1}{2}\pi i}\, e^{-(n+\frac{1}{2})t}}{\sqrt{\cos\theta(\cosh t - 1) + i\sin\theta\sinh t}}\,dt\right]. \tag{6.55}$$

We can apply Watson's lemma by substituting the expansion

$$\frac{1}{\sqrt{\cos\theta(\cosh t - 1) + i\sin\theta\sinh t}} = \sum_{k=1}^{\infty} c_k(\theta)\,t^{k-\frac{1}{2}}, \quad c_0(\theta) = \frac{e^{-\frac{1}{4}\pi i}}{\sqrt{\sin\theta}}. \tag{6.56}$$

Higher coefficients $c_k(\theta)$ are complicated expressions and can be computed by standard techniques. However, it is possible to modify this method, and to obtain an asymptotic expansion in which the coefficients are available in a simple closed form. It is easily verified that

$$\cos\theta(\cosh t - 1) + i\sin\theta\sinh t = iue^t\sin\theta(1-\lambda u), \qquad \lambda = \frac{ie^{-i\theta}}{2\sin\theta}.$$

Substituting this in (6.55), we obtain

$$P_n(\cos\theta) = \frac{\sqrt{2}}{\pi\sqrt{\sin\theta}}\Re\left[e^{-i[(n+\frac{1}{2})\theta-\frac{1}{4}\pi]}\int_0^1 u^{-\frac{1}{2}}(1-u)^n(1-\lambda u)^{-\frac{1}{2}}\,du\right].$$

That is, we have a new representation of the Legendre polynomials in terms of the hypergeometric function (see 4.4)):

$$P_n(\cos\theta) = \frac{4\,n!}{\pi\left(\frac{3}{2}\right)_n\sqrt{2\sin\theta}}\Re\left[e^{-i[(n+\frac{1}{2})\theta-\frac{1}{4}\pi]}\,F\left(\frac{1}{2},\frac{1}{2};n+\frac{3}{2};\lambda\right)\right].$$

We can expand the F–function when $|\lambda| < 1$. It follows that, if $2\sin\theta > 1$ on $[0,\pi]$, that is, if $\frac{1}{6}\pi < \theta < \frac{5}{6}\pi$, we have the convergent expansion

$$P_n(\cos\theta) = \frac{4\,n!}{\pi\left(\frac{3}{2}\right)_n}\sum_{k=0}^{\infty}\frac{\cos[(n+k+\frac{1}{2})\theta-(k+\frac{1}{2})\frac{\pi}{2}]}{(2\sin\theta)^{k+\frac{1}{2}}}\,\frac{(\frac{1}{2})_k(\frac{1}{2})_k}{k!\,(n+\frac{3}{2})_k}, \qquad (6.57)$$

Outside the θ–interval $[\frac{1}{6}\pi, \frac{5}{6}\pi]$ this expansion is divergent, but for all $\theta \in [(0,\pi)$ it has as an asymptotic character as $n \to \infty$. In fact we have

$$P_n(\cos\theta) \sim \frac{4\,n!}{\pi\left(\frac{3}{2}\right)_n}\sum_{k=0}^{\infty}\frac{\cos[(n+k+\frac{1}{2})\theta-(k+\frac{1}{2})\frac{\pi}{2}]}{(2\sin\theta)^{k+\frac{1}{2}}}\,\frac{(\frac{1}{2})_k(\frac{1}{2})_k}{k!\,(n+\frac{3}{2})_k},$$

as $n \to \infty$, uniformly for $\varepsilon < \theta < \pi - \varepsilon$, where ε is a fixed positive number, $0 < \varepsilon < \frac{1}{2}\pi$. This expansion is given by Stieltjes (1890), who also gave a simple upper bound for the remainder of this asymptotic expansion; see SZEGÖ (1975), where many other other asymptotic results for orthogonal polynomials are given.

The expansion fails when θ or $\pi - \theta$ is small. It is not possible to give an expansion in terms of elementary functions in which θ may approach 0 or π. In these cases a Bessel function is needed. To see this, we assume that in (6.54) θ is small. Then t is also small, and we have $\sqrt{2\cos t - 2\cos\theta} \sim \sqrt{\theta^2 - t^2}$. Hence,

$$P_n(\cos\theta) \sim \frac{1}{\pi}\int_{-\theta}^{\theta}\frac{\cos[(n+\frac{1}{2})t]}{\sqrt{\theta^2-t^2}}\,dt.$$

Comparing this with the first integral in Exercise 9.12, which reads for $\nu = 0$:

$$J_0(z) = \frac{1}{\pi}\int_{-1}^{1} \frac{\cos zt}{\sqrt{1-t^2}}\,dt,$$

we obtain

$$P_n(\cos\theta) \sim \sqrt{\frac{\theta}{\sin\theta}}\, J_0\left[(n+\tfrac{1}{2})\theta\right], \quad \theta = o(1).$$

The term in front of the Bessel function comes from the limiting form

$$\lim_{t\to\theta} \frac{\theta^2 - t^2}{2\cos t - 2\cos\theta} = \frac{\theta}{\sin\theta}.$$

Observe that we have not used that n is large. However, the error in the asymptotic relation for $P_n(\cos\theta)$ becomes small when n is large. In fact we can write

$$P_n(\cos\theta) = \sqrt{\frac{\theta}{\sin\theta}}\, J_0\left[\left(n+\frac{1}{2}\right)\theta\right] + \mathcal{O}\left(n^{-3/2}\right),$$

as $n \to \infty$, where the $\mathcal{O}-$term is uniform with respect to $\theta \in [0, \pi - \varepsilon]$, $\varepsilon > 0$. For this result we again refer to SZEGÖ (1975).

6.9. Expansions in Terms of Orthogonal Polynomials

We will not give the general theory, but discuss some aspects in connection with the Legendre and Chebyshev polynomials.

6.9.1. An Optimal Result in Connection with Legendre Polynomials

Let $L^2(-1,1)$ be the class of functions f which are square integrable on $[-1,1]$. That is:

$$f \in L^2(-1,1) \Longleftrightarrow \int_{-1}^{1} [f(x)]^2\,dx \quad \text{exists and is finite.}$$

When we want to approximate a function from $L^2(-1,1)$ by a polynomial, then the best choice (with respect to the L^2- norm, that is according to the criterion formulated by (6.60)) is an expansion in terms of Legendre polynomials. We assume that for a given $f \in L^2(-1,1)$ the expansion

$$f(x) = \sum_{n=0}^{\infty} c_n P_n(x) \tag{6.58}$$

uniformly converges on $[-1, 1]$. Then we can express the coefficients c_n in terms of the sum $f(x)$. To obtain the coefficients multiply (6.58) by $P_m(x)$ and integrate on $[-1, 1]$. On account of (6.47) we can write

$$c_n = (n + \tfrac{1}{2}) \int_{-1}^{1} f(x) P_n(x)\, dx. \tag{6.59}$$

Assume that we want to approximate f by a polynomial f_N; this polynomial can always be written in the form

$$f_N(x) = \sum_{n=0}^{N} a_n P_n(x).$$

We want to choose f_N such that the *mean quadratic deviation*

$$m_N = \int_{-1}^{1} \left[f(x) - \sum_{n=0}^{N} a_n P_n(x) \right]^2 dx \tag{6.60}$$

is as small as possible. The claim is that m_N is minimal for the choice $a_n = c_n$.

A proof follows from the orthogonality property of the Legendre polynomials:

$$\begin{aligned} m_N &= \int_{-1}^{1} f^2(x)\, dx - 2 \sum_{n=0}^{N} \frac{a_n c_n}{n + \frac{1}{2}} + \sum_{n=0}^{N} \frac{a_n^2}{n + \frac{1}{2}} \\ &= \int_{-1}^{1} f^2(x)\, dx - \sum_{n=0}^{N} \frac{c_n^2}{n + \frac{1}{2}} + \sum_{n=0}^{N} \frac{(a_n - c_n)^2}{n + \frac{1}{2}}. \end{aligned} \tag{6.61}$$

The final expression is minimal if $a_n = c_n$. Hence, we find for the best approximation in the L^2–norm

$$f_N(x) = \sum_{n=0}^{N} c_n P_n(x),$$

where c_n are given by (6.59). Observe that the polynomial f_N is just the first part of the infinite series expansion in terms of Legendre polynomials of the function f. As a side result we have from (6.61) *Bessel's inequality*

$$\sum_{n=0}^{\infty} \frac{c_n^2}{n + \frac{1}{2}} \le \int_{-1}^{1} f^2(x)\, dx \tag{6.62}$$

for the coefficients c_n of the series expansion in terms of Legendre polynomials of the function f. Compare this with the inequality having the same name in the theory of Fourier series; see ZYGMUND (1959).

In the following theorem we formulate the conditions on f in order that (6.58) is a convergent expansion.

Theorem 6.6.

(*i*) *Let f be continuous on $[-1,1]$, with the exception of a finite number of points of discontinuity.*

(*ii*) *On $[-1,1]$ let the derivative f' exist in the points where f is continuous, and let the left and right derivatives exist in the points where f is not continuous.*

(*iii*) *Let c_n be given by (6.59).*

Then the series in (6.58) converges and for $-1 < x < 1$ the series is equal to $f(x)$ in the points where f is continuous; the series is equal to the value $\frac{1}{2}[f(x+0)+f(x-0)]$ in the points where f is not continuous.

For a proof we refer to NIKIFOROV & UVAROV (1988), where a proof is given for more general orthogonal systems.

When f and all its derivatives exist on $[-1,1]$ formula (6.59) for the coefficient c_n can be modified. By applying the Rodrigues formula (6.20) one obtains, after integrating by parts n times,

$$c_n = \frac{n+\frac{1}{2}}{2^n n!}\int_{-1}^{1} f^{(n)}(x)\left(1-x^2\right)^n\,dx. \tag{6.63}$$

Example 6.2. Consider the function $f(x)=e^{i\alpha x}$, $-1\le x\le 1$, with $\alpha\in\mathbb{C}$. We determine the coefficients c_n in the expansion (6.58). The n–th derivative of f is given by $f^{(n)}(x)=(i\alpha)^n e^{i\alpha x}$. Hence, using (6.63) for the coefficients c_n we obtain

$$c_n = \frac{n+\frac{1}{2}}{2^n n!}(i\alpha)^n\int_{-1}^{1} e^{i\alpha x}\left(1-x^2\right)^n\,dx.$$

From the *Poisson integral* in Exercise 9.12 we derive that c_n can be written in terms of a Bessel function. So we find the result (BAUER (1859)

$$e^{i\alpha x} = \sqrt{\frac{\pi}{2\alpha}}\sum_{n=0}^{\infty}(2n+1)i^n J_{n+\frac{1}{2}}(\alpha)P_n(x). \tag{6.64}$$

An application of this result will be given in §10.3.2.

6.9.2. Numerical Aspects of Chebyshev Polynomials

As follows from formula (6.42), an expansion in terms of Chebyshev polynomials is in fact a Fourier series. That is, let f have the expansion

$$f(x)=\frac{1}{2}c_0+\sum_{n=1}^{\infty}c_nT_n(x),\quad -1\le x\le 1,\quad c_n=\frac{2}{\pi}\int_{-1}^{1}\frac{f(x)\,T_n(x)}{\sqrt{1-x^2}}\,dx. \tag{6.65}$$

Then we can also write

$$f(\cos\theta) = \frac{1}{2}c_0 + \sum_{n=1}^{\infty} c_n \cos n\theta, \quad 0 \le \theta \le \pi, \quad c_n = \frac{2}{\pi}\int_0^{\pi} f(\cos\theta)\,\cos n\theta\,d\theta. \tag{6.66}$$

Hence, several methods and techniques from Fourier theory, for instance, the *discrete Fourier transform* and the *fast Fourier transform*, can be used to evaluate the coefficients c_n. For details and useful information on Chebyshev expansions we refer to LUKE (1969), (1975), RIVLIN (1990), FOX & PARKER (1968), CLENSHAW (1962), CLENSHAW & PICKEN (1966), and NÉMETH (1992). For the discrete and fast techniques see, for instance, OPPENHEIM & SCHAFER (1975).

Chebyshev expansions are used very frequently in numerical algorithms for special functions. In general the convergence is very fast. When using the truncated expansion $f_n(x) = \frac{1}{2}c_0 + \sum_{k=1}^{n} c_k T_k(x)$, the error

$$\max_{x\in(-1,1)} |f(x) - f_n(x)|$$

is only slightly larger than the error in the *best approximation* (in the Chebyshev sense). For details see RIVLIN (1990). Various methods have been developed for computing the coefficients c_n in Chebyshev expansions. Quadrature methods are not the only tool for computing the coefficients (although the quadrature methods can be based on fast algorithms and may efficiently produce the coefficients). See CLENSHAW (1962) or LUKE (1969) for methods based on differential equations. Many special functions obey a linear (second) order differential equation with polynomial coefficients in which a Chebyshev expansion of the solution can be substituted. In this way recurrence relations for the coefficients can be obtained.

Example 6.3. Let $f(x) = 1/(x+a)$ with $a > 1$. Then $(x+a)f' = -f$. Considering the expansion in (6.65) we can write $f'(x) = \frac{1}{2}c'_0 + \sum_{n=1}^{\infty} c'_n T_n(x)$ where the relation between c_n and c'_n is given by $c'_{n-1} = c'_{n+1} + 2nc_n, (n \ge 1)$. Verify this, for instance by observing that $2\int T_n(x)\,dx = T_{n+1}(x)/(n+1) - T_{n-1}(x)/(n-1), (n > 1)$. From the differential equation (and the relation $xT_n(x) = \frac{1}{2}[T_{n-1}(x) + T_{n+1}(x)]$) we obtain the recurrence relation $-c_n = ac'_n + \frac{1}{2}(c'_{n-1} + c'_{n+1})$. Writing this relation with n replaced with $n+2$, and subtracting these two relations, the coefficients c'_n can be eliminated, and we obtain $c_{n+2} + 2ac_{n+1} + c_n = 0$. A solution of this recursion is obtained by writing $c_n = c\lambda^n$. It follows that $\lambda = \sqrt{a^2-1} - a$. The value of c can be obtained by observing that $T_n(1) = 1$, and that, hence, $1/(a+1) =$

$\frac{1}{2}c + c\sum_{n=1}^{\infty}\lambda^n$, giving $c = 2/\sqrt{a^2-1}$. It follows that

$$\frac{1}{x+a} = \frac{2}{\sqrt{a^2-1}}\left\{\frac{1}{2} + \sum_{n=1}^{\infty}\left[\sqrt{a^2-1}-a\right]^n T_n(x)\right\}, \quad a > 1, \ -1 \le x \le 1.$$

The expansion holds also for certain complex values of x and a, but we have to be careful with selecting the square root when a is complex.

6.10. Remarks and Comments for Further Reading

6.1. An excellent reference for this chapter is SZEGÖ (1974). Many new books have been written since this classic. For instance, FREUD (1971), RIVLIN (1990) and CHIHARA (1978). Interesting conference proceedings are BREZINSKI (1985) and NEVAI (1990).

6.2. Several computational aspects of orthogonal polynomials and of Gauss-type quadrature rules are treated in GAUTSCHI (1990), (1994). These papers also discuss methods for obtaining recursion coefficients a_n, b_n, c_n of (6.6) in the case of non-classical weight functions. The representations of these coefficients as given in Theorem 6.1 are not always suitable for obtaining numerical values. In §13.3, Example 9, the stability of the recurrence relation of the Jacobi polynomials (with respect to numerical evaluation of the polynomials) is discussed.

6.3. Many new results on asymptotic approximations for classical orthogonal polynomials have been obtained since Szegö's book. VAN ASSCHE (1987) considers general orthogonal polynomials. For uniform asymptotic estimates for classical orthogonal polynomials, see, for instance, FRENZEN & WONG (1985) for Jacobi polynomials (generalization of *Hilb's formula*), and FRENZEN & WONG (1988) and TEMME (1990) for Laguerre polynomials. TEMME (1989) gives estimates for the classical polynomials in terms of other classical polynomials. FRENZEN & WONG (1985) also gives estimates for the zeros of the Jacobi polynomials for large values of the order.

6.4. Present research in orthogonal polynomials is often in terms of q–orthogonal polynomials; see ASKEY & WILSON (1985) and GASPER & RAHMAN (1990). A few definitions of q–functions are given in §5.9.1 of the previous chapter with references to the literature.

6.11. Exercises and Further Examples

6.1. Determine the coefficients a_n, b_n, c_n of the recurrence relation for the Jacobi polynomials given in §6.4. *Warning*: the relations given after (6.6) are for orthonormal polynomials. It is better to proceed as follows. First observe that from the recurrence relation follows that a_n is given by

$$a_n = \lim_{x\to\infty} \frac{P_{n+1}^{(\alpha,\beta)}(x)}{P_n^{(\alpha,\beta)}(x)},$$

and use (6.36) to compute this limit. Next use the values of $P_n^{(\alpha,\beta)}(x)$ at $x = 1$ and $x = -1$ (see (6.19)) for determining b_n and c_n.

6.2. Verify with the help of Exercise 6.1 and (6.37) the recurrence relation

$$(n+1)p_{n+1}(x) = 2(n+\gamma)xp_n(x) - (n+2\gamma-1)p_{n-1}(x)$$

for the Gegenbauer polynomials. Derive from this the generating function (6.45) by introducing $G(x,t) = \sum_{n=0}^{\infty} p_n(x)t^n$ with $p_n(x) = C_n^\gamma(x)$. We have

$$\begin{aligned}\frac{\partial G}{\partial t} &= \sum_{n=0}^{\infty} np_n(x)t^{n-1} = \sum_{n=0}^{\infty}(n+1)p_{n+1}t^n \\ &= \sum_{n=0}^{\infty}[2(n+\gamma)xp_n(x) - (n+2\gamma-1)p_{n-1}(x)]t^n.\end{aligned}$$

Verify that this yields

$$\frac{\partial G}{\partial t} = 2\gamma xG + 2xt\frac{\partial G}{\partial t} - 2\gamma tG - t^2\frac{\partial G}{\partial t}.$$

This gives the differential equation

$$\frac{1}{G}\frac{\partial G}{\partial t} = \frac{2\gamma(x-t)}{1-2xt+t^2}.$$

With the initial conditions $G(x,0) = 1$, and we again arrive at (6.45).

6.3. Give a direct proof of (6.46) based on the Rodrigues formula (6.20). For that purpose write

$$\left(\frac{d}{dx}\right)^n\left(1-x^2\right)^n = \frac{n!}{2\pi i}\int \frac{(1-\zeta^2)^n}{(\zeta-x)^{n+1}}\,d\zeta,$$

where the contour of integration is a circle with center x.

6.4. Prove the contiguous relations

$$\begin{aligned}
\frac{d}{dx}P_n^{(\alpha,\beta)}(x) &= \tfrac{1}{2}(B_n+1)P_{n-1}^{(\alpha+1,\beta+1)}(x)\\
A_n\left(1-x^2\right)\frac{d}{dx}P_n^{(\alpha,\beta)}(x) &= n[\alpha-\beta-A_nx]P_n^{(\alpha,\beta)}(x)\\
&\quad+2(n+\alpha)(n+\beta)P_{n-1}^{(\alpha,\beta)}(x),\\
\frac{A_n+2}{B_n+1}\left(1-x^2\right)\frac{d}{dx}P_n^{(\alpha,\beta)}(x) &= [\alpha-\beta+(A_n+2)x]P_n^{(\alpha,\beta)}(x)\\
&\quad-2(n+1)P_{n+1}^{(\alpha,\beta)}(x),\\
\tfrac{1}{2}(1+x)P_{n-1}^{(\alpha+1,\beta+1)}(x) &= P_n^{(\alpha+1,\beta)}(x)-P_n^{(\alpha,\beta)}(x),\\
2P_n^{(\alpha,\beta)}(x) &= (1-x)P_n^{(\alpha+1,\beta)}(x)+(1+x)P_n^{(\alpha,\beta+1)}(x),
\end{aligned}$$

where

$$A_n = 2n+\alpha+\beta, \quad B_n = n+\alpha+\beta.$$

To prove the results you may use values at $x = \pm 1$, see (6.19); use k_n, given in Exercise 6.1, for determining the coefficient of the highest $x-$power. Also, show that for $x \in (-1,1)$

$$2n\int_x^1 (1-y)^\alpha(1+y)^\beta P_n^{(\alpha,\beta)}(y)\,dy = (1-x)^{\alpha+1}(1+x)^{\beta+1}P_{n-1}^{(\alpha+1,\beta+1)}(x).$$

For Laguerre polynomials we have

$$\frac{d}{dx}L_n^\alpha(x) = -L_{n-1}^{\alpha+1}(x) = \frac{1}{x}\left[nL_n^\alpha(x)-(n+\alpha)L_{n-1}^\alpha(x)\right]$$

6.5. Prove the orthogonality relation for the Laguerre polynomials

$$\int_0^\infty x^\alpha e^{-x}L_n^\alpha(x)L_m^\alpha(x)\,dx = \frac{\Gamma(n+\alpha+1)}{n!}\,\delta_{n,m}, \quad n,m=0,1,2,\ldots\,.$$

First show that, when $\Re\alpha>0,\ \Re\mu>0,\ n=0,1,2,\ldots,$

$$\int_0^\infty x^\alpha e^{-\mu x}L_n^\alpha(x)\,dx = \mu^{-\alpha-1}\frac{\Gamma(n+\alpha+1)}{n!}\left(\frac{\mu-1}{\mu}\right)^n,$$

by substituting (6.40). Next, multiply both sides of (6.43) by $x^\alpha e^{-x}L_m^\alpha(x)$, and integrate with respect to x. We give another proof. Consider the relation

$$\sum_{n=0}^{\infty}\sum_{m=0}^{\infty}\int_0^{\infty}\left[x^{\alpha}\,L_n^{\alpha}(x)\,L_m^{\alpha}(x)e^{-x}\,dx\right]\,s^n\,t^m$$
$$=\int_0^{\infty}\frac{x^{\alpha}\,e^{-sx/(1-s)-tx/(1-t)-x}}{(1-s)^{\alpha+1}(1-t)^{\alpha+1}}\,dx,$$

which follows from (6.43). The right-hand side equals $\Gamma(\alpha+1)(1-st)^{-\alpha-1}$.

6.6. From (6.43) it follows that we have the representation

$$L_n^{\alpha}(x)=\frac{1}{2\pi i}\int_C(1-t)^{-\alpha-1}e^{-tx/(1-t)}\,\frac{dt}{t^{n+1}},$$

where C is a circle around the origin with radius less than 1. Expand the exponential function in powers of x and verify formula (6.40).

6.7. Verify that

$$L_n^{\frac{1}{2}}(x)=\frac{(-1)^n}{\sqrt{x}\;n!\,2^{2n+1}}H_{2n+1}(\sqrt{x}\,),\quad L_n^{-\frac{1}{2}}(x)=\frac{(-1)^n}{n!\,2^{2n}}H_{2n}(\sqrt{x}\,).$$

A different way of writing this interesting relation between Laguerre and Hermite polynomials reads:

$$H_{2n}(x)=(-1)^n2^{2n}\,n!\,L_n^{-\frac{1}{2}}\left(x^2\right),\quad H_{2n+1}(x)=(-1)^n2^{2n+1}\,n!\,xL_n^{\frac{1}{2}}\left(x^2\right).$$

Verify with the help of Exercise 6.5 that

$$\int_{-\infty}^{\infty}e^{-x^2}H_n(x)H_m(x)\,dx=\sqrt{\pi}\,2^n n!\,\delta_{m,n}.$$

6.8. Prove that

$$H_n(x)=\frac{2^n}{\sqrt{\pi}}\int_{-\infty}^{\infty}(x+it)^n e^{-t^2}\,dt$$

by verifying that the right-hand side satisfies the recurrence relation for the Hermite polynomials. Then, with the help of (6.44), give a proof of the bilinear generating function for Hermite polynomials

$$\sum_{n=0}^{\infty}\frac{H_n(x)H_n(y)}{n!}(z/2)^n=\frac{1}{\sqrt{1-z^2}}\exp\left[\frac{2xyz-(x^2+y^2)z^2}{1-z^2}\right],\quad |z|<1.$$

6.9. Verify that, by applying the transformation that yields (4.8), the differential equation of Hermite (6.34) becomes

$$U''+(2n+1-x^2)U=0,\quad U(x)=e^{-x^2/2}H_n(x).$$

Denote the zeros of the n-th Hermite polynomial by $x_1 < x_1 < \cdots < x_n$. The function U' has inside the interval (x_1, x_n) exactly $n-1$ zeros. Since $U(x)$ tends to zero as $x \to \pm\infty$, U' has outside the interval (x_1, x_n) two extra zeros. Observe that

$$U'(x) = e^{-x^2/2}[H_n'(x) - xH_n(x)],$$

a polynomial of degree $n+1$ multiplied by an exponential function. Hence U' has exactly $n+1$ real zeros. By a similar reasoning it follows that $U''(x)$ has exactly $n+2$ real zeros, of which two zeros are located outside (x_1, x_n). However,

$$U''(x) = \left(x^2 - 2n - 1\right) e^{-x^2/2} H_n(x).$$

Verify that this leads to the conclusion that the zeros of $H_n(x)$ satisfy the inequalities

$$|x_k| < \sqrt{2n+1}, \quad k = 1, 2, \ldots, n.$$

6.10. Verify, by taking limits in each term of the polynomial on the left-hand side

$$\lim_{\gamma\to\infty} \gamma^{-n/2} C_n^\gamma(x/\sqrt{\gamma}) = \frac{1}{n!} H_n(x).$$

Other interesting results are:

$$\lim_{\beta\to\infty} P_n^{(\alpha,\beta)}(1 - 2x/\beta) = L_n^\alpha(x),$$

which follows from taking termwise limits in (6.35). In Chapter 7 a limiting process is discussed in which a ${}_1F_1$-function is obtained as a certain limit of a ${}_2F_1$- function. The above relation between the Jacobi polynomial and the Laguerre polynomial is closely connected with this more general process. The weight function of the Jacobi polynomial also transforms to that of the Laguerre polynomial. Let $x = 1 - 2t/\beta$; then, as $\beta \to \infty$,

$$(1-x)^\alpha (1+x)^\beta = \frac{2^{\alpha+\beta}}{\beta^\alpha} t^\alpha (1 - t/\beta)^\beta \sim \frac{2^{\alpha+\beta}}{\beta^\alpha} t^\alpha e^{-t}.$$

Verify the following interesting limits:

$$\lim_{\alpha\to\infty} \frac{P_n^{(\alpha,\beta)}(x)}{P_n^{(\alpha,\beta)}(1)} = \left(\frac{1+x}{2}\right)^n, \qquad \lim_{\beta\to\infty} \frac{P_n^{(\alpha,\beta)}(x)}{P_n^{(\alpha,\beta)}(-1)} = \left(\frac{1-x}{2}\right)^n,$$

showing that the zeros of the Jacobi polynomial tend to -1 when α is large, and to $+1$ when β is large. The zeros of $P_n^{(\alpha,\beta)}(x)$ tend to zero when both parameters α, β (with fixed ratio α/β) tend to infinity. Prove the special case for the Gegenbauer polynomials:

$$\lim_{\gamma\to\infty} \frac{C_n^\gamma(x)}{C_n^\gamma(1)} = x^n.$$

6.11. Prove the orthogonality relation for Jacobi polynomials

$$\int_{-1}^{1} P_n^{(\alpha,\beta)}(x) P_m^{(\alpha,\beta)}(x)\,(1-x)^\alpha (1+x)^\beta\,dx =$$

$$\frac{2^{\alpha+\beta+1}}{2n+\alpha+\beta+1}\,\frac{\Gamma(n+\alpha+1)\,\Gamma(n+\beta+1)}{\Gamma(n+1)\,\Gamma(n+\alpha+\beta+1)}\delta_{n,m}.$$

Base the proof on the Rodrigues formula by using (6.17) with $k = n$. Show that the left-hand side of (6.17) then equals

$$(-1)^n n! \int_{-1}^{1} w(x)\left(1-x^2\right)\,dx.$$

Use also k_n from Exercise 6.1.

6.12. Prove that

$$P_{2n}^{(\alpha,\alpha)}(x) = \frac{n!\,\Gamma(2n+\alpha+1)}{(2n)!\,\Gamma(n+\alpha+1)} P_n^{(\alpha,-1/2)}\left(2x^2-1\right),$$

$$P_{2n+1}^{(\alpha,\alpha)}(x) = \frac{n!\,\Gamma(2n+\alpha+2)}{(2n+1)!\,\Gamma(n+\alpha+1)} x P_n^{(\alpha,1/2)}\left(2x^2-1\right).$$

For a proof of the first relation it is sufficient to verify that

$$\int_{-1}^{1} P_n^{(\alpha,-1/2)}\left(2x^2-1\right) x^k \left(1-x^2\right)^\alpha dx = 0, \quad k = 0, 1, \ldots, 2n-1.$$

This is trivial when k is odd. Complete the proof for k even by first reducing the interval of integration to $[0,1]$ and then replacing x by $\sqrt{(t+1)/2}$. A similar method can be used for the second relation. The constants in front of the Jacobi polynomials can be derived from (6.19). For the second relation you can also use Exercise 5.8; an analogous proof of the first relation is possible when you first derive a similar quadratic transformation for the hypergeometric function.

6.13. Show that for $m = 0, 1, 2, \ldots$

$$x^m = m!2^{-m}\sqrt{\pi} \sum_{k=0}^{\lfloor m/2 \rfloor} \frac{m-2k+\frac{1}{2}}{k!\,\Gamma(m-k+\frac{3}{2})} P_{m-2k}(x).$$

6.14. Expand the function

$$f(x) = \begin{cases} 0, & \text{if } \ -1 \le x < 0; \\ 1, & \text{if } \quad 0 < x \le 1. \end{cases}$$

into a series of Legendre polynomials on the interval $[-1, 1]$. Determine the sum of the series at $x = 0$.

6.15. Expand the functions $f(x) = \sqrt{1-x}$ and $f(x) = \ln(1-x)$ on the interval $[-1, 1]$ into series of Legendre polynomials. Investigate the convergence of the series at $x = \pm 1$.

6.16. Let $\mathcal{P}_n$ be the class of all polynomials having degree n, with coefficient of x^n equal to unity. Show that

$$\int_{-1}^{1} P^2(x)\, dx, \quad P \in \mathcal{P}_n$$

is minimal for

$$P(x) = \frac{2^n (n!)^2}{(2n)!} P_n(x).$$

6.17. Prove the generating functions for Chebyshev polynomials

$$\sum_{n=0}^{\infty} \frac{T_n(x)}{n!} z^n = e^{xz} \cos z\sqrt{1-x^2}, \quad \sum_{n=0}^{\infty} \frac{U_n(x)}{(n+1)!} z^n = e^{xz} \frac{\sin z\sqrt{1-x^2}}{z\sqrt{1-x^2}}$$

by substituting $x = \cos\theta$; see (6.42). Both series converge and represent their sums for all complex values of x and z.

6.18. Verify the multiplication formula for Laguerre polynomials:

$$L_n^{\alpha}(\lambda x) = \sum_{k=0}^{n} \binom{n+\alpha}{n-k} \lambda^k (1-\lambda)^{n-k} L_n^{\alpha}(x)$$

by replacing $L_k^{\alpha}(x)$ by representation (6.40) and interchanging the two summations.

7

Confluent Hypergeometric Functions

The functions of this chapter are also called *Kummer functions* or *Whittaker functions*, and several kinds of notations are in fashion; a subclass can be denoted as ${}_1F_1$ hypergeometric functions. Anyhow, the confluent hypergeometric functions constitute an important class with many applications in physics and probability theory. Special cases are exponential integrals, error functions, incomplete gamma functions (chi-square probability functions), Fresnel integrals, Hermite and Laguerre polynomials, Coulomb wave functions, parabolic cylinder functions, and Bessel functions.

For instance, the Coulomb wave functions are the solutions of the non-relativistic Coulomb wave equation

$$\frac{d^2w}{d\rho^2} + \left[1 - \frac{2\eta}{\rho} - \frac{\lambda(\lambda+1)}{\rho^2}\right] w = 0,$$

and it describes the radial variation of the scattering states of the two charges interacting, with positive energy, by means of the Coulomb potential, Ze^2/r. The quantity $\eta = Z\alpha/\beta$ is the Sommerfeld parameter, which determines the strength of the interaction, Ze^2 is the charge product (of either sign), e is the charge on the positron, α is the fine-structure constant, and $\beta = v/c$ is the relative velocity of the charges in terms of the speed of light. The dimensionless independent variable ρ is kr, where the wave number k is given in terms of the reduced mass M, and Planck's constant $\hbar = h/(2\pi)$, as $k = Mv/\hbar$. The solution in nuclear and atomic physics usually requires solutions for real η , positive ρ and integer $\lambda = L$ (the angular momentum number). The equation has two solutions, denoted by $F_\lambda(\eta,\rho)$ and $G_\lambda(\eta,\rho)$, which are regular and irregular, respectively, at the origin $\rho = 0$, and which behave asymptotically as circular functions,

$$F_\lambda(\eta,\rho) \sim \sin\theta_\lambda, \quad G_\lambda(\eta,\rho) \sim \cos\theta_\lambda, \quad \rho \to \infty,$$

where

$$\theta_\lambda = \rho - \eta \ln(2\rho) - \frac{1}{2}\lambda\pi + \sigma_\lambda, \quad \sigma_\lambda = \arg\Gamma(\lambda + 1 + i\eta);$$

σ_λ is called the Coulomb phase shift.

In §7.3.2 we give the relations between the confluent hypergeometric functions and the Coulomb wave functions. These functions are (when we consider ρ, η and λ as general complex variables) no special cases, but equivalent to the confluent hypergeometric functions (or Kummer functions, or Whittaker functions).

7.1. The *M*–function

Let us recall the differential equation of the Gauss hypergeometric functions:

$$z(1-z)F'' + [c - (a+b+1)z]F' - abF = 0. \tag{7.1}$$

From the theory of Chapters 4 and 5 it follows that $z = 0, z = 1, z = \infty$ are regular singular points and the theory confirms the possibility of power series expansions around these points. The relevant indicial equation is $\mu(\mu - 1 + c) = 0$. Indeed, we find independent solutions, unless c is an integer.

The confluent hypergeometric function arises when two of the regular singular points of (7.1) are allowed to merge into one singular point. Formally this process runs as follows. The hypergeometric function $F(a, b; c; z/b)$ has a regular singular point at $z = b$. We define

$$\boxed{M(a, c, z) = \lim_{b\to\infty} F(a, b; c; z/b).} \tag{7.2}$$

Using the power series (5.3) of the F–function we can compute the limit termwise, since we know

$$\lim_{b\to\infty} \frac{(b)_n}{b^n} = 1.$$

The result is

$$M(a, c, z) = \sum_{n=0}^{\infty} \frac{(a)_n}{(c)_n} \frac{z^n}{n!}. \tag{7.3}$$

In the series a, c and z may assume any finite complex value, with the exception $c = 0, -1, -2, \ldots$.

Performing the same limit in (7.1) we obtain the differential equation

$$\boxed{zF'' + (c - z)F' - aF = 0.} \tag{7.4}$$

This equation is called the *Kummer differential equation*. The function in (7.3) is called a *Kummer function* (KUMMER (1836), (1837)). It is not difficult to verify that (7.3) satisfies (7.4).

Applying the above limiting process on (see (5.5))

$$F(a,b;c;z) = (1-z)^{-b} F\left(c-a,b;c;\frac{z}{z-1}\right),$$

we obtain the useful functional relation for the M–function

$$\boxed{M(a,c,z) = e^z M(c-a,c,-z).} \tag{7.5}$$

A second solution of (7.4), which will usually be independent of (7.3), can be obtained through the second solution in (5.9). Using the above limiting process we obtain a function of the form

$$z^{1-c} M(a-c+1, 2-c, z). \tag{7.6}$$

Finally, applying the limiting process on the integral representation (5.4), we obtain for $Rec > \Re a > 0$

$$\boxed{M(a,c,z) = \frac{\Gamma(c)}{\Gamma(a)\Gamma(c-a)} \int_0^1 e^{zt} t^{a-1}(1-t)^{c-a-1}\,dt.} \tag{7.7}$$

A more general variant of this integral representation is given in Exercise 4.7. Observe that the relation in (7.5) follows from this integral by letting $t \to 1-t$.

Considering Kummer's equation (7.4) in the light of the theory of Chapter 4, we distinguish two singular points: $z = 0$ and $z = \infty$. The indicial equation of the regular singular point $z = 0$ reads $\mu(\mu+c-1) = 0$. When c is an integer we obtain, just as for the Gauss functions, for one of the solutions logarithmic terms in the power series expansions near the origin. This time the singularity at $z = \infty$ is not regular. Hence we cannot expect convergent series in powers of $1/z$. Apparently, by the above limiting process described in (7.2) the two regular singular points in $z = b$ and $z = \infty$ are transferred into a singular point at ∞, which is not regular. For $z = 0$ the theory is applicable and does not give new perspectives.

From (7.3) it follows that $M(a,c,z)$ is a hypergeometric function (as introduced earlier in (5.29)) with $p = q = 1$. Hence

$$M(a,c,z) = {}_1F_1(a;c;z).$$

From this point of view the confluent hypergeometric functions can be introduced without reference to the limiting process in (7.2). The function in (7.3) is an entire function of z and has (7.7) as an integral representation. This can be derived directly from (7.7) by expanding the exponential function $\exp(zt)$ and by using the representation of the beta integral in (3.2). With the help of Theorem 2.2 from Chapter 2 (and possibly invoking the principle of analytic continuation) we can then confirm the relation between (7.3) and (7.7) for $\Re a > 0$, $\Re(c-a) > 0$.

Writing (7.7) in the form

$$M(a,c,z) = \frac{\Gamma(c)}{\Gamma(a)\Gamma(c-a)} e^z \int_0^1 e^{-zt} t^{c-a-1}(1-t)^{a-1}\, dt \tag{7.8}$$

we recognize a Laplace integral, on a finite interval. We can apply Watson's lemma (Theorem 2.3) to obtain the asymptotic behavior of the M–function for large values of $|z|$. We substitute in (7.8) the power series

$$(1-t)^{a-1} = \sum_{n=0}^{\infty} \frac{(1-a)_n}{n!} t^n$$

and we obtain the asymptotic expansion

$$\boxed{M(a,c,z) \sim \frac{\Gamma(c) e^z z^{a-c}}{\Gamma(a)} \sum_{n=0}^{\infty} \frac{(c-a)_n\,(1-a)_n}{n!} z^{-n}, \qquad z \to \infty,} \tag{7.9}$$

which is valid in the sector $|\arg z| < \frac{1}{2}\pi$. In Exercise 7.7 this limited range will be extended.

As remarked earlier (see (7.6)) we can write the general solution of the Kummer differential equation (7.4) in the form:

$$F(z) = AM(a,c,z) + Bz^{1-c} M(a-c+1, 2-c, z), \tag{7.10}$$

at least when c is not an integer. The behavior of F as $z \to 0$ is clear. As $z \to \infty$ it is more complicated. Using (7.9) we find

$$F(z) \sim \left[A \frac{\Gamma(c)}{\Gamma(a)} + B \frac{\Gamma(2-c)}{\Gamma(a-c+1)}\right] e^z z^{a-c} \sum_{n=0}^{\infty} \frac{(c-a)_n\,(1-a)_n}{n!\; z^n}, \qquad z \to \infty.$$

Surprisingly, a special choice of A and B (A, B may not depend on z) allows the expansion to completely vanish. This does not mean that, when A and B satisfy

$$A \frac{\Gamma(c)}{\Gamma(a)} + B \frac{\Gamma(2-c)}{\Gamma(a-c+1)} = 0, \tag{7.11}$$

the function F will vanish identically. When (7.11) holds we expect that F is a solution of lower order at $+\infty$. It does not behave as e^z times an algebraic function. In other words, when we take in (7.10) A and B as in (7.11), we arrive at a solution of (7.4) that is independent of both M–functions in (7.10). In the following section we will bring out this solution without reference to the previous considerations.

7.2. The *U*–function

We try to find a solution of (7.4) in the form

$$y(z) = \int_\alpha^\beta e^{-zt} v(t)\, dt$$

and we use the method of §4.4 to determine α, β and v. In the present case the operator M_t is given by $m_0 = -(ct + a)$, $m_1 = -t^2 - t$. The adjoint reads

$$M_t^*[v] = t(t+1)\frac{dv}{dt} + [(2-c)t + (1-a)]v.$$

A solution of the equation $M_t^*[v] = 0$ is given by

$$v(t) = t^{a-1}(t+1)^{c-a-1}.$$

This gives the occasion for introducing the following standard solution of the Kummer differential equation:

$$\boxed{U(a, c, z) = \frac{1}{\Gamma(a)} \int_0^\infty e^{-zt} t^{a-1} (1+t)^{c-a-1}\, dt,} \tag{7.12}$$

where we assume that $\Re a > 0$, $\Re z > 0$. The reciprocal gamma function in front of the integral is chosen for normalization and on account of convention.

An application of Watson's lemma indeed leads to the conclusion that $U(a, c, z)$ has the aforementioned asymptotic behavior of lower order. Expanding the integrand of (7.12) according to

$$(1+t)^{c-a-1} = \sum_{n=0}^\infty (-1)^n \frac{(a-c+1)_n}{n!} t^n$$

we obtain the asymptotic expansion for the U–function

$$\boxed{U(a, c, z) \sim z^{-a} \sum_{n=0}^\infty \frac{(a)_n (a-c+1)_n}{n!} (-z)^{-n}, \quad z \to \infty.} \tag{7.13}$$

According to Theorem 2.4, (7.13) holds for $|\arg z| < 3\pi/2$.

We verify that indeed the U–function can be written in the form (7.10). The proper values are obtained by taking into account (7.11) and the behavior at $z = 0$. Let $\Re c < 1$; then by (7.12)

$$A = U(a, c, 0) = \frac{1}{\Gamma(a)} \int_0^\infty t^{a-1}(1+t)^{c-a-1}\, dt,$$

giving (see Exercise 3.3)

$$A = \frac{\Gamma(1-c)}{\Gamma(a-c+1)}$$

and subsequently

$$B = \frac{\Gamma(c-1)}{\Gamma(a)}.$$

The result is

$$U(a,c,z) = \frac{\Gamma(1-c)}{\Gamma(a-c+1)} M(a,c,z) + \frac{\Gamma(c-1)}{\Gamma(a)} z^{1-c} M(a-c+1, 2-c, z). \quad (7.14)$$

This solution has a meaning for all values of z, a and c with the exception of the point $z = 0$; in general $U(a, c, z)$ is singular at $z = 0$. Observe that U is defined at the points $c = 0, -1, -2, \ldots$; see (7.12). The M–function itself is not defined at these c–values.

We have for $M(a, c, z)$ the alternative (7.6). Has the U–function a similar alternative? To answer the question we use the differential equation (7.4). It can be verified that $z^{1-c}U(a-c+1, 2-c, z)$ is also a solution. This should be a linear combination of two by now available solutions. In this linear combination no M–function can occur, because of the behavior of this function at infinity. The asymptotic behavior of the U–function given in (7.13) thus gives the remarkable functional relation for the U–function

$$\boxed{U(a,c,z) = z^{1-c} U(a-c+1, 2-c, z).} \quad (7.15)$$

This relation can be verified directly by using (7.14).

The right-hand side of (7.14) takes an undefined form when c is an integer number. A limiting process, in which both terms of the right-hand side in (7.14) should be involved, yields however a well-defined result. This is also confirmed by the theory of Chapter 4. When $c \in \mathbb{Z}$ the power series expansion of the U–function will contain logarithmic terms. We have for $n = 0, 1, 2, \ldots$

$$U(a, n+1, z) = \frac{(-1)^{n+1}}{n!\,\Gamma(a-n)} \Big[M(a, n+1, z) \ln z$$
$$+ \sum_{m=0}^{\infty} \frac{(a)_m\, z^m}{(n+1)_m\, m!} \times \{\psi(a+m) - \psi(1+m) - \psi(1+n+m)\} \Big]$$
$$+ \frac{(n-1)!}{\Gamma(a)} z^{-n} \sum_{m=0}^{n-1} \frac{(a-n)_m\, z^m}{(1-n)_m\, m!};$$

$(n-1)!$ is to be interpreted as zero when $n=0$. For negative values of n we can use (7.15) in the form

$$U(a,1-n,z)=z^nU(a+n,1+n,z).$$

Observe that (7.14) can be used for the analytic continuation with respect to z of the U–function. The M–functions in (7.14) are entire functions of z and the singularity of U comes from z^{1-c}. When it is required to use the U–function for negative values of $\Re z$ it is convenient to specify the phase of z. By (7.14) and (7.5),

$$\begin{aligned}e^zU(a,c,ze^{\pm\pi i}) &= \frac{\Gamma(1-c)}{\Gamma(a-c+1)}M(c-a,c,z)\\ &+\frac{\Gamma(c-1)}{\Gamma(a)}e^{\pm\pi i(1-c)}z^{1-c}M(1-a,2-c,z).\end{aligned}$$

It is also possible to express the M–function in terms of certain U–functions. Replacing in the above relation a with $c-a$ we obtain

$$\begin{aligned}e^zU(c-a,c,ze^{\pm\pi i}) &= \frac{\Gamma(1-c)}{\Gamma(1-a)}M(a,c,z)\\ &+\frac{\Gamma(c-1)}{\Gamma(c-a)}e^{\pm\pi i(1-c)}z^{1-c}M(1+a-c,2-c,z).\end{aligned}$$

Eliminating $M(1+a-c,2-c,z)$ from this relation and (7.14), we find after some algebra

$$\frac{1}{\Gamma(c)}M(a,c,z)=\frac{e^{\pm\pi i(c-a)}}{\Gamma(a)}e^zU\left(c-a,c,ze^{\pm\pi i}\right)+\frac{e^{\mp\pi ia}}{\Gamma(c-a)}U(a,c,z). \tag{7.16}$$

This formula can be used to extend the range of the asymptotic parameter z in (7.9); see Exercise 7.7.

7.3. Special Cases and Further Relations

Many well-known special functions can be expressed in terms of the confluent hypergeometric functions. An elementary example is the exponential function: $M(a,a,z)=e^z$. In this section we give the most important special cases. In Chapter 11 special attention will be paid to the error functions and incomplete gamma functions.

7.3.1. Whittaker Functions

In the literature an alternative pair for the confluent hypergeometric functions is given, called the *Whittaker functions*. The definitions are

$$\begin{aligned} M_{\kappa,\mu}(z) &= e^{-\frac{1}{2}z} z^{\frac{1}{2}+\mu} M\left(\frac{1}{2}+\mu-\kappa, 1+2\mu, z\right), \\ W_{\kappa,\mu}(z) &= e^{-\frac{1}{2}z} z^{\frac{1}{2}+\mu} U\left(\frac{1}{2}+\mu-\kappa, 1+2\mu, z\right). \end{aligned} \tag{7.17}$$

They satisfy the *Whittaker equation*

$$w'' + \left(-\frac{1}{4} + \frac{\kappa}{z} + \frac{\frac{1}{4}-\mu^2}{z^2}\right) w = 0. \tag{7.18}$$

This equation follows from the Kummer equation (7.4) by applying the transformation that yields (4.8). It is useful to know that a differential equation of the form

$$w'' + \frac{az^2+bz+c}{z^2} w = 0$$

has solutions which can be expressed in terms of Whittaker functions, and hence, by using the above relations, in terms of Kummer functions.

7.3.2. Coulomb Wave Functions

The differential equation

$$w'' + \left[1 - \frac{2\eta}{\rho} - \frac{\lambda(\lambda+1)}{\rho^2}\right] w = 0 \tag{7.19}$$

is a special form of (7.18) but deserves special attention. As we mentioned in the introduction to this chapter, it plays an important part in physics, in particular in quantum mechanics as a form of the Schrödinger equation in a central Coulomb field. The solutions of (7.19) are called *Coulomb wave functions*, and are usually denoted by $F_\lambda(\eta,\rho)$, $G_\lambda(\eta,\rho)$. We give the relations with the Kummer functions:

$$\begin{aligned} F_\lambda(\eta,\rho) &= A\,M(\lambda+1-i\eta, 2\lambda+2, 2i\rho), \\ G_\lambda(\eta,\rho) &= iF_\lambda(\eta,\rho) + iB\,U(\lambda+1-i\eta, 2\lambda+2, 2i\rho), \\ A &= \frac{|\Gamma(\lambda+1+i\eta)|e^{-\pi\eta/2-i\rho}(2\rho)^{\lambda+1}}{2\Gamma(2\lambda+2)}, \\ B &= e^{\pi\eta/2+\lambda\pi i-i\sigma_\lambda-i\rho}(2\rho)^{\lambda+1}, \\ \sigma_\lambda &= \arg\Gamma(\lambda+1+i\eta) \quad \text{(the Coulomb phase shift).} \end{aligned}$$

The functions $F_\lambda(\eta,\rho)$ and $G_\lambda(\eta,\rho)$ are real for real values of η, $\rho > 0, \lambda \geq 0$. This certainly does not follow directly from the above definitions, considering the many complex parameters in the definitions. For $F_\lambda(\eta,\rho)$, however, this result follows directly from (7.5).

The quantities A and B in the above definitions are chosen such that $F_\lambda(\eta,\rho)$, $G_\lambda(\eta,\rho)$ behave asymptotically as circular functions,

$$F_\lambda(\eta,\rho) \sim \sin\theta_\lambda, \quad G_\lambda(\eta,\rho) \sim \cos\theta_\lambda, \quad \rho\to\infty,$$

where

$$\theta_\lambda = \rho - \eta\ln(2\rho) - \frac{1}{2}\lambda\pi + \sigma_\lambda.$$

See Exercise 7.11.

7.3.3. Parabolic Cylinder Functions

The solutions of the differential equation

$$y'' + \left(z^2 + pz + q\right) y = 0$$

are called *parabolic cylinder functions* or *Weber parabolic cylinder functions.* Another standard form is

$$y'' - \left(a + \frac{1}{4}z^2\right) y = 0. \tag{7.20}$$

There are no finite singular points. Hence, all solutions are entire functions of z. It is straightforward to verify that the following even and odd solutions exist:

$$\begin{aligned} y_1 &= e^{-z^2/4} M\left(\frac{1}{2}a + \frac{1}{4}, \frac{1}{2}, \frac{1}{2}z^2\right) = e^{z^2/4} M\left(-\frac{1}{2}a + \frac{1}{4}, \frac{1}{2}, -\frac{1}{2}z^2\right), \\ y_2 &= z e^{-z^2/4} M\left(\frac{1}{2}a + \frac{3}{4}, \frac{3}{2}, \frac{1}{2}z^2\right) = z e^{z^2/4} M\left(-\frac{1}{2}a + \frac{3}{4}, \frac{3}{2}, -\frac{1}{2}z^2\right). \end{aligned}$$

Although the Wronskian of this pair equals 1, this is not a satisfactory pair (see Miller (1952b)). For instance, y_1 and y_2 have almost the same behavior at infinity. A better pair is defined by writing combinations of y_1, y_2:

$$\begin{aligned} U(a,z) &= \sqrt{\pi}\, 2^{-1/4-a/2} \left[\frac{y_1}{\Gamma\left(\frac{3}{4}+\frac{a}{2}\right)} - \frac{\sqrt{2}\, y_2}{\Gamma\left(\frac{1}{4}+\frac{a}{2}\right)} \right] \\ &= 2^{-3/4-a/2} e^{-\frac{1}{4}z^2}\, z\, U\left(\frac{3}{4}+\frac{a}{2}, \frac{3}{2}, \frac{1}{2}z^2\right), \\ V(a,z) &= \frac{1}{\pi}\Gamma\left(\frac{1}{2}+a\right) \left[\sin \pi a\, U(a,z) + U(a,-z)\right]. \end{aligned} \tag{7.21}$$

The Wronskian of the pair U, V equals $\sqrt{2/\pi}$. In the notation of Whittaker we have $D_\nu(z) = U\left(-\nu - \frac{1}{2}, z\right)$. When $a = -1/2, -3/2, -5/2, \ldots$ the Hermite polynomials arise:

$$H_n(z) = 2^{\frac{1}{2}n} e^{\frac{1}{2}z^2} U\left(-n-\frac{1}{2}, z\sqrt{2}\right) = 2^{\frac{1}{2}n} e^{\frac{1}{2}z^2} D_n(z\sqrt{2}). \tag{7.22}$$

In Exercise 7.9 integral representations for $U(a,z)$ are given.

7.3.4. Error Functions

The *error functions* are considered in more detail in Chapter 11, because of their importance in statistics and probability theory. The definitions are

$$\boxed{\operatorname{erf} z = \frac{2}{\sqrt{\pi}} \int_0^z e^{-t^2}\, dt, \quad \operatorname{erfc} z = 1 - \operatorname{erf} z = \frac{2}{\sqrt{\pi}} \int_z^\infty e^{-t^2}\, dt.} \tag{7.23}$$

The relations with the Kummer functions are

$$\operatorname{erf} z = zM\left(\frac{1}{2}, \frac{3}{2}, -z^2\right), \quad \operatorname{erfc} z = e^{-z^2} U\left(\frac{1}{2}, \frac{1}{2}, z^2\right).$$

7.3.5. Exponential Integrals

The *exponential integrals* are defined for $n = 1, 2, \ldots$ by

$$\boxed{E_n(z) = \int_1^\infty \frac{e^{-zt}}{t^n}\, dt, \quad \Re z > 0.} \tag{7.24}$$

The relation with the U–function is

$$E_n(z) = e^{-z} U(1, 2 - n, z) = z^{n-1} e^{-z} U(n, n, z),$$

which follows from (7.12) and (7.15). The latter gives

$$\boxed{E_n(z) = \frac{z^{n-1} e^{-z}}{\Gamma(n)} \int_0^\infty \frac{e^{-zt} t^{n-1}}{t+1}\, dt, \quad \Re z > 0.} \tag{7.25}$$

When $n = 1$ one usually writes

$$\boxed{E_1(z) = \int_z^\infty \frac{e^{-t}}{t}\, dt, \quad |\arg z| < \pi.}$$

This function is also written as $-Ei(-z)$. For real values of $z = x$ it is more convenient to define

$$\boxed{Ei(x) = \int_{-\infty}^x \frac{e^t}{t}\, dt,}$$

where for $x > 0$ the integral should be interpreted as a *Cauchy principal value integral*. The *logarithmic integral* follows from writing

$$\boxed{li(x) = \int_0^x \frac{dt}{\ln t} = Ei(\ln x).}$$

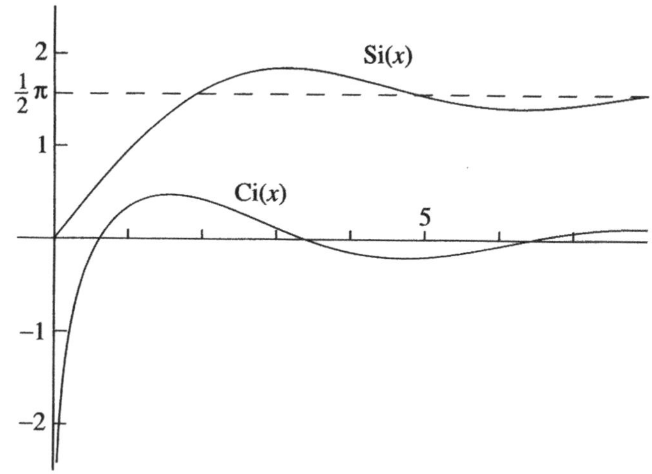

Figure 7.1. The sine and cosine integrals $\mathrm{Si}(x)$, $\mathrm{Ci}(x)$, $0 \le x \le 8$.

The function $li(x)$ is related to the *asymptotic distribution of prime numbers.*

For complex values of the argument it is more convenient to take as definition of the Ei–function:

$$Ei(z) = -e^z U(1, 1, -z), \quad |\arg(-z)| < \pi.$$

Obviously, the parameter n in (7.24) may assume any complex value, whereas (7.24) is valid only if $\Re n > 0$.

The *sine and cosine integrals* are defined by

$$\boxed{\mathrm{Si}(z) = \int_0^z \frac{\sin t}{t}\, dt, \quad \mathrm{Ci}(z) = \gamma + \ln z + \int_0^z \frac{\cos t - 1}{t}\, dt,}$$

where $\gamma = -\Gamma'(1)$ (Euler's constant; see also (3.8)). The integrals represent entire functions of z. In Exercises 7.1 and 7.2 more representations of these functions are given. The graphs of the sine and cosine integrals are given in Figure 7.1.

The sine integral can be used to describe the *Gibbs phenomenon.* To explain this consider the Fourier series

$$1 = \frac{4}{\pi}\left(\sin x + \frac{1}{3}\sin 3x + \frac{1}{5}\sin 5x + \ldots\right),$$

with $0 < x < \pi, n = 0, 1, 2, \ldots$. Denote the partial sum of the series by

$$S_n(x) = \frac{4}{\pi} \sum_{k=0}^{n-1} \frac{\sin(2k+1)x}{2k+1}.$$

It is not difficult to show (for example by using induction with respect to n) that

$$S_n(x) = \frac{2}{\pi} \int_0^x \frac{\sin 2nt}{\sin t}\, dt$$

which we write using the sine integral in the form

$$S_n(x) = \frac{2}{\pi} \mathrm{Si}(2nx) + R_n(x),$$

where

$$R_n(x) = \frac{2}{\pi} \int_0^x \sin 2nt \left(\frac{1}{\sin t} - \frac{1}{t} \right) dt.$$

For large values of n we have $R_n(x) = \mathcal{O}(1/n)$, uniformly with respect to x in closed intervals of $[0, \pi)$ (this can be shown by integrating by parts). As follows from Exercises 7.2–7.3, the quantity $\frac{2}{\pi}\mathrm{Si}(2nx)$ approaches 1 as $n \to \infty$ when x is a fixed positive number. Hence, the Fourier series converges, because the partial sums of the Fourier series approach 1 when n becomes large. But $\mathrm{Si}(x)$ has maxima at $x = \pi, 3\pi, 5\pi, \ldots$ and

$$\frac{2}{\pi} \mathrm{Si}(2nx) = \frac{2}{\pi} \int_0^\pi \frac{\sin t}{t}\, dt = 1.089490\ldots$$

when $x = \pi/(2n)$. So, when n is large, $S_n(x)$ is not uniformly close to 1 at the right of the origin. The maximal "overshooting" value of approximately $1.089490\ldots$ at $x = \pi/(2n)$, and smaller overshootings at

$$x = 3\pi/(2n), 5\pi/(2n), \ldots,$$

is the famous Gibbs phenomenon, and it occurs also in the Fourier expansions of other discontinuous functions; see Zygmund (1959). Since the $S_n(x)$ are odd functions of x, a similar situation occurs in the left-hand neighborhood of $x = 0$, where $S_n(x)$ tends to -1 as $n \to \infty$ with x fixed. In Figure 7.2 we give the details near the origin of the partial sum $S_n(x)$ with $n = 250$.

7.3.6. Fresnel Integrals

The *Fresnel integrals* are

$$\boxed{C(z) = \sqrt{\frac{2}{\pi}} \int_0^z \cos t^2\, dt, \quad S(z) = \sqrt{\frac{2}{\pi}} \int_0^z \sin t^2\, dt.} \tag{7.26}$$

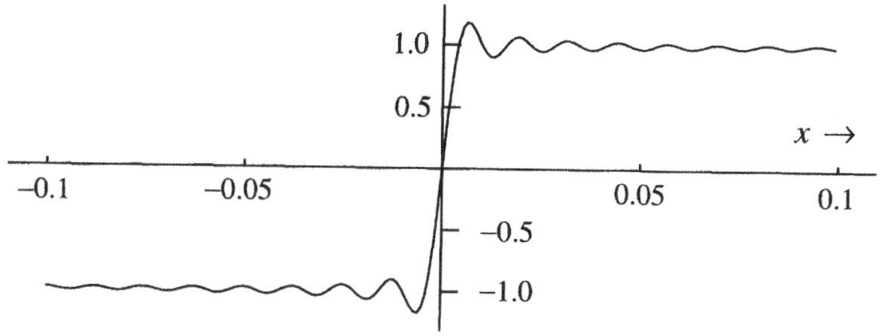

Figure 7.2. The graph of $S_n(x)$, $n = 250$, $-0.1 < x < 0.1$ showing the Gibbs phenomenon.

The t^2 in the circular functions suggests a relation with the error functions. Indeed we have:

$$C(z) + iS(z) = \frac{1+i}{2} \operatorname{erf} \frac{(1-i)z}{\sqrt{2}}.$$

Auxiliary functions are

$$f(z) = [\tfrac{1}{2} - S(z)] \cos z^2 - [\tfrac{1}{2} - C(z)] \sin z^2,$$

$$g(z) = [\tfrac{1}{2} - S(z)] \sin z^2 + [\tfrac{1}{2} - C(z)] \cos z^2.$$

On inverting these, we obtain

$$C(z) = \frac{1}{2} + f(z) \sin(z^2) - g(z) \cos(z^2),$$

$$S(z) = \frac{1}{2} - f(z) \cos(z^2) - g(z) \sin(z^2).$$

These representations of $S(z)$ and $C(z)$ describe precisely how $C(z)$ and $S(z)$ behave for large values of z. The fact is that the functions f and g are slowly varying and monotonic (when $z > 0$). To show this we first observe that $S(+\infty) = C(+\infty) = \frac{1}{2}$ (use Exercise 3.18). Then it easily follows that

$$G(z) := g(z) + if(z) = \sqrt{\frac{2}{\pi}}\, e^{-iz^2} \int_z^\infty e^{it^2}\, dt.$$

Observe that G vanishes at $+\infty$. Next consider the function

$$H(z) := \frac{\sqrt{2}}{\pi} \int_0^\infty \frac{e^{-z^2t^2}}{t^2 - i}\, dt, \quad \Re z^2 > 0.$$

This function satisfies the equation

$$\frac{d}{dz}\left[H(z)e^{iz^2}\right] = -\sqrt{\frac{2}{\pi}}\, e^{iz^2}$$

and H vanishes at $+\infty$ (verify this by using Watson's lemma). Hence $H(z) = G(z)$. Separating real and imaginary parts in the above integral representation of $H(z)$ we find for $z > 0$:

$$f(z) = \frac{1}{\pi\sqrt{2}} \int_0^\infty \frac{e^{-z^2 t}}{\sqrt{t}\,(t^2+1)}\, dt, \quad g(z) = \frac{1}{\pi\sqrt{2}} \int_0^\infty \frac{\sqrt{t}\, e^{-z^2 t}}{(t^2+1)}\, dt.$$

These representations hold for $\Re z^2 > 0$, and are useful for deriving asymptotic expansions for large values of z (Watson's lemma can be used). In diffraction theory one uses the Fresnel integral

$$F(z) = \sqrt{\frac{2}{\pi}} \int_z^\infty e^{it^2}\, dt.$$

From the above formulas we infer that

$$F(z) = [g(z) + if(z)]e^{iz^2}.$$

The Fresnel integrals $C(t), S(t), t \geq 0$ are known to form *Cornu's spiral*. Let the set $\{x(t), y(t), t\}$ be defined by

$$x(t) = C(t), \quad y(t) = S(t), \quad t \geq 0.$$

Then the set $\{x(t), y(t)\}$ is called Cornu's spiral, which is visualized in Figure 7.3. In fact Cornu's spiral is the projection of the cork-screw in the $\{x, y\}-$plane.

The spiral has a very special property. Let $P = P(x, y)$ be any point on the projected spiral. The curvature $K(x, y)$ at P, that is,

$$K(x, y) = \frac{d^2 y}{dx^2}\left[1 + \left(\frac{dy}{dx}\right)^2\right]^{-3/2}$$

is directly proportional to the arc length $L(x, y)$ between the origin and point P, that is,

$$L(x, y) = \int_0^x \sqrt{1 + (dy/d\xi)^2}\, d\xi.$$

We verify this by computing

$$\frac{dy}{dx} = \frac{dy}{dt}\frac{dt}{dx} = \tan t^2.$$

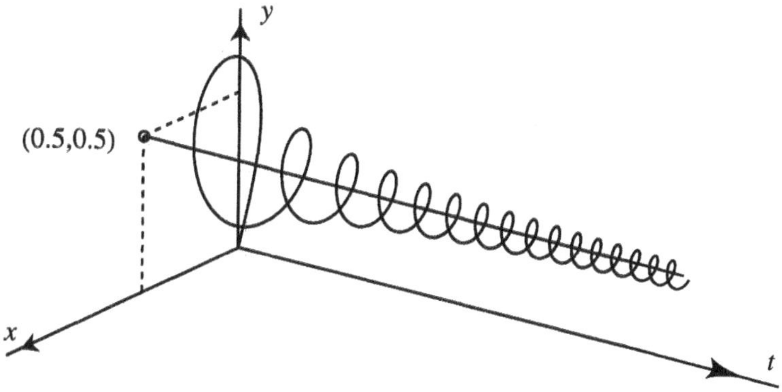

Figure 7.3. Cornu's spiral, formed from Fresnel integrals, is the set $\{C(t), S(t), t\}$, $t \geq 0$.

Hence, the arc length at P equals

$$L(x,y) = \int_0^x \frac{d\xi}{\cos t^2} = \sqrt{\frac{2}{\pi}} \int_0^t dt = t\sqrt{\frac{2}{\pi}}.$$

Next,

$$\frac{d^2 y}{d x^2} = \frac{d}{dx}\left[\frac{dy}{dx}\right] = \frac{d}{dt}\left[\tan t^2\right]\frac{dt}{dx} = \frac{t\sqrt{2\pi}}{\cos^3 t^2}.$$

Hence, the curvature at P equals

$$K(x,y) = t\sqrt{2\pi}.$$

It follows that the ratio $K(x,y)/L(x,y)$ at any point $P(x,y)$ of Cornu's spiral equals π.

7.3.7. Incomplete Gamma Functions

In Chapter 11 we give more details on incomplete gamma functions. The definitions are

$$\gamma(a,z) = \int_0^z t^{a-1} e^{-t}\, dt, \quad \Gamma(a,z) = \int_z^\infty t^{a-1} e^{-t}\, dt.$$

For $\gamma(a,z)$ we assume the condition $\Re a > 0$; with respect to z we assume $|\arg z| < \pi$. In probability theory these functions show up in connection with the gamma distribution. In this area of applications the normalizations

$$P(a,z) = \frac{\gamma(a,z)}{\Gamma(a)}, \quad Q(a,z) = \frac{\Gamma(a,z)}{\Gamma(a)}$$

are frequently used, which satisfy $P(a,z) + Q(a,z) = 1$.

The relations with the Kummer functions are as follows:

$$\boxed{\begin{aligned}\gamma(a,z) &= a^{-1}z^a e^{-z} M(1, a+1, z)\\ &= a^{-1}z^a M(a, a+1, -z),\\ \Gamma(a,z) &= z^a e^{-z} U(1, a+1, z)\\ &= e^{-z} U(1-a, 1-a, z).\end{aligned}}$$

7.3.8. Bessel Functions

Bessel functions arise when in $M(a,c,z)$ and $U(a,c,z)$ the parameters satisfy $c = 2a$. Bessel functions will be treated in Chapter 9. Two important relations are

$$J_\nu(z) = \frac{(z/2)^\nu}{\Gamma(\nu+1)} e^{iz} \, M\left(\nu + \frac{1}{2}, 2\nu+1, 2iz\right),$$

$$K_\nu(z) = \sqrt{\pi}\, e^{-z}\, (2z)^\nu \, U\left(\nu + \frac{1}{2}, 2\nu+1, 2z\right).$$

The latter is a modified Bessel function.

7.3.9. Orthogonal Polynomials

The Hermite and Laguerre polynomials introduced in the previous chapter are special cases of the confluent confluent hypergeometric functions. For the Laguerre polynomials, see Exercise 7.10.

7.4. Remarks and Comments for Further Reading

7.1. The books of BUCHHOLTZ (1969) and SLATER (1960) are exclusively devoted to the class of confluent hypergeometric functions or Whittaker functions. Especially in the first book many references are given to physical applications. Many properties of these functions follow from the general theory of hypergeometric functions, which is extensively discussed in LUKE (1969).

7.2. Rational approximations (based on the *Padé method*) and Chebyshev expansions of the $M-$ and $U-$functions are found in LUKE (1968), (1975).

7.3. OLVER (1959) gives a very detailed account on the asymptotics of the Weber parabolic cylinder functions. A discussion on the choice of standard solutions of Weber's equation (7.20) is given in MILLER (1952). In OLVER (1980) and DUNSTER (1989) uniform asymptotic expansions are given for the Whittaker functions (equivalently: for the confluent hypergeometric functions). The approach is based on the differential equation. In TEMME (1978) integral representations are used for deriving uniform expansions of the confluent hypergeometric functions.

7.5. Exercises and Further Examples

7.1. The exponential integral $E_1(z) = \Gamma(0, z)$ has the representation:

$$E_1(z) = -\gamma - \ln z + \int_0^z \frac{1 - e^{-t}}{t}\, dt.$$

Verify this by considering

$$\Gamma(a, z) = \Gamma(a) - \frac{1}{a}\, z^a + \int_0^z t^{a-1} \left(1 - e^{-t}\right) dt.$$

The integral is defined for $a = 0$. The two other terms give, by invoking l'Hôpital's theorem,

$$\lim_{a \to 0} \left[\frac{\Gamma(a+1) - z^a}{a}\right] = \Gamma'(1) - \ln z$$

The above representation of $E_1(z)$ shows clearly the singularity of the function, since the integral is an entire function of z.

7.2. Show that

$$\int_z^\infty \frac{\sin t}{t}\, dt = \frac{1}{2}\pi - \mathrm{Si}(z), \qquad \int_z^\infty \frac{\cos t}{t}\, dt = -\mathrm{Ci}(z).$$

Show that

$$E_1\left(z e^{\frac{1}{2}\pi i}\right) = -\gamma - \ln z - \mathrm{Ci}(z) + i\left[-\frac{1}{2}\pi + \mathrm{Si}(z)\right].$$

Show that, on the other hand,

$$E_1\left(z e^{\frac{1}{2}\pi i}\right) = \int_0^\infty \frac{\cos(z+t)}{z+t}\, dt - i \int_0^\infty \frac{\sin(z+t)}{z+t}\, dt,$$

and hence, that the sine and cosine integrals can be written as

$$\mathrm{Si}(z) = -f(z)\cos z - g(z)\sin z + \frac{1}{2}\pi,$$
$$\mathrm{Ci}(z) = +f(z)\sin z - g(z)\cos z,$$

where

$$f(z) = \int_0^\infty \frac{\sin t}{z+t}\, dt \quad g(z) = \int_0^\infty \frac{\cos t}{z+t}\, dt, \quad z \neq 0, \quad |\arg z| < \pi.$$

Show that for $\Re z > 0$ (cf. §7.3.6, where a similar procedure is used for the Fresnel integrals):

$$f(z) = \int_0^\infty \frac{e^{-zt}}{t^2+1}\,dt, \quad g(z) = \int_0^\infty \frac{te^{-zt}}{t^2+1}\,dt.$$

7.3. The asymptotic behavior of Si and Ci follows from the above representations in terms of the functions f and g. Show that for $N = 0, 1, 2, , \ldots$ and $\Re z > 0$

$$f(z) = \frac{1}{z}\sum_{n=0}^{N-1}(-1)^n\frac{(2n)!}{z^{2n}} + (-1)^N\int_0^\infty \frac{t^{2N}e^{-zt}}{t^2+1}\,dt,$$

$$g(z) = \frac{1}{z^2}\sum_{n=0}^{N-1}(-1)^n\frac{(2n+1)!}{z^{2n}} + (-1)^N\int_0^\infty \frac{t^{2N+1}e^{-zt}}{t^2+1}\,dt.$$

Bounds for the remainders in these expansions follow from replacing $1/(1+t^2)$ by unity.

7.4. Show that the error functions are special cases of the incomplete gamma functions:

$$\operatorname{erf} z = \frac{1}{\sqrt{\pi}}\gamma\left(\tfrac{1}{2}, z^2\right), \quad \operatorname{erfc} z = \frac{1}{\sqrt{\pi}}\Gamma\left(\tfrac{1}{2}, z^2\right).$$

7.5. Verify the following relations:

$$\frac{d^n}{dz^n}M(a,c,z) = \frac{(a)_n}{(c)_n}M(a+n, c+n, z),$$

$$\frac{d^n}{dz^n}U(a,c,z) = (-1)^n\,(a)_n\,U(a+n, c+n, z),$$

$$(c-a)M(a-1,c,z) + (2a-c+z)M(a,c,z) - aM(a+1,c,z) = 0,$$

$$c(c-1)M(a,c-1,z) + c(1-c-z)M(a,c,z) + z(c-a)M(a,c+1,z) = 0,$$

$$(1+a-c)M(a,c,z) - aM(a+1,c,z) + (c-1)M(a,c-1,z) = 0,$$

$$zM(a+1,c+1,z) + cM(a,c,z) - cM(a+1,c,z) = 0,$$

$$U(a-1,c,z) + (c-2a-z)U(a,c,z) + a(1+a-c)U(a+1,c,z) = 0,$$

$$(c-a-1)U(a,c-1,z) + (1-c-z)U(a,c,z) + zU(a,c+1,z) = 0,$$

$$U(a,c,z) - aU(a+1,c,z) - U(a,c-1,z) = 0,$$

$$zU(a+1,c+1,z) - U(a,c,z) - (c-a-1)U(a+1,c,z) = 0.$$

The recurrence relations for the M–functions can be derived from (7.3) by substituting the series and comparing equal powers of z. For the U–function the recursions follow from integrating by parts in the integral in (7.12). A remarkable feature is that U and M do not satisfy the same relations. This follows from the normalizations used for the functions, which are based on convention. Show that $M(a,c,z)/\Gamma(1+a-c)$ also satisfies the a–recursion of the U–function and that $[\Gamma(c-a)/\Gamma(c)]M(a,c,z)$ satisfies the c–recursion of the U–function.

7.6. The behavior of the U–function near $z=0$ follows from (7.14). A useful overview can be obtained by considering several c–values. Verify that

$$\begin{aligned}
U(a,c,z) &= \frac{\Gamma(c-1)}{\Gamma(a)} z^{1-c} + \mathcal{O}\left(|z|^{2-\Re c}\right), \quad \Re c \geq 2, c \neq 2,\\
&= \frac{\Gamma(c-1)}{\Gamma(a)} z^{1-c} + \mathcal{O}(|\ln z|), \quad c=2,\\
&= \frac{\Gamma(c-1)}{\Gamma(a)} z^{1-c} + \mathcal{O}(1), \quad 1<\Re c<2,\\
&= -\frac{1}{\Gamma(a)}[\ln z + \psi(a) + 2] + \mathcal{O}(|z\ln z|), \quad c=1, \ .\\
&= \frac{\Gamma(1-c)}{\Gamma(a+1-c)} + \mathcal{O}\left(|z|^{1-\Re c}\right), \quad 0<\Re c<1,\\
&= \frac{1}{\Gamma(a+1)} + \mathcal{O}(|z\ln z|), \quad c=0,\\
&= \frac{\Gamma(1-c)}{\Gamma(a+1-c)} + \mathcal{O}(|z|), \quad \Re c \leq 0, \quad c \neq 0.
\end{aligned}$$

7.7. By combining (7.13) and (7.16) verify that

$$\begin{aligned}
\frac{1}{\Gamma(c)} M(a,c,z) \sim{}& \frac{e^z z^{a-c}}{\Gamma(a)} \sum_{n=0}^{\infty} \frac{(c-a)_n (1-a)_n}{n!} z^{-n}\\
&+ \frac{e^{\pm \pi i a} z^{-a}}{\Gamma(c-a)} \sum_{n=0}^{\infty} \frac{(a)_n (1+a-c)_n}{n!} (-z)^{-n},
\end{aligned}$$

where the upper sign is taken if $-\frac{1}{2}\pi < \arg z < \frac{3}{2}\pi$ and the lower sign if $-\frac{3}{2}\pi < \arg z < \frac{1}{2}\pi$. The first part is dominant when $\Re z > 0$ and corresponds with (7.9); the second part becomes dominant when z enters the half plane $\Re z < 0$.

7.8. Use the methods of §5.6 for deriving the following Mellin-Barnes integral for the M–function:

$$M(a,c,-z) = \frac{1}{2\pi i} \frac{\Gamma(c)}{\Gamma(a)} \int_{\mathcal{L}} \frac{\Gamma(a+s)\Gamma(-s)}{\Gamma(c+s)} z^s \, ds, \quad |\arg z| < \pi/2,$$

where $\mathcal{L}$ runs from $-i\infty$ to $i\infty$, and separates the poles of $\Gamma(-s)\,\Gamma(1-c-s)$ from those of $\Gamma(a+s)$. For the U–function take the integral (7.12) and use the result of Exercise 5.6 to derive the Mellin-Barnes representation

$$U(a,c,z) = \frac{z^{-a}}{2\pi i}\int_{\mathcal{L}} \frac{\Gamma(a+s)\Gamma(1+a-c+s)\Gamma(-s)z^s}{\Gamma(a)\Gamma(1+a-c)}\,ds, \quad |\arg z| < 3\pi/2.$$

The path $\mathcal{L}$ separates the poles of $\Gamma(-s)$ from those of $\Gamma(a+s)\Gamma(1+a-c+s)$.

7.9. Show that a solution of the differential equation

$$y'' + zy' + (\tfrac{1}{2} - a)y = 0 \qquad (1)$$

is given by

$$y(z) = e^{-\frac{1}{4}z^2}U(a,z),$$

where $U(a,z)$ is the parabolic cylinder function introduced in §7.3.3. Verify that for equation (1) the operators M_t and M_t^* introduced in §4.4 (with corresponding kernel $K(z,t) = \exp(-zt)$) read

$$M_t = t^2 + t\frac{d}{dt} + (\tfrac{1}{2} - a), \quad M_t^* = t^2 - \frac{1}{2} - a - t\frac{d}{dt}.$$

A solution of $M_t^* v(t) = 0$ is $v(t) = t^{-1/2-a}\exp(\frac{1}{2}t^2)$. Show that

$$y(z) = \int_{c-i\infty}^{c+i\infty} e^{\frac{1}{2}s^2 - zs} s^{-\frac{1}{2}-a}\,ds, \quad c > 0,$$

is a solution of (1) with

$$y(0) = 2\pi i\,\frac{2^{-3/4-a/2}}{\Gamma\left(\frac{3}{4}+\frac{a}{2}\right)}, \quad y'(0) = -2\pi i\,\frac{2^{-1/4-a/2}}{\Gamma\left(\frac{1}{4}+\frac{a}{2}\right)}.$$

Show with the help of (7.21) for determining $U(a,0)$, $U'(a,0)$ that

$$U(a,z) = \frac{1}{i\sqrt{2\pi}} e^{\frac{1}{4}z^2} \int_{c-i\infty}^{c+i\infty} e^{\frac{1}{2}s^2 - zs} s^{-\frac{1}{2}-a}\,ds, \quad c > 0.$$

7.10. When in (7.14) a is a negative integer, the second term on the right-hand side and the first M–function become polynomials. Verify that

$$\boxed{L_n^\alpha(z) = \frac{(-1)^n}{n!}U(-n,\alpha+1,z) = \binom{n+\alpha}{n}M(-n,\alpha+1,z),}$$

where $L_n^\alpha(z)$ is the Laguerre polynomial introduced in (5.40).

7.11. Verify that the Coulomb wave functions introduced in §7.3.2 have the following behavior as $\rho \to \infty$:

$$F_\lambda(\eta,\rho) \sim \sin\theta_\lambda, \quad G_\lambda(\eta,\rho) \sim \cos\theta_\lambda, \quad \rho \to \infty,$$

where

$$\theta_\lambda = \rho - \eta\ln(2\rho) - \frac{1}{2}\lambda\pi + \sigma_\lambda.$$

Use the expansion of Exercise 7.7 for the M–function and expansion (7.13) for the U–function; observe that σ_λ satisfies

$$e^{i\sigma_\lambda} = \frac{\Gamma(\lambda+1+i\eta)}{|\Gamma(\lambda+1+i\eta)|}.$$

7.12. Verify the Laplace transformation of the M–function

$$\int_0^\infty t^{c-1}e^{-st}\,M(a,c,t)\,dt = \Gamma(c)\,s^{a-c}\,(s-1)^{-a}, \quad \Re c > 0, \quad \Re s > 1,$$

by substituting expansion (7.3). Verify the inversion formula

$$M(a,c,t) = \frac{t^{1-c}\Gamma(c)}{2\pi i}\int_{\mathcal{L}} e^{st}\,s^{a-c}\,(s-1)^{-a}\,ds,$$

where $\mathcal{L}$ is a vertical line in the half plane $\Re s > 1$. Initially we need the condition $\Re c > 0$. By deforming $\mathcal{L}$, for instance when $|\arg t| < \frac{1}{2}\pi$ into the Hankel contour for the reciprocal gamma function in Figure 3.4, we see that the condition on c can be dropped (except for the usual condition $c \neq 0, -1, -2, \ldots$).

8

Legendre Functions

Legendre functions have as a subclass the Legendre polynomials introduced in Chapter 6. Legendre functions are of great importance in physics and arise in several branches of the physical sciences. One of the interpretations of the generating function (6.46) for the Legendre polynomials comes from *potential theory*. The expression $1/\sqrt{a^2 - 2ar\cos\gamma + r^2}$ represents the potential at a point P of a source situated at A when r and a are the distances respectively of P and A from a point O, and γ is the angle subtended by PA at O. In terms of the present parameters (6.46) reads

$$\frac{1}{\sqrt{a^2 - 2ar\cos\gamma + r^2}} = \frac{1}{a}\sum_{n=0}^{\infty} P_n(\cos\gamma)\left(\frac{r}{a}\right)^n, \quad 0 < r < a.$$

In this expansion of the potential the parameters r and $\cos\gamma$ are separated, and the coefficients $P_n(\cos\gamma)$ are the Legendre polynomials introduced by Legendre in 1784 (LEGENDRE (1785)).

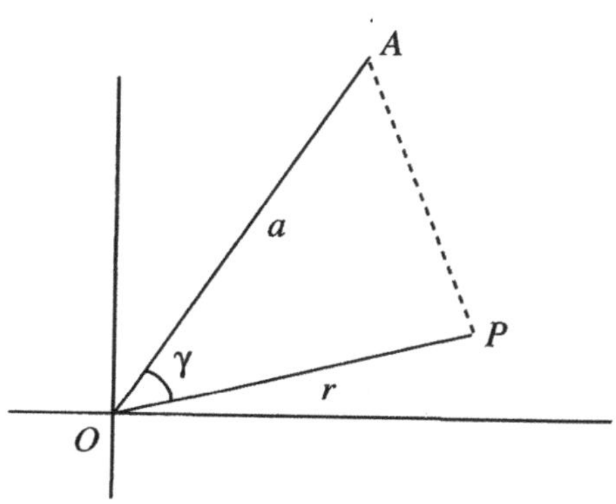

Figure 8.1. The potential at A due to a source at P is given by $1/\sqrt{a^2 - 2ar\cos\gamma + r^2}$.

When Laplace's equation $\Delta V = 0$ (the potential equation) is discussed in spherical polar coordinates the Legendre functions arise. This will become clear in Chapter 10, when we consider several coordinate systems, and separate the variables. Because of the connection with spherical coordinate systems Legendre functions are also called *spherical harmonics*.

8.1. The Legendre Differential Equation

Our starting point is the differential equation

$$\boxed{\left(1-z^2\right)y'' - 2zy' + \left[\nu(\nu+1) - \frac{\mu^2}{1-z^2}\right]y = 0,} \tag{8.1}$$

which, when $\mu = 0$ and $\nu = n = 0, 1, 2, \ldots$, indeed reduces to equation (5.32) for the Legendre polynomials. When $\mu \neq 0$ (8.1) is called the associated equation and the solutions are called the *associated Legendre functions*.

The singularities of (8.1) are located at $z = -1$, 1 and ∞, each singularity being regular, as is easily verified through Definition 4.1 of Chapter 4. In Riemann's notation (8.1) reads (see (5.21))

$$y = \mathcal{P}\left\{\begin{matrix} 1 & \infty & -1 & \\ \frac{1}{2}\mu & \nu+1 & \frac{1}{2}\mu & z \\ -\frac{1}{2}\mu & -\nu & -\frac{1}{2}\mu & \end{matrix}\right\}. \tag{8.2}$$

By using elementary transformations this differential equation can be written as a hypergeometric equation:

$$y = \frac{(z-1)^{\mu/2}}{(z+1)^{\mu/2}}\,\mathcal{P}\left\{\begin{matrix} 0 & \infty & 1 & \\ 0 & \nu+1 & \mu & \frac{1}{2}-\frac{1}{2}z \\ -\mu & -\nu & 0 & \end{matrix}\right\}. \tag{8.3}$$

From (5.23) it follows that hypergeometric functions $F(a, b; c; \zeta)$ are involved if

$$1 - c = -\mu, \quad c - a - b = \mu, \quad a = \nu + 1, \quad b = -\nu, \quad \zeta = \frac{1}{2} - \frac{1}{2}z.$$

In §5.8 we have mentioned that for this type of hypergeometric function quadratic transformations exist.

8.2. Ordinary Legendre Functions

When $\mu = 0$, (8.1) reduces to

$$\boxed{\left(1-z^2\right)y'' - 2zy' + \nu(\nu+1)y = 0.} \tag{8.4}$$

We define

$$\boxed{P_\nu(z) = F\left(-\nu, \nu+1; 1; \frac{1}{2} - \frac{1}{2}z\right).} \tag{8.5}$$

as the *Legendre function of the first kind.* This is a straightforward generalization of the Legendre polynomial. From the symmetry relation $F(a, b; c; z) = F(b, a; c; z)$ we derive the important property

$$\boxed{P_{-\nu-1}(z) = P_\nu(z).} \tag{8.6}$$

The right-hand side of (8.5) has a regular singular point at $z = -1$. From the properties of the hypergeometric functions (see for instance the conditions for (5.4)) it follows that we can consider $P_\nu(z)$ as an analytic function in the complex $z-$plane, with a branch cut along $(-\infty, -1]$.

The function of the second kind does not follow directly from (5.9), since $c = 1$. However, from transformation (2) in Exercise 5.8 it follows that

$$\boxed{P_\nu(z) = F\left(-\frac{1}{2}\nu, \frac{1}{2}\nu + \frac{1}{2}; 1; 1 - z^2\right)}$$

in a domain that contains the point $z = 1$. (*Warning*: from the above representation it does *not* follow that the function $P_\nu(z)$ is even with respect to z.) The last function will now be transformed by using (5.12). The result is:

$$\begin{aligned} P_\nu(z) = &\frac{z^\nu\,\Gamma(\nu+\frac{1}{2})}{\Gamma(\frac{1}{2}\nu+\frac{1}{2})\,\Gamma(1+\frac{1}{2}\nu)} F\left(-\frac{1}{2}\nu, \frac{1}{2} - \frac{1}{2}\nu; -\nu + \frac{1}{2}; z^{-2}\right) + \\ &\frac{z^{-\nu-1}\,\Gamma(-\nu-\frac{1}{2})}{\Gamma(-\frac{1}{2}\nu)\,\Gamma(\frac{1}{2}-\frac{1}{2}\nu)} F\left(\frac{1}{2}\nu + \frac{1}{2}, \frac{1}{2}\nu + 1; \nu + \frac{3}{2}; z^{-2}\right). \end{aligned} \tag{8.7}$$

Both terms on the right-hand side are solutions of (8.4). We now concentrate on the second term and we define

$$\boxed{Q_\nu(z) = \frac{\sqrt{\pi}\,\Gamma(\nu+1)}{\Gamma(\nu+\frac{3}{2})\,(2z)^{\nu+1}} F\left(\frac{1}{2}\nu + 1, \frac{1}{2}\nu + \frac{1}{2}; \nu + \frac{3}{2}; z^{-2}\right)} \tag{8.8}$$

as the *Legendre function of the second kind.* For the $Q-$function the values $\nu = -1, -2, \ldots$ are always excluded. The $F-$function represents an analytic function in the domain $|z| > 1$. In the same domain the $Q-$function is analytic and is single valued for $|\arg z| < \pi$; the many-valuedness outside this domain is due to the factor $z^{-\nu-1}$. The analytic continuation of the $Q-$function follows from that of the $F-$function, for instance by using (5.5) and (5.10)–(5.13). We conclude that the $Q-$function is an analytic function in the entire complex plane with the exception of the points $z = \pm 1$ and with a branch cut along $(-\infty, 1]$. Especially, the point $z = 0$ is a regular point, although in (8.8) the factor $z^{-\nu-1}$ may suggest differently. The two domains for $P_\nu(z)$ and $Q_\nu(z)$ are shown in Figure 8.2. It is clear that $P_\nu(1) = 1$, whereas $Q_\nu(z)$ is not defined at $z = \pm 1$ (see (5.6)).

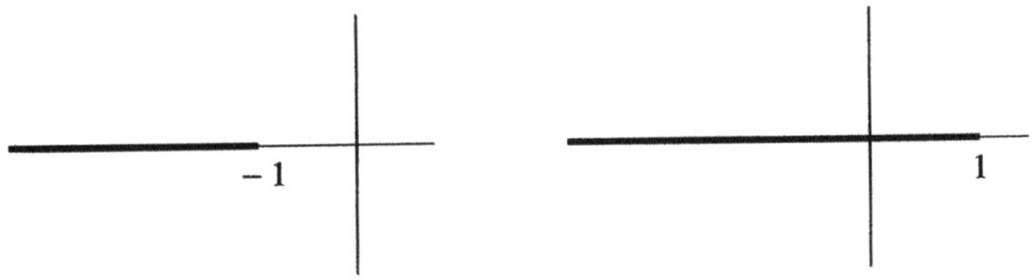

Figure 8.2. Domains with branch cuts where $P_\nu(z)$ (left) and $Q_\nu(z)$ are analytic.

8.3. Other Solutions of the Differential Equation

The functions $P_\nu(-z)$, $Q_\nu(-z)$, $P_{-\nu-1}(\pm z)$, $Q_{-\nu-1}(\pm z)$ are also solutions of (8.4). For $P_{-\nu-1}(z)$ this has been already mentioned in (8.6). A similar symmetry does not exist for the Q–function. We know that a linear relation should exist between $P_\nu(z)$, $Q_\nu(z)$, $Q_{-\nu-1}(z)$. This relation reads

$$\boxed{Q_\nu(z) - Q_{-\nu-1}(z) = \pi \cot \nu\pi P_\nu(z).} \tag{8.9}$$

A hint for a proof of this relation is given in Exercise 8.1. Equation (8.9) holds for all z in the complex plane, with a branch cut along $(-\infty, 1]$, since in this domain both sides of (8.9) are analytic. Also, formula (8.9) holds for all complex values of ν, with the exception of $\nu = 0, \pm 1, \pm 2, \ldots$. A special case of (8.9) is

$$Q_{n-\frac{1}{2}}(z) = Q_{-n-\frac{1}{2}}(z), \quad n = 0, \pm 1, \pm 2, \ldots .$$

Next we explain the relations between $P_\nu(-z)$, $Q_\nu(-z)$ and $P_\nu(z)$, $Q_\nu(z)$. Due to the many-valuedness of the functions it is convenient to indicate the meaning of the minus sign. This can be done by writing $e^{\pm i\pi}$. Let z be a point in the complex plane with $|\arg z| < \pi$. Then from (8.8) it follows that

$$Q_\nu\left(ze^{mi\pi}\right) = -e^{-m\nu\pi i} Q_\nu(z), \quad m \in \mathbb{Z}, \quad \nu \neq -1, -2, \ldots . \tag{8.10}$$

Initially (8.8) allows $|z| > 1$ only; by using the principle of analytic continuation we conclude that (8.10) holds for all z in the complex plane with a branch cut along $(-\infty, 1]$. Combining (8.9) and (8.10) yields

$$-\sin \nu\pi \left[e^{\mp \nu\pi i} Q_\nu(z) + e^{\pm \nu\pi i} Q_{-\nu-1}(z)\right] = \pi \cos \nu\pi P_\nu\left(ze^{\pm i\pi}\right),$$

where the upper (lower) sign in the left side corresponds to the upper (lower) sign on the right side. We can use (8.9) to eliminate $Q_{-\nu-1}(z)$. The result is

$$\frac{2\sin\nu\pi}{\pi}Q_\nu(z) = P_\nu(z)e^{\pm\nu\pi i} - P_\nu\left(ze^{\pm i\pi}\right), \tag{8.11}$$

where again $\nu \neq -1, -2, \ldots$, although the factor $\sin(\nu\pi)$ in front of the Q–function makes the left-hand side defined for those ν–values.

A different way of writing (8.11) is:

$$\frac{2\sin\nu\pi}{\pi}Q_\nu(z) = P_\nu(z)e^{\pm\nu\pi i} - P_\nu(-z), \tag{8.12}$$

where the upper sign is chosen when $\Im z < 0$ and the lower when $\Im z > 0$. To see this, take z with $\Im z < 0$ outside the unit disc. Turning z in the positive direction over π radians, that is writing $ze^{i\pi}$, we avoid the branch line of the Q–function; hence we take the upper signs in (8.11). Similarly for $\Im z > 0$, in which case we must take the lower signs. By the principle of analytic continuation, the restriction $|z| > 1$ can be dropped (on the unit circle only the points ± 1 are singular).

We now wish to interpret (8.11) and (8.12) when $z = x$ real and in the interval $(-1, 1)$. We know that the two values $P_\nu(xe^{\pm i\pi})$ are real (see (8.5)) and equal; hence we can write $P_\nu(xe^{\pm i\pi}) = P_\nu(-x)$, without ambiguity. For the Q–function the situation is different. $Q_\nu(z)$ is real on $(1, \infty)$. On the cut from $-\infty$ to 1 there are two possible values for $Q_\nu(z)$, depending on whether the cut is approached from the upper or lower side. We denote the two values by $Q_\nu(x + i0)$ and $Q_\nu(x - i0)$. With this notation, we derive from (8.12) for $-1 < x < 1$ the relations

$$\begin{aligned} \frac{2\sin\nu\pi}{\pi}Q_\nu(x+i0) &= P_\nu(x)e^{-\nu\pi i} - P_\nu(-x), \\ \frac{2\sin\nu\pi}{\pi}Q_\nu(x-i0) &= P_\nu(x)e^{+\nu\pi i} - P_\nu(-x). \end{aligned} \tag{8.13}$$

Hence

$$Q_\nu(x+i0) - Q_\nu(x-i0) = -i\pi P_\nu(x), \quad -1 < x < 1,$$

which clearly shows that indeed the values $Q_\nu(x+i0)$ and $Q_\nu(x-i0)$ are different and that the interval $[-1, 1]$ is part of the branch cut for the Q–function, as indicated in Figure 8.2.

In practical problems it is very convenient to have available two solutions of (8.4) which are real on $(-1, 1)$. Besides $P_\nu(x)$ we define

$$Q_\nu(x) = \frac{1}{2}[Q_\nu(x+i0) + Q_\nu(x-i0)], \quad -1 < x < 1. \tag{8.14}$$

Indeed, it is disturbing that the symbol Q_ν on the left-hand side has a different meaning from that on the right-hand side. But we use the convention that $Q_\nu(x)$ always denotes the function as defined in (8.14), whereas $Q_\nu(z)$ denotes the function from (8.8). For the P–function we do not need an agreement like this.

So, the general solution of (8.4) with $z = x \in (-1, 1)$ can be written in the form

$$y = AP_\nu(x) + BQ_\nu(x).$$

Remark 8.1. For the sake of clarity, the functions $P_\nu(z)$, $Q_\nu(z)$ are sometimes denoted in the literature by Gothic symbols: $\mathfrak{P}_\nu(z)$, $\mathfrak{Q}_\nu(z)$ when z is outside the interval $(-1, 1)$. For the functions on the cut $(-1, 1)$ the argument often is written in terms of a cosine: $P_\nu(\cos\theta)$, $Q_\nu(\cos\theta)$, $\theta \in (0, \pi)$.

8.4. A Few More Series Expansions

After manipulations with the gamma functions, we write (8.7) in the following form:

$$\begin{aligned} P_\nu(z) =& \frac{\Gamma(\nu+\frac{1}{2})}{\sqrt{\pi}\,\Gamma(\nu+1)}(2z)^\nu F\left(-\frac{1}{2}\nu, \frac{1}{2}-\frac{1}{2}\nu; -\nu+\frac{1}{2}; z^{-2}\right) + \\ & \frac{\Gamma(-\nu-\frac{1}{2})}{\sqrt{\pi}\,\Gamma(-\nu)}(2z)^{-\nu-1} F\left(\frac{1}{2}\nu+\frac{1}{2}, \frac{1}{2}\nu+1; \nu+\frac{3}{2}; z^{-2}\right). \end{aligned} \tag{8.15}$$

This representation is useful for large values of z. Again we assume that $|\arg z| < \pi$; with respect to ν we exclude the values $\pm\frac{1}{2}, \pm\frac{3}{2}, \ldots$, although a proper limiting process is available to define the right-hand side of (8.15) for these cases; see the discussion at the end of §4.3.

We next consider an expansion of $P_\nu(z)$ at the regular singular point $z = -1$. Combining (8.5) and (5.14) we obtain

$$\begin{aligned} P_\nu(z) &= \frac{1}{\Gamma(-\nu)\Gamma(\nu+1)} \sum_{n=0}^{\infty} \frac{(-\nu)_n\,(\nu+1)_n}{n!\; n!} C_n \left[\frac{1}{2}(1+z)\right]^n, \\ C_n &= 2\psi(n+1) - \psi(n-\nu) - \psi(n+\nu+1) - \ln\frac{1}{2}(1+z), \end{aligned} \tag{8.16}$$

with conditions: $|1+z| < 2$, $|\arg(1+z)| < \pi$. A similar result holds for the function of the second kind and can be obtained by combining (8.8) and (5.14).

From a practical point of view, it is interesting to have an expansion of $P_\nu(z)$ in powers of z. Observe that (8.5) yields an expansion in powers of

$z-1$. To start with, we compute the values $P_\nu(z)$, $P'_\nu(z)$ at $z=0$. Applying the first formula of (5.5) on (8.5) we obtain

$$P_\nu(z) = 2^{-\nu}(z+1)^\nu F\left(-\nu, -\nu; 1; \frac{z-1}{z+1}\right).$$

When $z=0$ this gives, with the help of Exercise 5.5,

$$P_\nu(0) = \frac{\sqrt{\pi}}{\Gamma\left(1+\frac{1}{2}\nu\right)\Gamma\left(\frac{1}{2}-\frac{1}{2}\nu\right)}. \tag{8.17}$$

From relation (3) in Exercise 8.3 it follows that $P'_\nu(0) = \nu P_{\nu-1}(0)$. Hence

$$P'_\nu(0) = \frac{\nu\sqrt{\pi}}{\Gamma\left(1-\frac{1}{2}\nu\right)\Gamma\left(\frac{1}{2}+\frac{1}{2}\nu\right)}. \tag{8.18}$$

To obtain the desired power series for $P_\nu(z)$ we remark that the substitution $t=z^2$ brings equation (8.4) in the form

$$t(1-t)\ddot{y} + \frac{1}{2}(1-3t)\dot{y} + \frac{1}{4}\nu(\nu+1)y = 0, \tag{8.19}$$

where the dots in $\dot{y}$, $\ddot{y}$ denote differentiation with respect to t. This is a special case of the hypergeometric differential equation (5.8) with

$$a = -\frac{1}{2}\nu, \quad b = \frac{1}{2}(\nu+1), \quad c = \frac{1}{2}.$$

So the general solution of (8.19) can be written in the form (5.9). In our case we obtain (observe that with the present $c-$value two independent solutions occur in (5.9))

$$P_\nu(z) = P_\nu(0)F\left(-\frac{1}{2}\nu, \frac{1}{2}\nu+\frac{1}{2}; \frac{1}{2}; z^2\right) + P'_\nu(0)zF\left(\frac{1}{2}-\frac{1}{2}\nu, 1+\frac{1}{2}\nu; \frac{3}{2}; z^2\right), \tag{8.20}$$

valid for each complex value of ν and initially for $|z|<1$. Analytic continuation with respect to z gives the domain of validity in the cut plane indicated in Figure 8.2. For the function of the second kind we obtain a similar representation:

$$Q_\nu(z) = Q_\nu(0)F\left(-\frac{1}{2}\nu, \frac{1}{2}\nu+\frac{1}{2}; \frac{1}{2}; z^2\right) + Q'_\nu(0)zF\left(\frac{1}{2}-\frac{1}{2}\nu, 1+\frac{1}{2}\nu; \frac{3}{2}; z^2\right). \tag{8.21}$$

The values of $Q_\nu(0)$, $Q'_\nu(0)$ are still to be computed. They depend on the sign of $\Im z$. After a few manipulations we find from (8.13), (8.17) and (8.18) that

$$Q_\nu(\pm i0) = \pm\frac{\sqrt{\pi}\,\Gamma\left(\frac{1}{2}\nu+\frac{1}{2}\right)}{2i\Gamma\left(\frac{1}{2}\nu+1\right)}\,e^{\mp\frac{1}{2}\nu\pi i}, \quad Q'_\nu(\pm i0) = \frac{\sqrt{\pi}\,\Gamma\left(1+\frac{1}{2}\nu\right)}{2i\Gamma\left(\frac{1}{2}\nu+\frac{1}{2}\right)}\,e^{\mp\frac{1}{2}\nu\pi i}. \tag{8.22}$$

Just as, or even more, important is the representation of the function introduced in (8.14):

$$Q_\nu(x) = \frac{\sqrt{\pi}\,\Gamma\left(\frac{1}{2}\nu+1\right)\cos\frac{1}{2}\nu\pi}{\Gamma\left(\frac{1}{2}\nu+\frac{1}{2}\right)} xF\left(\frac{1}{2}-\frac{1}{2}\nu, \frac{1}{2}\nu+1; \frac{3}{2}; x^2\right) - \frac{\sqrt{\pi}\,\Gamma\left(\frac{1}{2}\nu+\frac{1}{2}\right)\sin\frac{1}{2}\nu\pi}{2\Gamma\left(\frac{1}{2}\nu+1\right)} F\left(\frac{1}{2}+\frac{1}{2}\nu, -\frac{1}{2}\nu; \frac{1}{2}; x^2\right), \tag{8.23}$$

valid for $-1 < x < 1$ and, as usual, $\nu \neq 0, -1, -2, \ldots$. In Exercise 8.12 you are invited to prove this formula.

8.5. The function $Q_n(z)$

It is of importance to examine the Legendre function of the second kind for non-negative integer values of the parameter ν. From (8.8) it follows by summing the hypergeometric series (see also the list at the end of §5.1)

$$\boxed{Q_0(z) = \frac{1}{2}\ln\frac{z+1}{z-1}, \quad Q_1(z) = \frac{1}{2}z\ln\frac{z+1}{z-1} - 1.} \tag{8.24}$$

These representations show clearly the many-valuedness of the functions. We again consider the domain shown in Figure 8.2. The other Q_n–functions can be obtained by using the recurrence relations of Exercise 8.3. When doing so it becomes clear that the following representation arises:

$$Q_n(z) = P_n(z)Q_0(z) - W_{n-1}(z), \quad n = 0, 1, 2, \ldots \tag{8.25}$$

where W_{n-1} is a polynomial of degree $n-1$. That is, we have $W_{-1}(z) = 0$, $W_0(z) = 1$ and $W_{n-1}(z)$ satisfies recursion (1) of Exercise 8.3, since $P_n(z)$ and $Q_n(z)$ satisfy this relation too. Substituting $W_{n-1}(z)$ in the differential equation (8.4) we obtain

$$\left(1-z^2\right)W''_{n-1}(z) - 2zW'_{n-1}(z) + n(n+1)W_{n-1}(z) = 2P'_n(z). \tag{8.26}$$

This enables us to derive an explicit representation for $W_{n-1}(z)$. Let

$$W_{n-1}(z) = \sum_{k=0}^{n-1} a_k P_k(z). \tag{8.27}$$

Substitution of the left-hand side of (8.26) yields

$$\sum_{k=0}^{n-1} (n+k+1)(n-k)a_k P_k(z). \tag{8.28}$$

Expansion of the right-hand side gives

$$2P_n'(z) = 2\sum_{k=0}^{n-1} b_k P_k(z), \tag{8.29}$$

where b_k follow from elementary relations for the Legendre polynomials, such as (5.47). For, we see after integrating by parts that b_k can be written as

$$\frac{2}{2k+1} b_k = 1 - (-1)^{k+n} - \int_{-1}^{1} P_n(z) P_k'(z)\, dz$$

where the integral vanishes since $k < n$. Combining (8.28) and (8.29) we have finally

$$Q_n(z) = P_n(z) Q_0(z) - \sum_{k=0}^{n-1} \frac{(2k+1)[1-(-1)^{n+k}]}{(n+k+1)(n-k)} P_k(z). \tag{8.30}$$

For the real function $Q_n(x)$ defined in (8.14) and $-1 < x < 1$ the formulas become

$$\boxed{Q_0(x) = \frac{1}{2}\ln\frac{1+x}{1-x}, \quad Q_1(x) = \frac{1}{2} x \ln\frac{1+x}{1-x} - 1,} \tag{8.31}$$

$$Q_n(x) = P_n(x) Q_0(x) - \sum_{k=0}^{n-1} \frac{(2k+1)[1-(-1)^{n+k}]}{(n+k+1)(n-k)} P_k(x). \tag{8.32}$$

The Legendre polynomials and the Legendre functions of the second kind are connected through *Cauchy integrals* and *Hilbert transformations* on the interval $(-1, 1)$. We show that the function $Q_n(z)$ is the Hilbert transform of the Legendre polynomial $P_n(z)$ on the interval $(-1, 1)$. That is,

$$\boxed{Q_n(z) = \frac{1}{2}\int_{1}^{1} \frac{P_n(t)}{z-t}\, dt, \quad n = 0, 1, 2, \ldots .} \tag{8.33}$$

This is *Neumann's integral for* $Q_n(z)$ (1848). We consider $z-$values outside the interval $(-1, 1)$. To prove (8.33) we show that it satisfies (8.4). Denote the integral of the right-hand side of (8.33) by $g(z)$. Then we have

$$\begin{aligned}
\left(1 - z^2\right) g'(z) &= -\int_{-1}^{1} \frac{1-z^2}{(z-t)^2} P_n(t)\, dt \\
&= \int_{-1}^{1} P_n(t)\, dt - \int_{-1}^{1} \frac{t^2 - 2tz + 1}{(z-t)^2} P_n(t)\, dt \\
&= C - \int_{-1}^{1} P_n(t) \frac{d}{dt}\left(\frac{1-t^2}{z-t}\right) dt \\
&= C + \int_{-1}^{1} \frac{1-t^2}{z-t} P_\nu'(t)\, dt,
\end{aligned}$$

where C is a constant ($C = 2$ when $n = 0$ and $C = 0$ when $n \geq 1$). Next we verify

$$\frac{d}{dz}\left[\left(1-z^2\right)g'(z)\right] = \int_{-1}^{1} \frac{d}{dt}[(1-t^2)P_\nu'(t)]\,\frac{dt}{z-t} = -n(\nu+1)\int_{-1}^{1}\frac{P_n(t)}{z-t}\,dt,$$

where in the final step we have used (8.4). It follows that the function $g(z)$ satisfies (8.4). Hence it is a linear combination of $P_n(z)$ and $Q_n(z)$. For $n = 0$ and $n = 1$ the corresponding functions $g(z)$ can easily be evaluated. The validity for $n = 2, 3, \ldots$ then follows from the recurrence relation (1) of Exercise 8.3. We can also verify that the linear combination $P_n(z)$ and $Q_n(z)$ only contains $Q_n(z)$; for instance, by using the behavior of the functions as $z \to \infty$. We conclude that (8.33) is verified. When $z = x \in (-1, 1)$ the formula reads

$$\boxed{Q_n(x) = \frac{1}{2}\fint_{-1}^{1}\frac{P_n(t)}{x-t}\,dt, \quad n = 0, 1, 2, \ldots,} \tag{8.34}$$

where the integral should be interpreted as a Cauchy principal value integral.

From (8.33) it follows that

$$Q_n(z) = P_n(z)Q_0(z) - \frac{1}{2}\int_{1}^{1}\frac{P_n(z)-P_n(t)}{z-t}\,dt.$$

The integrand can be expanded in powers of z. A polynomial of degree $n-1$ remains. This again leads to formula (8.30).

8.6. Integral Representations

From Exercise 5.12 it follows that

$$P_\nu(z) = \frac{2}{\pi}\int_0^\pi F\left(-\nu, \nu+1; \frac{1}{2}; \frac{1}{2}(1-z)\sin^2\phi\right)\,d\phi,$$

where the hypergeometric function is an elementary function:

$$\begin{aligned} f_\nu(w) &= F\left(-\nu, \nu+1; \frac{1}{2}; -w\right) \\ &= \frac{\left(\sqrt{1+w}+\sqrt{w}\right)^{2\nu+1} + \left(\sqrt{1+w}+\sqrt{w}\right)^{-2\nu-1}}{2\sqrt{1+w}}. \end{aligned} \tag{8.35}$$

To prove this, use Exercise 5.4, formula (2), and substitute $w = \sinh^2(t)$. We obtain

$$P_\nu(z) = \frac{2}{\pi}\int_0^\pi f_\nu\left[\frac{1}{2}(z-1)\sin^2\phi\right]\,d\phi. \tag{8.36}$$

The right-hand side is an analytic function for z in the cut plane of Figure 8.2, that is for $|\arg(1+z)| < \pi$. Restrictions on $\nu \in \mathbb{C}$ are not needed.

Now let $z = \cosh\alpha$, with $\alpha > 0$ and introduce in (8.36) the following transformation:

$$\sin\phi = \frac{\sinh\frac{1}{2}\theta}{\sinh\frac{1}{2}\alpha}.$$

Then (8.36) becomes

$$\boxed{P_\nu(\cosh\alpha) = \frac{2}{\pi}\int_0^\alpha \frac{\cosh\left(\nu+\frac{1}{2}\right)\theta}{\sqrt{2\cosh\alpha - 2\cosh\theta}}\,d\theta.} \tag{8.37}$$

Writing this in the form

$$P_\nu(\cosh\alpha) = \frac{1}{\pi}\int_{-\alpha}^\alpha \frac{e^{-(\nu+\frac{1}{2})\theta}}{\sqrt{2\cosh\alpha - 2\cosh\theta}}\,d\theta$$

and putting $e^\theta = \cosh\alpha + \sinh\alpha\,\cos\psi$, we obtain

$$\boxed{P_\nu(\cosh\alpha) = \frac{1}{\pi}\int_0^\pi \frac{d\psi}{(\cosh\alpha + \sinh\alpha\,\cos\psi)^{\nu+1}}.} \tag{8.38}$$

From (8.6) we have the alternative representation

$$\boxed{P_\nu(\cosh\alpha) = \frac{1}{\pi}\int_0^\pi (\cosh\alpha + \sinh\alpha\,\cos\psi)^\nu\,d\psi.} \tag{8.39}$$

The above representations hold for any $\nu \in \mathbb{C}$ and are derived under the condition $z > 0$. Analytic continuation with respect to z gives a much larger domain, viz. $\Re\cosh\alpha > 0$.

Representations for $-1 < x < 1$ follow from (8.35) via

$$x = \cos\beta \quad (0 < \beta < \pi), \quad \sin\phi = \frac{\sin\frac{1}{2}\theta}{\sin\frac{1}{2}\beta}.$$

Then (8.35) becomes the Mehler-Dirichlet formula for Legendre functions

$$\boxed{P_\nu(\cos\beta) = \frac{2}{\pi}\int_0^\beta \frac{\cos\left(\nu+\frac{1}{2}\right)\theta}{\sqrt{2\cos\theta - 2\cos\beta}}\,d\theta.} \tag{8.40}$$

Integrating this on the interval $[-\beta, \beta]$ and writing $\cos(\nu+\frac{1}{2})\theta$ in exponential form, the substitution

$$e^{i\theta} = \cos\beta + i\sin\beta\cos\psi$$

gives the analog of (8.38), that is,

$$\boxed{P_\nu(\cos\beta) = \frac{1}{\pi}\int_0^\pi \frac{d\psi}{(\cos\beta + i\sin\beta\,\cos\psi)^{\nu+1}},} \tag{8.41}$$

with the alternative

$$\boxed{P_\nu(\cos\beta) = \frac{1}{\pi}\int_0^\pi (\cos\beta + i\sin\beta\,\cos\psi)^\nu\,d\psi.} \tag{8.42}$$

When $\nu = n$, (8.42) is called Laplace's formula for the Legendre polynomial, which usually is derived by using the Rodrigues formula (6.20). Namely, from (6.20) it follows that

$$\boxed{P_n(z) = \frac{2^{-n}}{2\pi i}\oint \frac{\left(t^2-1\right)^n}{(t-z)^{n+1}}\,dt,} \tag{8.43}$$

only condition is that the contour encircles the point z in positive direction. Using the substitution $t = \cos\beta + i\sin\beta\,e^{i\psi}$, $z = \cos\beta$ (8.43) becomes (8.42) with $\nu = n$. Writing z for $\cos\beta$ we obtain

$$\boxed{P_n(z) = \frac{1}{\pi}\int_0^\pi \left(z + \sqrt{z^2-1}\,\cos\psi\right)^n d\psi,}$$

which in this case is valid for all $z \in \mathbb{C}$. The transition from (8.43) (with $\nu = n$) to the last integral can also be described as integrating in (8.44) over the circle with center $t = z$ and radius $\sqrt{|z^2-1|}$.

A similar approach for general values of ν cannot be set up via (6.20), since Rodrigues' formula permits no generalization to complex values of n. However, with some extra care when handling the many-valued functions, we can generalize (8.43) in the form of *Schläfli's formula*

$$P_\nu(z) = \frac{2^{-\nu}}{2\pi i}\oint_C \frac{\left(t^2-1\right)^\nu}{(t-z)^{\nu+1}}\,dt, \tag{8.44}$$

where the contour encircles both the points $t = 1$ and $t = z$ in the positive sense, but *does not* encircle the point $t = -1$ (see Figure 8.3). From this we again obtain (8.42). Usually, for complex values of z, this formula is written in the form

$$\boxed{P_\nu(z) = \frac{1}{\pi}\int_0^\pi \left(z + \sqrt{z^2-1}\,\cos\psi\right)^\nu d\psi.} \tag{8.45}$$

The conditions will be formulated later.

A very important aspect in the theory of integral representations of Legendre functions is that the differential equation (8.4) is solved by the function defined by the integral

$$F(z) = \frac{2^{-\nu}}{2\pi i} \int_C \frac{(t^2-1)^\nu}{(t-z)^{\nu+1}}\, dt. \tag{8.46}$$

A solution like this can be found by the methods of §4.4. We give a direct verification however. The contour of integration in (8.46) avoids the singularities at $t = \pm 1$, $t = z$ and satisfies the condition that the function

$$f(t) = (t^2-1)^{\nu+1}(t-z)^{-\nu-2}$$

assumes its original value when t runs along through the contour. This remark is related to the many-valuedness of the function f. A proof that (8.46) is indeed a solution of (8.4) is quite simple. By substituting $F(z)$ into (8.4), one obtains

$$\begin{aligned}
&\left(1-z^2\right) F'' - 2zF' + \nu(\nu+1)F \\
&= \frac{(\nu+1)2^{-\nu}}{2\pi i} \int_C \frac{\left(t^2-1\right)^\nu}{(t-z)^{\nu+3}} \left[-(\nu+2)\left(t^2-1\right) + 2(\nu+1)t(t-z)\right] dt\,. \\
&= \frac{(\nu+1)2^{-\nu}}{2\pi i} \int_C \frac{d}{dt} f(t)\, dt
\end{aligned}$$

Hence the integral vanishes when C satisfies the above mentioned condition. Since the contour avoids the singularities of f, the function $F(z)$ is a solution for any $\nu \in \mathbb{C}$. If $\nu = n$ then (8.46) yields a special case with $F(z) = P_n(z)$. We now take two special choices of C. The first choice will lead to (8.44) and the other one to an integral representation for the function of the second kind.

We concentrate on contours as given in Figure 8.3 with the properties described after (8.46). First we remark that, after a full circuit around the point $t = 1$, the function

$$f(t) = \left(t^2-1\right)^{\nu+1} (t-z)^{-\nu-2}$$

assumes its original value multiplied by the factor $e^{2\pi i(\nu+1)}$; after a full circuit around the point $t = z$ the change equals $e^{2\pi i(-\nu-2)}$. Hence, when C is a contour that encircles once the points $t = 1$ and $t = z$, but does not encircle the point $t = -1$, then the function $f(t)$ assumes its original value when t runs along the contour. To point out the sense of the contour we write

$$P_\nu(z) = \frac{2^{-\nu}}{2\pi i} \int_A^{(1^+, z^+)} \frac{(t^2-1)^\nu}{(t-z)^{\nu+1}}\, dt, \tag{8.47}$$

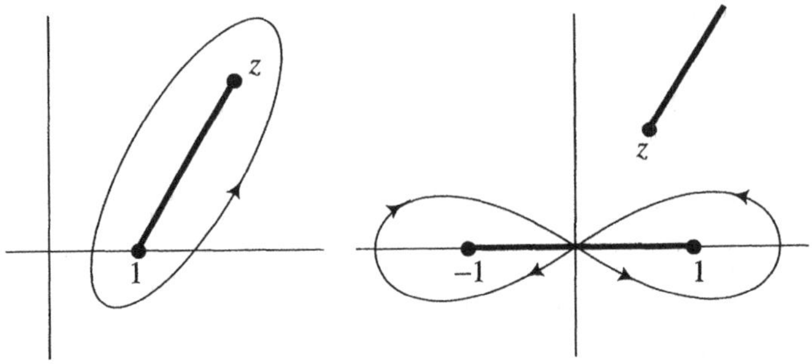

Figure 8.3. Contours of integration and branch cuts for (8.47) (left) and for (8.49).

where A is a point on the real axis larger than unity (and larger than z if z is real and larger than unity); see Figure 8.3. We need to specify the many-valued functions in the integrand of (8.47). At the point A we assume that $\arg(t-1) = \arg(t+1) = 0$ and $|\arg(t-z)| < \pi$. With this condition we know that $F(1) = 1$. Since the function of the second kind is not bounded at $z = 1$ we conclude that $F(z) = P_\nu(z)$ and that (8.44) is verified.

A proof of (8.45) now follows by integrating in (8.47) over a circle with center $t = z$ and radius $\sqrt{|z^2-1|}$. If $\Re z > 0$ the point $t = 1$ is within the circle, but $t = -1$ lies outside the circle. Next we substitute in (8.47) $t = z + \sqrt{z^2-1}\,e^{i\psi}$, $-\pi < \psi < \pi$. Indeed we arrive at (8.45), with the conditions $\nu \in \mathbb{C}$, $|\arg z| < \frac{1}{2}\pi$ and $\arg[z + \sqrt{z^2-1}\,\cos\psi]$ equals $\arg z$ if $\psi = \frac{1}{2}\pi$. With similar conditions the variant of (8.41) holds:

$$\boxed{P_\nu(z) = \frac{1}{\pi}\int_0^\pi \frac{d\psi}{\left(z + \sqrt{z^2-1}\,\cos\psi\right)^{\nu+1}}.} \tag{8.48}$$

Formulas (8.45) and (8.48) are called *Laplace's first integral* and *Laplace's second integral*, respectively.

An integral for $Q_\nu(z)$ by choosing a contour $\mathcal{D}$, as depicted in Figure 8.3: a 'figure of eight', where we assume that z does not belong to the interval $[-1, 1]$. Moreover the contour should encircle the points $t = +1$ and $t = -1$; the first point in the positive, while the second is in the negative direction. The point $t = z$ lies outside the contour. The contour cuts the real t−axis at a point $A > 1$, where $\arg(t+1) = \arg(t-1) = 0$. Furthermore, $|\arg z| < \pi$.

Reasoning similarly as earlier we infer that (8.46), with $\mathcal{C}$ replaced by $\mathcal{D}$, solves equation (8.4). We claim that, when $\nu \notin \mathbb{Z}$,

$$Q_\nu(z) = \frac{2^{-\nu}}{4i \sin \nu\pi} \int_{\mathcal{D}} \frac{(t^2-1)^\nu}{(z-t)^{\nu+1}}\, dt. \tag{8.49}$$

We prove this for $\Re\nu > -1$. In that case we can push the contour along the interval $(-1,1)$. Taking into account the phases of $(t \pm 1)^\nu$, we obtain

$$Q_\nu(z) = \frac{2^{-\nu}\left(e^{i\nu\pi} - e^{-i\nu\pi}\right)}{4i \sin \nu\pi} \int_{-1}^{1} \frac{(1-t^2)^\nu}{(z-t)^{\nu+1}}\, dt$$

with $\arg(1-t) = \arg(1+t) = 0$. Hence, for $\Re\nu > -1$, $|\arg z| < \pi$, $z \notin [-1,1]$

$$\boxed{Q_\nu(z) = 2^{-\nu-1} \int_{-1}^{1} \frac{(1-t^2)^\nu}{(z-t)^{\nu+1}}\, dt.} \tag{8.50}$$

When $\nu = n$ (integer) we can integrate by parts n times. With Rodrigues' formula (5.20) Neumann's integral (8.33) then appears. The identification with earlier definitions of the function of the second kind follows from the behavior of the above integral as $z \to \infty$:

$$2^{-\nu-1} \int_{-1}^{1} \frac{(1-t^2)^\nu}{(z-t)^{\nu+1}}\, dt \sim \frac{2}{(2z)^{\nu+1}} \int_0^1 \left(1-t^2\right)^\nu dt = \frac{\Gamma(\nu+1)\sqrt{\pi}}{(2z)^{\nu+1}\Gamma\left(\nu+\frac{3}{2}\right)}.$$

From (8.15) it follows that $P_\nu(z)$ cannot be present in the right-hand side of (8.50). Hence, the identification with (8.8) is verified.

Next we derive Laplace's integral for the Q–function that corresponds with (8.48). The starting point is (8.50). Assume that $z > 1$. Introduce the new variable of integration ϕ by writing

$$t = \frac{e^\phi\sqrt{z+1} - \sqrt{z-1}}{e^\phi\sqrt{z+1} + \sqrt{z-1}}. \tag{8.51}$$

The range $(-1,1)$ of real t–values corresponds to the range $(-\infty,\infty)$ of real ϕ–values. After some simple manipulations we obtain

$$\begin{aligned} Q_\nu(z) &= \tfrac{1}{2} \int_{-\infty}^{\infty} \left[z + \sqrt{z^2-1}\cosh\phi\right]^{-\nu-1} d\phi \\ &= \int_0^{\infty} \left[z + \sqrt{z^2-1}\cosh\phi\right]^{-\nu-1} d\phi. \end{aligned} \tag{8.52}$$

This is the desired Laplace integral for the function of the second kind.

To obtain a variant - and a variant for (8.37) as well - we introduce the variable of integration θ by putting

$$e^{\theta} = z + \sqrt{z^2 - 1}\ \cosh\phi.$$

Then

$$\boxed{Q_\nu(\cosh\alpha) = \int_\alpha^\infty \frac{e^{-(\nu+\frac{1}{2})\theta}}{\sqrt{2\cosh\theta - 2\cosh\alpha}}\, d\theta, \qquad z = \cosh\alpha.} \tag{8.53}$$

Both results hold when $\Re\nu > -1$. By using the principle of analytic continuation we infer that the restriction with respect to real values of z can be relaxed: (8.52) and (8.53) hold in the familiar z–domain for the Q–function as shown in Figure 8.2; the function $\sqrt{z^2-1}$ also assumes in that domain the principal branch (it is real if $z > 1$ and continuous in the mentioned domain).

Finally we present the following representations:

$$P_\nu(\cosh\alpha) = \frac{2}{\pi}\cos\left(\nu + \frac{1}{2}\right)\pi \int_0^\infty \frac{\cosh\left(\nu+\frac{1}{2}\right)\theta}{\sqrt{2\cosh\theta + 2\cosh\alpha}}\, d\theta, \tag{8.54}$$

$$P_\nu(\cosh\alpha) = \frac{2}{\pi}\cot\left(\nu + \frac{1}{2}\right)\pi \int_\alpha^\infty \frac{\sinh\left(\nu+\frac{1}{2}\right)\theta}{\sqrt{2\cosh\theta - 2\cosh\alpha}}\, d\theta, \tag{8.55}$$

valid when $\alpha > 0, -1 < \Re\nu < 0$. The reader is invited to prove this by using the hints given in Exercise 8.2.

8.7. Associated Legendre Functions

From the theory of hypergeometric functions (see Chapter 5) it can be verified that the following two functions are solutions of (8.1):

$$P_\nu^\mu(z) = \frac{1}{\Gamma(1-\mu)}\left(\frac{z+1}{z-1}\right)^{\frac{1}{2}\mu} F\left(-\nu, \nu+1; 1-\mu; \frac{1}{2} - \frac{1}{2}z\right), \tag{8.56}$$

$$\begin{aligned} Q_\nu^\mu(z) = {} & \frac{\sqrt{\pi}\, e^{\mu\pi i}\,\Gamma(\nu+\mu+1)\,\left(z^2-1\right)^{\frac{1}{2}\mu}}{2^{\nu+1}\, z^{\nu+\mu+1}\,\Gamma\left(\nu+\frac{3}{2}\right)} \\ & \times F\left(\frac{1}{2}\nu + \frac{1}{2}\mu + 1, \frac{1}{2}\nu + \frac{1}{2}\mu + \frac{1}{2}; \nu + \frac{3}{2}; z^{-2}\right). \end{aligned} \tag{8.57}$$

These functions are called the *associated Legendre function of the first and second kinds*, respectively. Other representations are also available. They are all generated through the well-known relations of the hypergeometric functions. An example is given in Exercise 8.19.

The domain of definition of both functions with respect to z is (after analytic continuation of the above relations) the complex plane with a branch cut along $(-\infty, 1]$. In (8.56) we assume that $\arg[(z+1)/(z-1)] = 0$ if z is real and greater than 1. In (8.57) we assume that $\arg(z^2-1) = 0$ when z is real and greater than 1, and that $\arg z = 0$ when z is real and greater than 0.

When $\mu = 0$ the above functions reduce to the definitions in (8.5) and (8.8); μ is called the *order* and ν the *degree*. The word 'degree' corresponds to the polynomial case $P_\nu^\mu(z)$ when $\mu = 0$, $\nu = n = 0, 1, 2, \ldots$. We shall not give an extensive treatment of the associated functions. A few basic properties will be mentioned; for more information we refer to the literature.

When $\mu = m$ (a positive integer) the hypergeometric function in (8.56) is not defined. The reciprocal gamma function in front of the F–function vanishes, however, and a meaningful expression remains when $\mu = m = 1, 2, 3\ldots$. The fact is that

$$\begin{aligned} P_\nu^m(z) &= (z^2-1)^{\frac{1}{2}m}\frac{(-\nu)_m(\nu+1)_m}{m!}\left(-\frac{1}{2}\right)^m \\ &\times F\left(-\nu+m, \nu+1+m; m+1; \frac{1}{2}-\frac{1}{2}z\right), \end{aligned} \tag{8.58}$$

as follows immediately from the definition (5.3) of the F–functions. From Exercise 5.1 the fundamental relation

$$\boxed{P_\nu^m(z) = \left(z^2-1\right)^{\frac{1}{2}m}\frac{d^m}{dz^m}P_\nu(z),} \tag{8.59}$$

can then be derived, which is often used as the definition of the associated function (if $\mu = m$), since it is quite important for practical purposes. For the function of the second kind we have the analog

$$\boxed{Q_\nu^m(z) = \left(z^2-1\right)^{\frac{1}{2}m}\frac{d^m}{dz^m}Q_\nu(z).} \tag{8.60}$$

In fact this can be derived from (8.57), but we give a direct proof, because it gives a better insight into the nature of equation (8.1). Introduce in (8.1), if $\mu = m$, the substitution $y = (z^2-1)^{m/2}w$. Then, after a few calculations (8.1) becomes

$$\left(1-z^2\right)w'' - 2(m+1)zw' + (\nu-m)(\nu+m+1)w = 0. \tag{8.61}$$

Consider now the Legendre differential equation (8.4). Differentiate this equation m times with respect to z and put $d^m y/dz^m = v$. With the help of

Leibniz' formula we derive

$$\frac{d^m}{dz^m}\left[\left(1-z^2\right)y''\right] = \sum_{k=0}^{m}\binom{m}{k}\left(\frac{d^{m-k}(y'')}{dz^{m-k}}\right)\left[\frac{d^k\left(1-z^2\right)}{dz^k}\right]$$
$$= \left(1-z^2\right)v'' - 2mzv' - m(m-1)v$$

and also

$$\frac{d^m}{dz^m}(2zy') = 2zv' + 2mv.$$

Hence, after m–fold differentiation of (8.4), we again find for w equation (8.61). On the one hand the functions

$$\frac{d^m P_\nu(z)}{dz^m}, \quad \frac{d^m Q_\nu(z)}{dz^m}$$

are solutions of this equation; on the other hand both left-hand sides of (8.59) and (8.60), multiplied by $(z^2-1)^{-m/2}$ are also solutions. The verification of (8.59) has already been carried out. The verification of (8.60) now follows from the behavior of both sides as $z \to \infty$, by using (8.57).

From (8.58) and the third formula of (5.5) it follows, with the help of a few relations for the gamma function, that

$$\boxed{P_\nu^{-m}(z) = \frac{\Gamma(1+\nu-m)}{\Gamma(1+\nu+m)} P_\nu^m(z).} \tag{8.62}$$

As we will see later (cf. (8.67), for general order)

$$\boxed{Q_\nu^{-m}(z) = \frac{\Gamma(1+\nu-m)}{\Gamma(1+\nu+m)} Q_\nu^m(z).} \tag{8.63}$$

Both formulas give a useful alternative to the result for $P_\nu^{-m}(z)$, $Q_\nu^{-m}(z)$ given in Exercise 8.17, which gives an interpretation of (8.59) and (8.60) when m is a negative integer.

Equation (8.1) remains the same when we replace μ with $-\mu$, z with $-z$ or ν with $-\nu-1$. Hence,

$$P_\nu^{\pm\mu}(\pm z), \quad Q_\nu^{\pm\mu}(\pm z), \quad P_{-\nu-1}^{\pm\mu}(\pm z), \quad Q_{-\nu-1}^{\pm\mu}(\pm z)$$

all satisfy the differential equation (8.1). There must exist several relations between these functions. As in (8.6) we have

$$P_{-\nu-1}^{\mu}(z) = P_\nu^\mu(z), \tag{8.64}$$

which follows immediately from (8.56). Furthermore, we have the analog of (8.9):

$$Q^{\mu}_{-\nu-1}(z) = \frac{-\pi\ e^{\mu\pi i} \cos\nu\pi\ P^{\mu}_{\nu}(z) + \sin\pi(\nu+\mu)\ Q^{\mu}_{\nu}(z)}{\sin\pi(\nu-\mu)}. \tag{8.65}$$

The proof runs as for (8.9) itself; see Exercise 8.13. Next we consider the reflection with respect to μ. Special cases are already given in (8.62) and (8.63). For general μ we have

$$P^{-\mu}_{\nu}(z) = \frac{\Gamma(\nu-\mu+1)}{\Gamma(\nu+\mu+1)}\left[P^{\mu}_{\nu}(z) - \frac{2}{\pi}e^{-\mu\pi i}\sin\mu\pi\ Q^{\mu}_{\nu}(z)\right], \tag{8.66}$$

$$Q^{-\mu}_{\nu}(z) = e^{-2\mu\pi i}\frac{\Gamma(\nu-\mu+1)}{\Gamma(\nu+\mu+1)}Q^{\mu}_{\nu}(z). \tag{8.67}$$

The last formula follows directly from (8.57) in combination with the third formula in (5.5). Formula (8.66) can be verified in the same way as (8.65) can be proved.

The analogs of (8.47) and (8.49) for the associated functions (see Exercise 8.14) read

$$P^{\mu}_{\nu}(z) = \frac{e^{-\mu\pi i}\Gamma(-\nu)}{2^{\nu+1}\pi i\Gamma(-\mu-\nu)}\left(z^2-1\right)^{-\frac{1}{2}\mu}\int_{\mathcal{C}}\frac{\left(t^2-1\right)^{\nu}}{(t-z)^{\nu-\mu+1}}\,dt, \tag{8.68}$$

$$Q^{\mu}_{\nu}(z) = \frac{e^{\mu\pi i}\Gamma(\nu+\mu+1)\Gamma(-\nu)}{2^{\nu+2}\pi i}\left(z^2-1\right)^{\frac{1}{2}\mu}\int_{\mathcal{D}}\frac{\left(t^2-1\right)^{\nu}}{(t-z)^{\nu+\mu+1}}\,dt. \tag{8.69}$$

The contour $\mathcal{D}$ is the same as in Figure 8.3; for the contour $\mathcal{C}$ the real point A should be equal to $+\infty$. Since now the contour extends to infinity we need $\Re(\nu+\mu) < 0$.

From (8.68) we again derive, as in (8.45), a Laplace integral. For $m = 0, 1, 2, \ldots$ we have

$$P^{m}_{\nu}(z) = \frac{\Gamma(\nu+m+1)}{\pi\Gamma(\nu+1)}\int_0^{\pi}\left(z+\sqrt{z^2-1}\ \cos\psi\right)^{\nu}\cos m\psi\ d\psi. \tag{8.70}$$

For the function of the second kind we mention the generalization of (8.50):

$$Q^{\mu}_{\nu}(z) = \frac{e^{\mu\pi i}}{2^{\nu+1}}\frac{\Gamma(\nu+\mu+1)}{\Gamma(\nu+1)}(z^2-1)^{\frac{1}{2}\mu}\int_{-1}^{1}\frac{\left(1-t^2\right)^{\nu}}{(z-t)^{\nu+\mu+1}}\,dt, \quad \Re\nu > -1,\quad |\arg z| < \pi,\quad z \notin [-1,1]. \tag{8.71}$$

The associated functions satisfy a number of recurrence relations. See Exercise 8.16. The first pair is with varying order, the second pair with varying

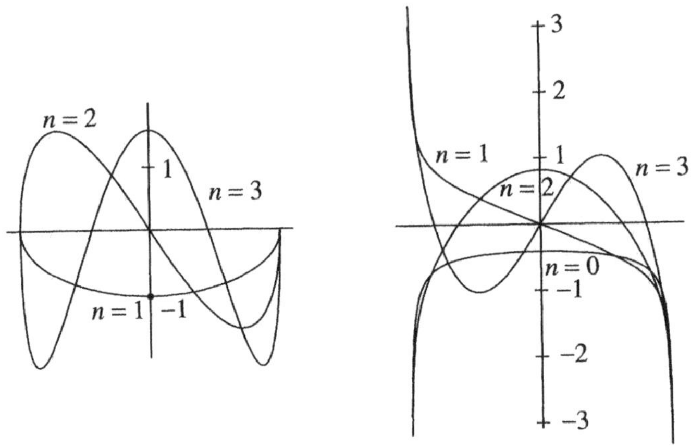

Figure 8.4. Legendre functions on the interval $[-1,1]$; $P_n^1(x), n = 1,2,3$ (left) and $Q_n^1(x), n = 0,1,2,3$.

degree. The proofs of the recurrences follow from the integral representations in (8.68) and (8.69). See Exercise 8.3 for a demonstration of the method.

In Figure 8.4 we show graphs of Legendre functions on the interval $(-1,1)$. In fact, for real values of $z = x \in (-1,1)$ we can define real solutions. as we did in (8.14). In the associated case we need a definition for the P–functions too. We introduce the functions

$$\begin{aligned} P_\nu^\mu(x) &= \frac{1}{2}\left[e^{\frac{1}{2}\mu\pi i}P_\nu^\mu(x+i0) + e^{-\frac{1}{2}\mu\pi i}P_\nu^\mu(x-i0)\right], \\ Q_\nu^\mu(x) &= \frac{1}{2}e^{-\mu\pi i}\left[e^{-\frac{1}{2}\mu\pi i}Q_\nu^\mu(x+i0) + e^{\frac{1}{2}\mu\pi i}Q_\nu^\mu(x-i0)\right]. \end{aligned} \qquad -1 < x < 1. \tag{8.72}$$

When $\mu = m$ (a positive integer) we have

$$\boxed{P_\nu^m(x) = (-1)^m\left(1-x^2\right)^{\frac{1}{2}m}\frac{d^m}{dx^m}P_\nu(x),} \tag{8.73}$$

$$\boxed{Q_\nu^m(x) = (-1)^m\left(1-x^2\right)^{\frac{1}{2}m}\frac{d^m}{dx^m}Q_\nu(x).} \tag{8.74}$$

Formulas for $P_\nu^\mu(x)$, $Q_\nu^\mu(x)$ with $x \in (-1,1)$ are usually obtained from results for $P_\nu^\mu(z)$, $Q_\nu^\mu(z)$ by writing $z = x \pm i0$ and replacing $z-1$ and z^2-1 with $(1-x)e^{\pm\pi i}$ and $(1-x^2)e^{\pm\pi i}$, respectively, and $z+1$ with $x+1$. Apart from (8.72) it is convenient to have explicit representations in the form of hypergeometric functions. See Exercise 8.8 for expansions in powers of x. The following representation is the analog of (8.5):

$$P_\nu^\mu(x) = \frac{1}{\Gamma(1-\mu)} \left(\frac{1+x}{1-x}\right)^{\frac{1}{2}\mu} F\left(-\nu, \nu+1; 1-\mu; \frac{1}{2} - \frac{1}{2}x\right). \tag{8.75}$$

You are asked to prove this in Exercise 8.15, together with the analog of (8.66), that is,

$$P_\nu^{-\mu}(x) = \frac{\Gamma(\nu-\mu+1)}{\Gamma(\nu+\mu+1)} \left[\cos\mu\pi \; P_\nu^\mu(x) - \frac{2}{\pi}\sin\mu\pi \; Q_\nu^\mu(x)\right]. \tag{8.76}$$

8.8. Remarks and Comments for Further Reading

8.1. A very extensive treatment of Legendre functions and associated functions is given in ROBIN (1957–1959). An overview of the functions with many formulas can be found in the BATEMAN PROJECT (1953, Vol. I, Ch. 3), ABRAMOWITZ & STEGUN (1964, Ch. 8) and MAGNUS, OBERHETTINGER & SONI (1966, Ch. 4). In the classic HOBSON (1931, pp. 183–200, 236–243, 266) integral representations as loop integrals are discussed quite extensively, and in a very lucid style.

8.2. The numerical stability of the recurrence relations of the Legendre functions is discussed in §13.3.1, Examples 2 and 3; more details are given in GAUTSCHI (1967). See also OLVER & SMITH (1983). For computing a single value the representations and transformations of the hypergeometric functions may be exploited. However, when the parameters μ, ν are large, computations based on the power series expansions may suffer from severe instabilities.

8.3. Asymptotic expansions with respect to large parameters μ, ν can be found in the references given above. In some cases these expansions are just the power series expansions of the hypergeometric functions. More recent are the more powerful uniform expansions, in which more than one of the quantities z, μ, ν may tend to infinity. See OLVER (1974), where uniform expansions are given in terms of Bessel functions, and OLVER (1975) for expansions in terms of parabolic cylinder functions. In succession to Olver's pioneering work, we mention BOYD & DUNSTER (1986), DUNSTER (1990), (1991); these results are based on the Legendre differential equation. In URSELL (1984), SHIVAKUMAR & WONG (1988), FRENZEN (1990) integral representations are the starting points for obtaining uniform expansions of the Legendre functions.

8.9. Exercises and Further Examples

8.1. Prove that

$$Q_\nu(z) - Q_{-\nu-1}(z) = \pi \cot \nu\pi \; P_\nu(z)$$

by substituting (8.8) and a similar form for $Q_{-\nu-1}(z)$. The result can also be transformed into (8.7), by using several relations for the gamma function.

8.2. Prove the integral representation (8.54) by expanding

$$\frac{1}{\sqrt{2\cosh\theta + 2\cosh\alpha}} = \frac{1}{2\cosh\frac{1}{2}\theta\,\sqrt{1+\frac{1}{2}(z-1)/\cosh^2\frac{1}{2}\theta}}, \qquad z = \cosh\alpha$$

in powers of $(z-1)$. Formulas (3.4) and (3.5), Exercise 3.12 and definition (8.5) should also be used. Prove (8.55) by using (8.9) and (8.53).

8.3. Verify that the Legendre functions satisfy the following recurrence relations:

$$(\nu+1)P_{\nu+1}(z) = (2\nu+1)zP_\nu(z) - \nu P_{\nu-1}(z), \tag{1}$$

$$P'_{\nu+1}(z) - P'_{\nu-1}(z) = (2\nu+1)P_\nu(z), \tag{2}$$

$$P'_{\nu+1}(z) - zP'_\nu(z) = (\nu+1)P_\nu(z), \tag{3}$$

$$zP'_\nu(z) - P'_{\nu-1}(z) = \nu P_\nu(z), \tag{4}$$

$$\left(1-z^2\right)P'_\nu(z) = \nu P_{\nu-1}(z) - \nu z P_\nu(z). \tag{5}$$

The functions $Q_\nu(z)$ satisfy the same relations. (*Warning*: when in (3) the Q–function is used with $\nu = -1$, we cannot replace the right-hand side by 0; we now from (8.8) that $\lim_{\nu\to-1}(\nu+1)Q_\nu(z) = 1$; from the first relation in Exercise 5.1 and (8.8) it follows that $\lim_{\nu\to-1} Q'_\nu(z)$ exists.) The above recurrence relations also hold for the Legendre polynomials; for instance, for (1) see §6.4. The recurrence relations can be verified by substituting the series expansions that follow from (8.5) and (8.8). Much more elegant proofs follow from (8.47). For example,

$$\frac{d}{dt}\frac{\left(t^2-1\right)^{\nu+1}}{(t-z)^{\nu+1}} = \frac{2(\nu+1)t\left(t^2-1\right)^\nu}{(t-z)^{\nu+1}} - \frac{(\nu+1)\left(t^2-1\right)^{\nu+1}}{(t-z)^{\nu+2}}.$$

Hence, upon integrating,

$$0 = 2\int_C \frac{t\left(t^2-1\right)^\nu}{(t-z)^{\nu+1}}\,dt - \int_C \frac{\left(t^2-1\right)^{\nu+1}}{(t-z)^{\nu+2}}\,dt.$$

By writing t from the numerator of the first integral in the form $(t-z)+z$ we obtain

$$P_{\nu+1}(z) - zP_\nu(z) = \frac{1}{2^{\nu+1}\pi i}\int_C \frac{(t^2-1)^\nu}{(t-z)^\nu}\,dt.$$

Differentiating this we obtain (3).

8.4. Verify using Abel's identity (4.14) that, for a pair u, v that solves (8.1), the Wronskian $W[u,v](z) = uv' - u'v$ satisfies

$$W[u,v](z) = C/(z^2-1),$$

where C may depend on ν and μ. Verify the following cases:

$$u(z) = P_\nu^\mu(z),\ v(z) = Q_\nu^\mu(z),\quad W[u,v](z) = e^{\mu\pi i}\frac{\Gamma(1+\nu+\mu)}{\Gamma(1+\nu-\mu)}\frac{1}{1-z^2};$$

$$u(z) = P_\nu^{-\mu}(z),\ v(z) = Q_\nu^\mu(z),\quad W[u,v](z) = e^{\mu\pi i}\frac{1}{1-z^2};$$

$$u(x) = P_\nu^\mu(x),\ v(x) = Q_\nu^\mu(x),\quad W[u,v](x) = \frac{\Gamma(1+\nu+\mu)}{\Gamma(1+\nu-\mu)}\frac{1}{1-x^2};$$

$$u(x) = P_\nu^{-\mu}(x),\ v(x) = Q_\nu^\mu(x),\quad W[u,v](x) = \frac{\cos\mu\pi}{1-x^2}.$$

In the last two cases we assume that $x \in (-1,1)$. The second case is especially of interest, since W cannot vanish for any choice of z, ν, μ.

8.5. Show that the associated equation (8.1) with $\mu = m = 0,1,2,\ldots$ admits a solution that is finite at $z = \pm 1$ if and only if $\nu = n$ (integer) with $n \geq m$. This solution is given by $P_n^m(z)$.

8.6. Show that the associated functions $\{P_n^m(x)\}$ given in (8.73) with fixed m and $n = m, m+1, m+2, \ldots$ constitute an *orthogonal system* on the interval $[-1,1]$ with *weight function* equal to unity, and that

$$\int_{-1}^{1} P_n^m(x)P_k^m(x)\,dx = \frac{2}{2n+1}\frac{(n+m)!}{(n-m)!}\delta_{n,k}.$$

Whittaker & Watson (1927, p. 324)

8.7. Show that

$$\int_{-1}^{1} P_n^m(x)Q_k^m(x)\,dx = \frac{\left[1-(-1)^{k+n}\right](n+m)!}{(n-k)(k+n+1)(n-m)!},$$

where k, m, n are non-negative integers.

8.8. Verify the representations, valid for $-1 < x < 1$,

$$P_\nu^\mu(x) = \frac{2^\mu \sqrt{\pi}}{(1-x^2)^{\frac{1}{2}\mu}} \left[A - 2xB\right].$$

$$Q_\nu^\mu(x) = \frac{2^\mu \pi \sqrt{\pi}}{(1-x^2)^{\frac{1}{2}\mu}} \left[\cot \frac{1}{2}\pi(\nu+\mu)\, xB - \frac{1}{2} \tan \frac{1}{2}\pi(\nu+\mu)\, A\right],$$

where

$$A = \frac{F\left(-\frac{1}{2}\mu - \frac{1}{2}\nu, \frac{1}{2} - \frac{1}{2}\mu + \frac{1}{2}\nu; \frac{1}{2}; x^2\right)}{\Gamma\left(\frac{1}{2} - \frac{1}{2}\nu - \frac{1}{2}\mu\right)\Gamma\left(1 + \frac{1}{2}\nu - \frac{1}{2}\mu\right)},$$

$$B = \frac{F\left(\frac{1}{2} - \frac{1}{2}\mu - \frac{1}{2}\nu, 1 + \frac{1}{2}\nu - \frac{1}{2}\mu; \frac{3}{2}; x^2\right)}{\Gamma\left(\frac{1}{2} + \frac{1}{2}\nu - \frac{1}{2}\mu\right)\Gamma\left(-\frac{1}{2}\nu - \frac{1}{2}\mu\right)}.$$

8.9. Verify Barnes' integral (1908)

$$\int_0^1 x^\sigma \left(1 - x^2\right)^{-\frac{1}{2}\mu} P_\nu^\mu(x)\, dx = \frac{2^{\mu-1}\Gamma\left(\frac{\sigma+1}{2}\right)\Gamma\left(\frac{\sigma+2}{2}\right)}{\Gamma\left(\frac{\sigma-\nu-\mu+2}{2}\right)\Gamma\left(\frac{\sigma+\nu-\mu+3}{2}\right)}.$$

8.10. Show that for non-negative integer values of n and m:

$$Q_n^m(z) = (-1)^m P_n^m(z) \frac{(n+m)!}{(n-m)!} \int_z^\infty \frac{dt}{(t^2-1)\left[P_n^m(t)\right]^2}, \quad n \geq m,$$

where the path of integration avoids the interval $[-1, 1]$.

8.11. Show that the Legendre function $P_\nu(z)$ with $z > 1$ and complex degree of the form $\nu = -\frac{1}{2} + i\tau$, with $\tau \in \mathbb{R}$, is real. Legendre functions of this type arise in *boundary value problems inside a cone.* They also arise in integral transforms known as the *Mehler-Fock transformation.* The function $P_{-\frac{1}{2}+i\tau}(z)$ is also called a *Mehler function* or *conical function.* Show that this function satisfies the integral equation

$$P_{-\frac{1}{2}+i\tau}(z) = \frac{\cosh \pi\tau}{\pi} \int_1^\infty \frac{P_{-\frac{1}{2}+i\tau}(\zeta)}{z+\zeta}\, d\zeta, \quad z \geq 1.$$

Hint: take (8.37) and substitute this in the integral; interchange the order of integration; the resulting integral looks like (8.54).

8.12. Verify formula (8.23).

8.13. Prove the relations in (8.65) and (8.66).

8.14. Prove the relations in (8.68) and (8.69). First show that the function $w(z) = (z^2-1)^{\mu/2} P_\nu^\mu(z)$ satisfies the equation

$$(z^2-1)w'' + 2(1-\mu)zw' - (\nu+\mu)(\nu-\mu+1)w = 0.$$

Next substitute $v(z) = \int_C (t^2-1)^\nu (t-z)^{\mu-\nu-1}\, dt$ in the equation and verify that this time the right-hand side equals $(\nu-\mu+1)u(z)$, where

$$\begin{aligned} u(z) &= \int_C \frac{(t^2-1)^\nu}{(t-z)^{\nu-\mu+3}} [(\nu-\mu+2)\left(t^2-1\right) - 2(\nu+1)t(t-z)]\, dt \\ &= -\int_C \frac{d}{dt} \frac{(t^2-1)^{\nu+1}}{(t-z)^{\nu-\mu+2}}\, dt. \end{aligned}$$

Verify that this expression indeed vanishes, when C is as in (8.68). As $z \to 1$, $w(z) \sim 2^\mu \Gamma(-\mu)/[\Gamma(1+\nu-\mu)\Gamma(-\nu-\mu)]$ when $\Re\mu < 0$. Use this behavior and compute $v(1)$ by using Exercise 3.13, to verify that the terms in front of the integral in (8.68) are as shown. The condition $\Re\mu < 0$ can be dropped, by using analytic continuation. [Olver (1974, p. 174.)]

8.15. Prove the relations in (8.75) and (8.76).

8.16. Show that the associated Legendre functions $P_\nu^\mu(z)\, Q_\nu^\mu(z)$ both satisfy the following recurrence relations:

$$P_\nu^{\mu+1}(z) + \frac{2\mu z}{\sqrt{z^2-1}} P_\nu^\mu(z) = (\nu-\mu+1)(\nu+\mu)P_\nu^{\mu-1}(z), \qquad (1)$$

$$\left(z^2-1\right)\frac{dP_\nu^\mu(z)}{dz} = \mu z P_\nu^\mu(z) + \sqrt{z^2-1}\, P_\nu^{\mu+1}(z), \qquad (2)$$

$$(2\nu+1)zP_\nu^\mu(z) = (\nu-\mu+1)P_{\nu+1}^\mu(z) + (\nu+\mu)P_{\nu-1}^\mu(z), \qquad (3)$$

$$\left(z^2-1\right)\frac{dP_\nu^\mu(z)}{dz} = (\nu-\mu+1)P_{\nu+1}^\mu(z) - (\nu+1)zP_\nu^\mu(z), \qquad (4)$$

$$P_{\nu-1}^\mu(z) - P_{\nu+1}^\mu(z) = -(2\nu+1)\sqrt{z^2-1}\, P_\nu^{\mu-1}(z). \qquad (5)$$

A similar warning as in Exercise 8.3 is effective when the Q–function is used; for instance, one needs to be careful when (3) is used with $\nu+\mu=0$.

8.17. Formulas (8.59) and (8.60) suggest that, when using them with negative values of m and interpreting d^m/dz^m as an $|m|$–fold integration,

$$\left.\begin{aligned} P_\nu^{-m}(z) &= \left(z^2-1\right)^{-\frac{1}{2}m} \int_1^z \cdots \int_1^z P_\nu(z)\,(dz)^m, \\ Q_\nu^{-m}(z) &= (-1)^m\left(z^2-1\right)^{-\frac{1}{2}m} \int_z^\infty \cdots \int_z^\infty Q_\nu(z)\,(dz)^m, \end{aligned}\right\} \quad m = 1,2,3,\ldots$$

where the choice of the limits of integration is based on the behavior of the functions at $z = 1$ and $z = \infty$, respectively; this behavior follows from (8.56) and (8.57). Give a direct proof of the above formulas.

Hint: differentiate the results; use induction with respect to m; use recursion (2) of the previous exercise.

8.18. Show that, when $\mu = \pm\frac{1}{2}$, the associated functions are elementary functions. In fact we have:

$$P_\nu^{\frac{1}{2}}(\cos\theta) = \sqrt{\frac{2}{\pi}}\,\frac{\cos\left(\nu+\frac{1}{2}\right)\theta}{\sqrt{\sin\theta}}, \quad P_\nu^{-\frac{1}{2}}(\cos\theta) = \sqrt{\frac{2}{\pi}}\,\frac{\sin\left(\nu+\frac{1}{2}\right)\theta}{\left(\nu+\frac{1}{2}\right)\sqrt{\sin\theta}}$$

$$Q_\nu^{\frac{1}{2}}(\cos\theta) = -\sqrt{\frac{\pi}{2}}\,\frac{\sin\left(\nu+\frac{1}{2}\right)\theta}{\sqrt{\sin\theta}}, \quad Q_\nu^{-\frac{1}{2}}(\cos\theta) = -\sqrt{\frac{\pi}{2}}\,\frac{\cos\left(\nu+\frac{1}{2}\right)\theta}{\left(\nu+\frac{1}{2}\right)\sqrt{\sin\theta}}.$$

Another elementary case for the P–functions gives $\mu = -\nu$. Show that

$$P_\nu^{-\nu}(z) = \frac{2^{-\nu}\left(z^2-1\right)^{\frac{1}{2}\nu}}{\Gamma(\nu+1)}, \quad P_\nu^{-\nu}(\cos\theta) = \frac{2^{-\nu}(\sin\theta)^\nu}{\Gamma(\nu+1)}.$$

9

Bessel Functions

9.1. Introduction

Bessel functions show up in many problems of physics and engineering, in Fourier theory and abstract harmonic analysis, in statistics and probability theory. Most frequently they occur in connection with differential equations. In BESSEL (1824) the earliest systematic study of the functions was made in a problem connected with planetary motions; see WATSON (1944) (Chapter 1: Bessel functions before 1826) and DUTKA (1995) for information on earlier occurrences of special cases of Bessel functions and the early history of Bessel functions.

In the theory of probability functions, the *modified Bessel function* $I_\mu(z)$ plays a role in the *non-central* χ^2- *distribution*, which can be defined by the integral

$$Q_\mu(x,y) = e^{-x} \int_y^\infty (z/x)^{\frac{1}{2}(\mu-1)}\, e^{-z} I_{\mu-1}\left(2\sqrt{xz}\right)\, dz.$$

The function $Q_\mu(x,y)$ plays also a role in physics, for instance in problems on radar communications, where it is called the generalized *Marcum* $Q-$*function.* In MARCUM (1960) the function is considered with $\mu = 1$. The parameter μ is related with the degrees of freedom and y with the non-centrality. The occurrence of the Bessel function in the integral for $Q_\mu(x,y)$ enables us to derive many interesting properties of Marcum's function, by using particular properties of the Bessel function. Series expansions in terms of the (incomplete gamma functions) follow from substituting the power series expansion of the Bessel function (see (9.28)), giving

$$Q_\mu(x,y) = e^{-x} \sum_{n=0}^{\infty} \frac{x^n}{n!} Q(\mu+n,y) = 1 - e^{-x} \sum_{n=0}^{\infty} \frac{x^n}{n!} P(\mu+n,y),$$

where $P(a,z)$ and $Q(a,z)$ are the incomplete gamma functions defined in §7.3.7, and which will be considered in more detail in Chapter 11.

In mathematical physics Bessel functions are associated most commonly with the partial differential equations of the potential, wave motion or diffusion, in cylindrical or spherical coordinates. By separating the variables with respect to these coordinates in the *time independent wave equation* (or *Helmholtz equation*) $\Delta v + k^2 v = 0$ several differential equations are obtained (for details we refer to Chapter 10), and one of them can be put in the form of the *Bessel differential equation*

$$\boxed{L_z[y] = z^2 y'' + z y' + \left(z^2 - \nu^2\right) y = 0.} \tag{9.1}$$

The *Bessel functions* are defined as solutions of this equation. Proper normalizations and combinations give the standard Bessel functions.

The point $z = 0$ is, according to the theory of Chapter 4, see Example 4.2, a regular singular point. The power series method yields a solution as given in (4.22). This is one of the starting points to treat and clarify the available wealth of information on Bessel functions. Here we choose the approach based on integral representations. Also in this method several starting points lead to the desired results. In our view the approach described in the following section produces pretty and flexible integral representations, which are of Schläfli type and related to the Sommerfeld integral representations of Bessel functions. As an extra benefit of this approach we mention the optimal domain of validity with respect to complex parameters for the representations. A different approach, considered extensively in WATSON (1944), will be mentioned in §9.6; see also Remark 9.3.

9.2. Integral Representations

We try to find a solution of (9.1) by using the method of §4.4:

$$y(z) = \int_\alpha^\beta K(z,t)v(t)\,dt,$$

with $K(z,t) = \exp(-z\sinh t)$. Then we have

$$L_z K(z,t) = \left[z^2\sinh^2 t - z\sinh t + (z^2 - \nu^2)\right] e^{-z\sinh t} = M_t e^{-z\sinh t}$$

with $M_t = d^2/dt^2 - \nu^2 = M_t^*$. In this case the function P of formula (4.28) is given by $P(u,v) = vu' - uv'$. A solution of $M_t^* v(t) = 0$ is $v(t) = \exp(\nu t)$. The required solution of (9.1) now reads

$$y(z) = \int_\alpha^\beta e^{-z\sinh t + \nu t}\,dt,$$

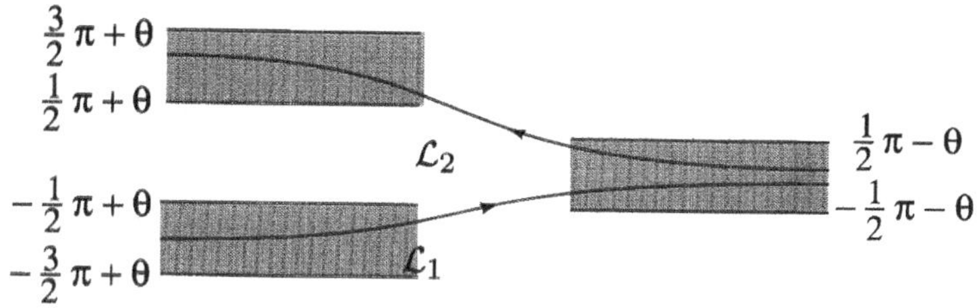

Figure 9.1. The contours of integration $\mathcal{L}_j$ for (9.3).

with α and β chosen in the complex t-plane such that

$$P\left(e^{-z\sinh t}, e^{\nu t}\right)\Big|_{\alpha}^{\beta} = 0. \tag{9.2}$$

When $z > 0$ we can choose $\beta = +\infty$; however $\alpha = -\infty$ is not possible. It is better to choose the path $\mathcal{L}$ such that for $\Re t \to -\infty$ the function $\sinh t$ approaches $+\infty$. Because of the periodicity of the hyperbolic function, this is possible in several strips running parallel to the real axis in the complex t-plane

Putting $z = x + iy = re^{i\theta}$ and $t = u + iv$ we take for $\mathcal{L}$ in particular the paths depicted in Figure 9.1; the variable u runs through all real values. Convergence of the integral and the relation in (9.2) then requires the following:

- When $u \to +\infty$ we must have $\cos(\theta + v) > 0$; we choose

$$-\frac{1}{2}\pi - \theta < v < \frac{1}{2}\pi - \theta.$$

- When $u \to -\infty$ we must have $\cos(\theta - v) < 0$; we choose

$$-\frac{3}{2}\pi + \theta < v < -\frac{1}{2}\pi + \theta$$

or

$$\frac{1}{2}\pi + \theta < v < \frac{3}{2}\pi + \theta.$$

This leads to the following definitions of the so called *Hankel functions*

$$\boxed{H_\nu^{(j)}(z) = \frac{1}{\pi i}\int_{\mathcal{L}_j} e^{-z\sinh t + \nu t}\, dt, \quad j = 1, 2;} \tag{9.3}$$

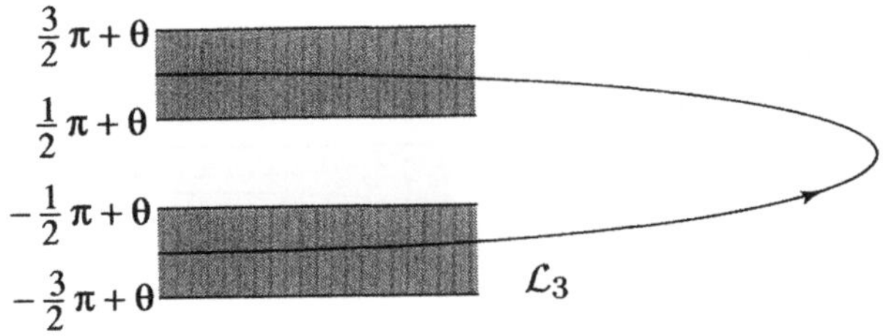

Figure 9.2. The path of integration $\mathcal{L}_3$ for (9.6).

ν is called the *order*. The integrals in (9.3) are called *Sommerfeld integrals.* The Hankel functions are entire functions with respect to the order (for any complex $z \neq 0$); considered as function of z they are analytic in $\mathbb{C}\setminus\{0\}$. From (9.3) it follows that

$$H^{(1)}_{-\nu}(z) = e^{i\nu\pi} H^{(1)}_\nu(z), \quad H^{(2)}_{-\nu}(z) = e^{-i\nu\pi} H^{(2)}_\nu(z). \tag{9.4}$$

The first relation is proved by substituting $t = -w - i\pi$ in (9.3) (with ν replaced by $-\nu$). Similarly for the second relation.

It will certainly support your insight in the theory of Bessel functions by attributing to the Hankel functions $H^{(1)}_\nu(z)$, $H^{(2)}_\nu(z)$ the roles the exponential functions e^{iz} and e^{-iz} play in the theory of circular functions. As will become clear in §9.6, the asymptotic expansions of the Hankel functions, as $z \to \infty$, contain the exponential functions as crucial components. When ν is real the Hankel functions are complex conjugates of each other. That is,

$$H^{(1)}_\nu(\overline{z}) = \overline{H^{(2)}_\nu(z)}, \quad H^{(2)}_\nu(\overline{z}) = \overline{H^{(1)}_\nu(z)}.$$

When ν is real and $z > 0$ we can split these functions into real and imaginary parts. This gives the analogs of the sine and cosine functions, namely, the *ordinary Bessel functions*

$$\boxed{H^{(1)}_\nu(z) = J_\nu(z) + iY_\nu(z), \quad H^{(2)}_\nu(z) = J_\nu(z) - iY_\nu(z).} \tag{9.5}$$

This can be written as

$$\boxed{\begin{aligned} J_\nu(z) &= \frac{1}{2}\left[H^{(1)}_\nu(z) + H^{(2)}_\nu(z)\right] = \frac{1}{2\pi i}\int_{\mathcal{L}_3} e^{-z\sinh t + \nu t}\, dt, \\ Y_\nu(z) &= \frac{1}{2i}\left[H^{(1)}_\nu(z) - H^{(2)}_\nu(z)\right] = \frac{1}{2\pi i}\int_{\mathcal{L}_4} e^{-z\sinh t + \nu t}\, dt, \end{aligned}} \tag{9.6}$$

which are now taken as definitions of $J_\nu(z)$ and $Y_\nu(z)$ for general complex values of z and ν ($z \neq 0$)†. The contour of integration $\mathcal{L}_3$ is depicted in Figure 9.2, whereas $\mathcal{L}_4$ is just the union of $\mathcal{L}_1$ and $\mathcal{L}_2$, with a different direction of integration on $\mathcal{L}_2$.

In Figures 9.3–9.6 we give the graphs of the Bessel functions $J_\nu(x), Y_\nu(x)$ for non-negative values of x and ν.

9.3. Cylinder Functions

Solutions of equation (9.1) are also called *cylinder functions*, because they occur as solutions in boundary value problems with cylindrical symmetry; see Chapter 10. A general notation for cylinder functions is $C_\nu(z)$, denoting an arbitrary solution of (9.1). Bessel functions satisfying (9.1) are called *ordinary Bessel functions*, as distinct from the *modified Bessel functions* introduced in §9.5. The functions $J_{\pm\nu}(z)$ are called the *Bessel functions of the first kind*, $Y_\nu(z)$ (also called the *Neumann function*) is the function of the second kind, and $H_\nu^{(1)}(z)$, $H_\nu^{(2)}(z)$ are the functions of the third kind.

We return to the Bessel differential equation (9.1). The equation remains the same when we replace ν by $-\nu$. Hence, the functions

$$H_{-\nu}^{(1)}(z),\ H_{-\nu}^{(2)}(z),\ J_{-\nu}(z),\ Y_{-\nu}(z)$$

are also solutions of (9.1). The relations between the Hankel functions

$$H_{-\nu}^{(1)}(z),\ H_{-\nu}^{(2)}(z) \quad \text{and} \quad H_\nu^{(1)}(z),\ H_\nu^{(2)}(z)$$

are given in (9.4). For the other Bessel functions the relations follow from (9.4) and (9.5):

$$\begin{aligned} 2i\,\sin\nu\pi\, H_\nu^{(1)}(z) &= -e^{-i\nu\pi} J_\nu(z) + J_{-\nu}(z), \\ 2i\,\sin\nu\pi\, H_\nu^{(2)}(z) &= \ \ e^{i\nu\pi} J_\nu(z) - J_{-\nu}(z), \end{aligned} \qquad (9.7)$$

When $\nu = n = 0, 1, 2, \ldots$ we conclude that

$$\boxed{J_{-n}(z) = (-1)^n J_n(z).} \qquad (9.8)$$

† In the physics literature the function $Y_\nu(z)$ is sometimes denoted by $N_\nu(z)$.

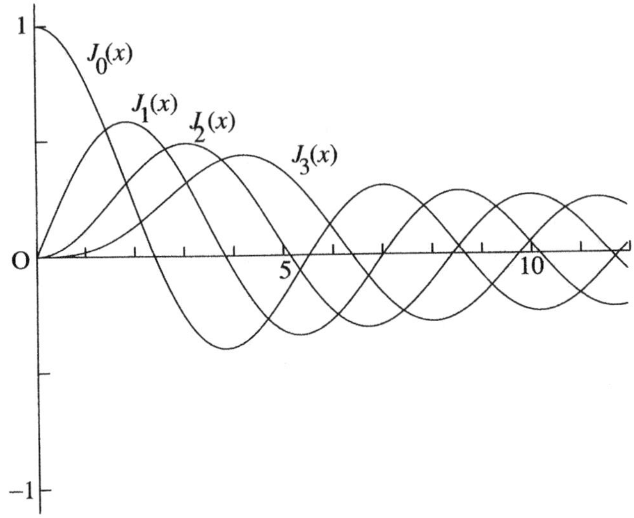

Figure 9.3. Bessel functions $J_\nu(x)$, $\nu = 0, 1, 2, 3$.

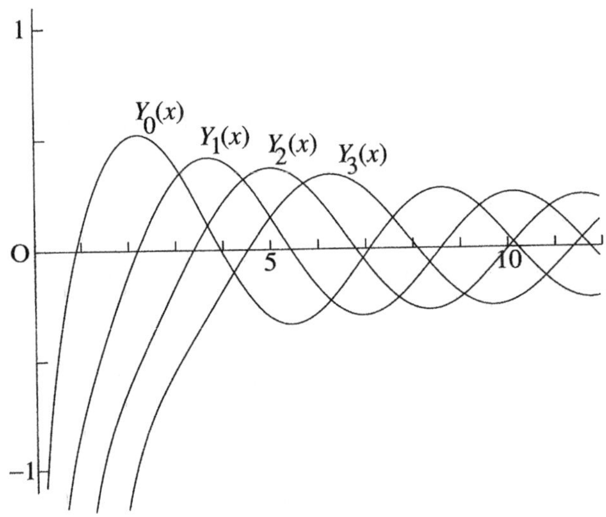

Figure 9.4. Bessel functions $Y_\nu(x)$, $\nu = 0, 1, 2, 3$.

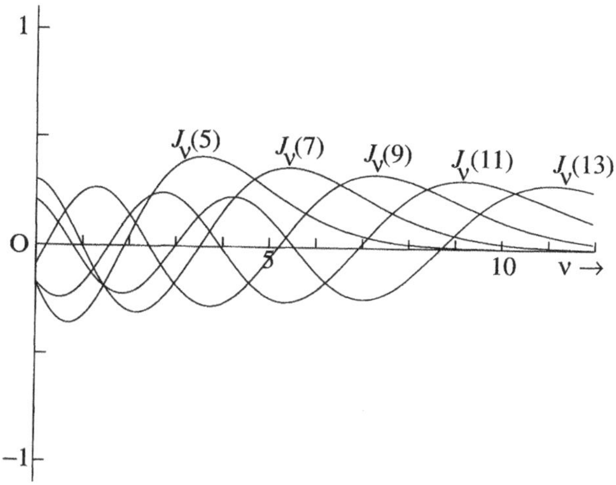

Figure 9.5. Bessel functions $J_\nu(x)$, $x = 5, 7, 9, 11, 13$, as functions of their order ν.

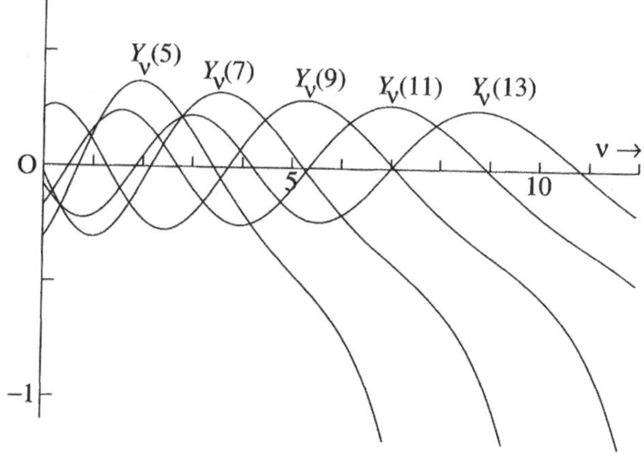

Figure 9.6. Bessel functions $Y_\nu(x)$, $x = 5, 7, 9, 11, 13$, as functions of their order ν.

The pair $\{J_\nu(z), Y_\nu(z)\}$ act, under all circumstances, as an independent pair of solutions of the Bessel equation (9.1). The same for the Hankel functions $\{H_\nu^{(1)}(z), H_\nu^{(2)}(z)\}$. In the terminology of Chapter 4, the pairs $\{J_\nu(z), Y_\nu(z)\}$ and $\{H_\nu^{(1)}(z), H_\nu^{(2)}(z)\}$ constitute fundamental pairs of solutions of (9.1). The solutions $J_\nu(z)$ and $J_{-\nu}(z)$ are only independent if ν is different from an integer. When $\nu \in \mathbb{Z}$, see (9.8). In Exercise 9.1 an overview of Wronskians is given, from which the independence of pairs of solutions follows.

From (9.5) and (9.7) follows the fundamental relation for the Neumann function

$$\boxed{Y_\nu(z) = \frac{\cos\nu\pi\ J_\nu(z) - J_{-\nu}(z)}{\sin\nu\pi},} \tag{9.9}$$

which often is used as the definition of the function of the second kind.

Remark 9.1. So far we have considered four cylinder functions (and also functions with order $-\nu$). As mentioned above

$$\{J_\nu(z),\ Y_\nu(z)\} \quad \text{and} \quad \{H_\nu^{(1)}(z),\ H_\nu^{(2)}(z)\}$$

constitute linearly independent pairs of solutions of the Bessel differential equation. In fact, one pair is sufficient to describe all interesting properties of the Bessel functions. However, in physical problems it is very important to have available all four functions. First because $J_\nu(z)$ and $Y_\nu(z)$ are real for positive z and real ν. Second, the Hankel function $H_\nu^{(1)}(z)$ is exponentially small as $\Im z \to +\infty$, whereas $H_\nu^{(2)}(z)$ is exponentially small as $\Im z \to -\infty$ (see §9.7). Therefore, the Hankel functions are important to describe solutions of physical problems with complex parameters.

Remark 9.2. In numerical and physical applications it is also important to have a *numerically satisfactory pair* of solutions. For instance, the pair $\{e^z, e^{-z}\}$ is a linearly independent pair of solutions of the equation $d^2w/dz^2 = w$, as is the pair $\{\sinh z, \cosh z\}$. However, with the latter pair severe cancellation takes place when we compute $e^{-z} = \cosh z - \sinh z$ for large positive values of $\Re z$. We infer that the pair $\{\sinh z, \cosh z\}$ is not a numerically satisfactory pair of the equation $d^2w/dz^2 = w$ for large positive values of $\Re z$, but the pair is numerically satisfactory for small values of $|z|$. A similar situation occurs in the case of the Bessel functions. The Hankel functions constitute a numerically satisfactory pair of the Bessel equation for large values of $|z|$ in $|\arg z| \le \pi$. They are numerically unsatisfactory for small $|z|$, in which case a better pair is $\{J_\nu(z), H_\nu^{(1)}(z)\}$ for $0 \le \arg z \le \pi$, and $\{J_\nu(z), H_\nu^{(2)}(z)\}$

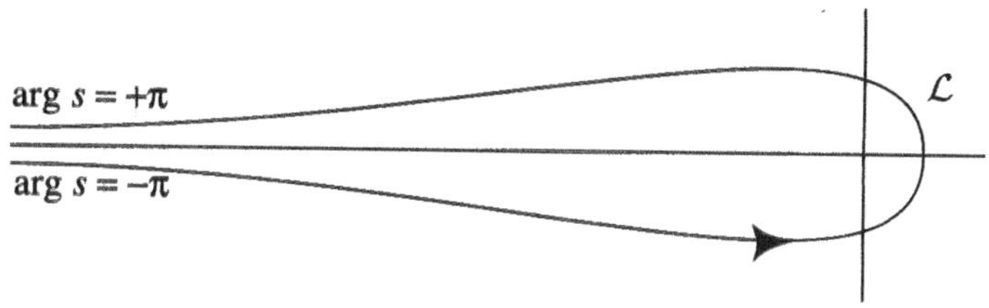

Figure 9.7. Contour $\mathcal{L}$ for (9.10).

for $-\pi \le \arg z \le 0$. For more discussions on this point we refer to MILLER (1950) and OLVER (1974).

9.4. Further Properties

We give an interpretation of the first formula in (9.6) in terms of the inversion formula for the Laplace transformation. Take s as new variable of integration by putting $e^{-t} = 2s/z$. First take $z > 0$. Then the transform of $\mathcal{L}_3$ can be taken as the vertical line

$$\mathcal{L} = \{s \mid \Re s = \text{constant} > 0, \ -\infty < \Im s < \infty\}.$$

Hence

$$J_\nu(z) = \left(\tfrac{1}{2}z\right)^\nu \frac{1}{2\pi i}\int_{\mathcal{L}} e^{s-z^2/(4s)} s^{-\nu-1}\, ds. \tag{9.10}$$

When $\mathcal{L}$ is a vertical line we have to assume that $\Re\nu > -1$. Analytic continuation with respect to ν is possible by deforming the vertical line at $\pm i\infty$ into the left half plane; see Figure 9.7. Also, the restriction with respect to $z > 0$ can be relaxed: (9.10) holds for $z \in \mathbb{C} \setminus \{0\}$. In (9.10) take $z = 2\sqrt{t}$ and change $s \to ts$. Applying Laplace inversion on the resulting integral, we obtain

$$\int_0^\infty t^{\frac{1}{2}\nu} J_\nu\left(2\sqrt{t}\right) e^{-st}\, dt = s^{-\nu-1} e^{-1/s}, \quad \Re\nu > -1, \quad \Re t > 0. \tag{9.11}$$

Other Laplace transforms of $J_\nu(z)$ are given in Exercises 9.8 and 9.9.

Expanding the factor $\exp(-\frac{1}{4}z^2/s)$ in the integrand of (9.10) into a power series, we obtain after interchanging the order of integration and summation

$$J_\nu(z) = \left(\tfrac{1}{2}z\right)^\nu \sum_{n=0}^{\infty} \frac{(-1)^n}{\Gamma(n+\nu+1)\, n!}\left(\tfrac{1}{2}z\right)^{2n}, \tag{9.12}$$

the well-known power series for the Bessel function of the first kind. In the proof of (9.12) we have used Hankel's integral (3.6); in (9.10) we take the contour as in Figure 9.7. Also Theorem 2.1 from Chapter 2 may be used. Recall that (9.12) directly follows from the power series method applied to the differential equation (9.1); see (4.22).

By using (9.7) and (9.9) we can obtain compound power series expansions for the the Hankel functions and the Neumann function. The dominant terms for $\Re\nu > 0$ are

$$\begin{aligned} J_\nu(z) &\sim \frac{(\frac{1}{2}z)^\nu}{\Gamma(\nu+1)}, & Y_\nu(z) &\sim -\frac{1}{\pi}\Gamma(\nu)\left(\frac{2}{z}\right)^\nu, \\ H_\nu^{(1)}(z) &\sim \frac{1}{\pi i}\Gamma(\nu)\left(\frac{2}{z}\right)^\nu, & H_\nu^{(2)}(z) &\sim -\frac{1}{\pi i}\Gamma(\nu)\left(\frac{2}{z}\right)^\nu. \end{aligned} \tag{9.13}$$

These relations hold as $z \to 0$. They also are *asymptotic relations for the Bessel functions as* $\Re\nu \to +\infty$, with z fixed. This easily follows from the series (9.12). It has an asymptotic character as $\nu \to \infty$.

When $\nu = 0$ the expansion in (9.12) reads

$$\boxed{J_0(z) = 1 - \left(\frac{1}{2}z\right)^2 + \frac{1}{2!\,2!}\left(\frac{1}{2}z\right)^4 - \frac{1}{3!\,3!}\left(\frac{1}{2}z\right)^6 + \ldots .} \tag{9.14}$$

When ν assumes integer values, the expansion of $Y_\nu(z)$ has logarithmic terms. This is in accordance with the theory of Chapter 4. Applying l'Hôpital's rule to (9.9) we obtain in the first instance

$$Y_n(z) = \frac{1}{\pi}\frac{\partial J_\nu(z)}{\partial \nu}\bigg|_{\nu=n} + \frac{(-1)^n}{\pi}\frac{\partial J_\nu(z)}{\partial \nu}\bigg|_{\nu=-n}, \quad n \text{ integer.}$$

Hence

$$Y_{-n}(z) = (-1)^n Y_n(z).$$

Next we assume that $n \geq 0$. Straightforward analysis eventually gives (when $n = 0, 1, 2, \ldots$)

$$\boxed{\begin{aligned} Y_n(z) = &\frac{2}{\pi}J_n(z)\ln\left(\frac{1}{2}z\right) - \frac{\left(\frac{1}{2}z\right)^{-n}}{\pi}\sum_{k=0}^{n-1}\frac{(n-k-1)!}{k!}\left(\frac{1}{2}z\right)^{2k} - \\ &\frac{\left(\frac{1}{2}z\right)^n}{\pi}\sum_{k=0}^{\infty}[\psi(k+1)+\psi(n+k+1)]\frac{(-1)^k\left(\frac{1}{2}z\right)^{2k}}{k!\,(n+k)!}, \end{aligned}} \tag{9.15}$$

where ψ is the logarithmic derivative of the gamma function, which is defined in §3.4.

Other interesting special cases occur when ν equals $n+\frac{1}{2}$. In Exercise 9.3 you can verify that in this case the Bessel functions reduce to elementary functions.

For the Bessel functions with consecutive indices $\nu-1, \nu, \nu+1$ a number of relations exist, which easily follow from (9.3). We have the recurrence relations for the Bessel functions:

$$\begin{aligned}
C_{\nu-1}(z)+C_{\nu+1}(z) &= \frac{2\nu}{z}C_\nu(z),\\
C_{\nu-1}(z)-C_{\nu+1}(z) &= 2C'_\nu(z),\\
C'_\nu(z) &= C_{\nu-1}(z)-\frac{\nu}{z}C_\nu(z),\\
C'_\nu(z) &= -C_{\nu+1}(z)+\frac{\nu}{z}C_\nu(z).
\end{aligned} \tag{9.16}$$

where $C_\nu(z)$ denotes one of the functions $H_\nu^{(1)}(z)$, $H_\nu^{(2)}(z)$, $J_\nu(z)$, $Y_\nu(z)$, the *cylinder functions*; see §9.2.

When we take $C_\nu(z) = H_\nu^{(j)}(z)$, the proof of the first relation runs as follows:

$$\begin{aligned}
H_{\nu-1}^{(j)}(z)+H_{\nu+1}^{(j)}(z) &= \frac{2}{\pi i}\int_{\mathcal{L}_j} e^{-z\sinh t+\nu t}\cosh t\,dt\\
&= -\frac{2}{z\pi i}\int_{\mathcal{L}_j} e^{\nu t}\,d\,e^{-z\sinh t} = \frac{2\nu}{z\pi i}\int_{\mathcal{L}_j} e^{-z\sinh t+\nu t}\,dt.
\end{aligned}$$

The other relations follow in a similar way; for $J_\nu(z)$, $Y_\nu(z)$ the first parts of (9.6) can be used. Next we derive from (9.16)

$$\begin{aligned}
\frac{1}{z}\frac{d}{dz}\left[z^\nu C_\nu(z)\right] &= z^{\nu-1}C_{\nu-1}(z),\\
\frac{1}{z}\frac{d}{dz}\left[z^{-\nu} C_\nu(z)\right] &= -z^{-\nu-1}C_{\nu+1}(z).
\end{aligned}$$

Applying the operator of the left-hand side repeatedly we obtain the nice results

$$\boxed{\begin{aligned}
&\left[\frac{1}{z}\frac{d}{dz}\right]^k \left[z^\nu C_\nu(z)\right] = z^{\nu-k}C_{\nu-k}(z),\\
&\qquad\qquad\qquad\qquad\qquad\qquad k=0,1,2,\ldots\,.\\
&\left[\frac{1}{z}\frac{d}{dz}\right]^k \left[z^{-\nu} C_\nu(z)\right] = (-1)^k z^{-\nu-k}C_{\nu+k}(z).
\end{aligned}} \tag{9.17}$$

We mention the following special cases, which occur frequently in practical problems,

$$\boxed{J_1(z) = -J'_0(z), \quad Y_1(z) = -Y'_0(z).} \tag{9.18}$$

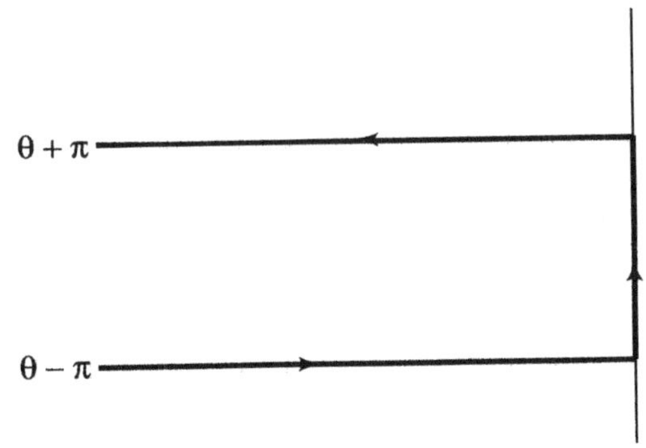

Figure 9.8. Special choice of the contour $\mathcal{L}_3$ of (9.6).

When $\nu \in \mathbb{Z}$ we can derive from (9.6) the well-known *Bessel integral representation*:

$$\boxed{\begin{aligned} J_n(z) &= \frac{1}{2\pi}\int_{-\pi}^{\pi} e^{-iz\sin v}\, e^{inv}\, dv, \\ &= \frac{1}{\pi}\int_0^{\pi} \cos(z\sin v - nv)\, dv \end{aligned} \qquad z \in \mathbb{C}, \quad n \in \mathbb{Z}.} \tag{9.19}$$

To prove this we write in (9.6) $t = u + iv$, $z = re^{i\theta}$ and deform the path $\mathcal{L}_3$ into a path constituted by the following straight lines:

$$\begin{aligned} &\text{(a)} \quad -\infty < u < 0, \quad v = i(\theta - \pi), \\ &\text{(b)} \quad u = 0, \quad i(\theta - \pi) < v < i(\theta + \pi), \\ &\text{(c)} \quad -\infty < u < 0, \quad v = i(\theta + \pi). \end{aligned}$$

Since $e^{i\nu(\theta-\pi)} = e^{i\nu(\theta+\pi)}$ when $\nu = n$ the contributions from (a) and (c) cancel each other. All that remains is the right-hand side of (9.19), with limits of integration $\theta - \pi$ and $\theta + \pi$. Since the integrand is a periodic function of period 2π, and the integral runs over a full period, we can change the interval into $[-\pi, \pi]$. In Figure 9.8 we show the special contour that is used in the proof of (9.19).

When $|\arg z| < \frac{1}{2}\pi$ we can take in (a) and (c) $v = -\pi$ and $v = \pi$, respectively. Then the above method leads to *Schläfli's integrals*

$$\begin{aligned} J_\nu(z) &= \frac{1}{\pi}\int_0^{\pi} \cos(\nu\theta - z\sin\theta)\, d\theta - \frac{\sin\nu\pi}{\pi}\int_0^{\infty} e^{-\nu t - z\sinh t}\, dt, \\ Y_\nu(z) &= \frac{1}{\pi}\int_0^{\pi} \sin(z\sin\theta - \nu\theta)\, d\theta - \int_0^{\infty} \left(e^{\nu t} + e^{-\nu t}\cos\nu\pi\right) e^{-z\sinh t}\, dt, \end{aligned}$$

which are valid for all complex values of ν. Observe that, when $\nu = n$ (integer), the first one is equivalent to the integrals in (9.19). Apparently, interpreting the first integral in (9.19) as a *Fourier coefficient*, we have

$$\boxed{e^{-iz\sin t} = \sum_{n=-\infty}^{\infty} e^{-int} J_n(z).} \tag{9.20}$$

This Fourier series can be considered as a generating function for the Bessel functions of integer order, that is, for the *Bessel coefficients*. From the *Parseval relation* for Fourier series it follows that

$$\boxed{\sum_{n=-\infty}^{\infty} J_n^2(z) = 1, \quad z \in \mathbb{C}.} \tag{9.21}$$

Several variants of the expansion (9.20) exist, for instance,

$$\begin{aligned}
\cos(z\sin\theta) &= J_0(z) + 2\sum_{n=1}^{\infty} J_{2n}(z)\cos 2n\theta,\\
\sin(z\sin\theta) &= \phantom{J_0(z) +{}} 2\sum_{n=1}^{\infty} J_{2n+1}(z)\sin(2n+1)\theta,\\
\cos(z\cos\theta) &= J_0(z) + 2\sum_{n=1}^{\infty} (-1)^n J_{2n}(z)\cos 2n\theta,\\
\sin(z\cos\theta) &= \phantom{J_0(z) +{}} 2\sum_{n=1}^{\infty} (-1)^n J_{2n+1}(z)\cos(2n+1)\theta.
\end{aligned}$$

We obtain from the second series, by differentiating with respect to θ and taking $\theta = 0$:

$$z = 2\sum_{n=0}^{\infty} (2n+1) J_{2n+1}(z).$$

In a similar way expansions can be derived for higher powers of z. In Exercise 9.4 a few other aspects of generating functions will be mentioned, in particular the expansion of the exponential function and the sine and cosine in terms of Chebyshev polynomials.

The Bessel functions introduced thus far have, for general values of ν, an algebraic singularity at $z = 0$; when $\nu \in \mathbb{Z}$ the J-function is regular at the origin. When $\nu = n$ the other functions have a pole and a logarithmic singularity at $z = 0$. For the Neumann function $Y_\nu(z)$ this is described in (9.15). Usually we consider the many-valued functions in the sector $|\arg z| <$

π, but we can consider the analytic continuation of the Bessel functions outside this sector. The differential equation (9.1) does not change when z is replace by $-z$. Hence, the functions $J_\nu(-z)$, and so on, can be written as linear combinations of the other functions. For the $J-$function this easily follows from (9.12). Since the series is an even entire function of z, we find

$$\boxed{J_\nu(ze^{m\pi i}) = e^{m\nu\pi i} J_\nu(z), \quad m \in \mathbb{Z}.} \tag{9.22}$$

For the $Y-$function we find using (9.9)

$$\boxed{Y_\nu(ze^{m\pi i}) = e^{-m\nu\pi i} Y_\nu(z) + 2i \sin m\nu\pi \ \cot \nu\pi \ J_\nu(z)} \tag{9.23}$$

and for the Hankel functions the relations follow from (9.5). We mention the simple special cases:

$$\begin{aligned}
H_\nu^{(1)}\left(ze^{+\pi i}\right) &= -e^{-\nu\pi i} H_\nu^{(2)}(z), \\
H_\nu^{(1)}\left(ze^{-\pi i}\right) &= 2\cos \nu\pi H_\nu^{(1)}(z) + e^{-\nu\pi i} H_\nu^{(2)}(z), \\
H_\nu^{(2)}\left(ze^{-\pi i}\right) &= -e^{\nu\pi i} H_\nu^{(1)}(z), \\
H_\nu^{(2)}\left(ze^{+\pi i}\right) &= 2\cos \nu\pi H_\nu^{(2)}(z) + e^{+\nu\pi i} H_\nu^{(1)}(z).
\end{aligned} \tag{9.24}$$

Bessel functions play an important role in the theory of integral transforms. We mention the *Hankel transform* pair

$$g(y) = \int_0^\infty \sqrt{xy}\, J_\nu(xy) f(x)\, dx, \quad f(x) = \int_0^\infty \sqrt{xy}\, J_\nu(xy) g(y)\, dy.$$

Other Bessel functions also occur as kernels in integral transforms. A good introduction to this topic is SNEDDON (1972). Tables can be found in the BATEMANN PROJECT (1953), in Oberhettinger's tables and in PRUDNIKOV (1986).

9.5. Modified Bessel Functions

The Bessel functions with argument $\pm iz$ are called *modified Bessel functions.* When z is replaced by iz, equation (9.1) becomes

$$\boxed{z^2 y'' + zy' - \left(z^2 + \nu^2\right) y = 0.} \tag{9.25}$$

The modified Bessel functions are the solutions of this equation. When z is positive and ν is real this equation has real solutions. The pair $\{I_\nu(z), K_\nu(z)\}$

constitute an attractive and conveniently chosen pair, their Wronskian being equal to $-1/z$ (see Exercise 9.1). They are defined by

$$\begin{aligned}
I_\nu(z) &= e^{-\frac{1}{2}\nu\pi i} J_\nu\left(z e^{\frac{1}{2}\pi i}\right), && -\pi < \arg z \le \frac{1}{2}\pi, \\
I_\nu(z) &= e^{\frac{3}{2}\nu\pi i} J_\nu\left(z e^{-\frac{3}{2}\pi i}\right), && \frac{1}{2}\pi < \arg z \le \pi, \\
K_\nu(z) &= \frac{1}{2}\pi i e^{\frac{1}{2}\nu\pi i} H_\nu^{(1)}\left(z e^{\frac{1}{2}\pi i}\right), && -\pi < \arg z \le \frac{1}{2}\pi, \\
K_\nu(z) &= -\frac{1}{2}\pi i e^{-\frac{1}{2}\nu\pi i} H_\nu^{(2)}\left(z e^{-\frac{1}{2}\pi i}\right), && -\frac{1}{2}\pi < \arg z \le \pi.
\end{aligned} \tag{9.26}$$

When $\nu = 0$ this leads to $J_0(\pm iz) = I_0(z)$. For the Y-function we have

$$Y_\nu\left(z e^{\frac{1}{2}\pi i}\right) = e^{\frac{1}{2}(\nu+1)\pi i} I_\nu(z) - \frac{2}{\pi} e^{-\frac{1}{2}\nu\pi i} K_\nu(z), \qquad -\pi < \arg z \le \frac{1}{2}\pi. \tag{9.27}$$

It is easily verified by means of (9.12) that

$$\boxed{I_\nu(z) = \left(\frac{1}{2}z\right)^\nu \sum_{n=0}^\infty \frac{\left(\frac{1}{2}z\right)^{2n}}{\Gamma(n+\nu+1)\, n!}.} \tag{9.28}$$

Furthermore, from the first formula in (9.7) and (9.26) it follows that

$$\boxed{K_\nu(z) = \frac{\pi}{2} \frac{I_{-\nu}(z) - I_\nu(z)}{\sin \nu\pi}, \qquad -\pi < \arg z < \pi} \tag{9.29}$$

where the right-hand side should be determined by a limiting process when ν assumes integer values. In that case we have, when $n = 0, 1, 2, \ldots$, the analog of (9.15):

$$\begin{aligned}
K_n(z) = (-1)^{n+1} I_n(z) \ln\left(\frac{z}{2}\right) + \frac{1}{2}\left(\frac{z}{2}\right)^{-n} \sum_{k=0}^{n-1} \frac{(n-k-1)!}{k!}(-1)^k \left(\frac{z}{2}\right)^{2k} + \\
(-1)^n \frac{1}{2}\left(\frac{z}{2}\right)^n \sum_{k=0}^\infty [\psi(k+1) + \psi(n+k+1)] \frac{\left(\frac{z}{2}\right)^{2k}}{k!\,(n+k)!}.
\end{aligned} \tag{9.30}$$

From (9.8) and (9.29) we derive the following properties:

$$\boxed{I_{-n}(z) = I_n(z), \quad K_{-\nu}(z) = K_\nu(z).} \tag{9.31}$$

The K-function is an even function with respect to ν. The analytic continuation with respect to z is described by

$$\begin{aligned}
I_\nu\left(z e^{m\pi i}\right) &= e^{m\nu\pi i} I_\nu(z), \\
K_\nu\left(z e^{m\pi i}\right) &= e^{-m\nu\pi i} K_\nu(z) - \pi i \frac{\sin m\nu\pi}{\sin \nu\pi} I_\nu(z).
\end{aligned} \qquad m \in \mathbb{Z} \tag{9.32}$$

The analog of (9.20) for the I–functions reads:

$$e^{z\cos t} = \sum_{n=-\infty}^{\infty} \cos nt \; I_n(z) = I_0(z) + 2\sum_{n=1}^{\infty} \cos nt \; I_n(z); \tag{9.33}$$

see also Exercise 9.4. The modified Bessel functions $I_\nu(z)$, $K_\nu(z)$ satisfy recurrence relations, which follow from (9.16) and (9.26). They read as follows:

$$\begin{aligned} I_{\nu-1}(z) - I_{\nu+1}(z) &= \frac{2\nu}{z} I_\nu(z), & K_{\nu+1}(z) - K_{\nu-1}(z) &= \frac{2\nu}{z} K_\nu(z) \\ I_{\nu-1}(z) + I_{\nu+1}(z) &= 2I'_\nu(z), & K_{\nu-1}(z) + K_{\nu+1}(z) &= -2K'_\nu(z), \\ I'_\nu(z) &= I_{\nu-1}(z) - \frac{\nu}{z} I_\nu(z), & K'_\nu(z) &= -K_{\nu-1}(z) - \frac{\nu}{z} K_\nu(z), \\ I'_\nu(z) &= I_{\nu+1}(z) + \frac{\nu}{z} I_\nu(z), & K'_\nu(z) &= -K_{\nu+1}(z) + \frac{\nu}{z} K_\nu(z). \end{aligned} \tag{9.34}$$

Unfortunately, $I_\nu(z)$, $K_\nu(z)$ do not satisfy the same relations; observe that the relation for $I_\nu(z)$ is also satisfied by $e^{\nu\pi i} K_\nu(z)$.

Next we have

$$\begin{aligned} \left[\frac{1}{z}\frac{d}{dz}\right]^k \left[z^\nu \mathcal{Z}_\nu(z)\right] &= z^{\nu-k} \mathcal{Z}_{\nu-k}(z), \\ \left[\frac{1}{z}\frac{d}{dz}\right]^k \left[z^{-\nu} \mathcal{Z}_\nu(z)\right] &= z^{-\nu-k} \mathcal{Z}_{\nu+k}(z). \end{aligned} \qquad k = 0, 1, 2, \ldots, \tag{9.35}$$

where $\mathcal{Z}_\nu(z)$ means the function $I_\nu(z)$ or $e^{\nu\pi i} K_\nu(z)$. Special cases of (9.35) are

$$I'_0(z) = I_1(z), \quad K'_0(z) = -K_1(z). \tag{9.36}$$

The following asymptotic relations as $z \to 0$ are important:

$$\begin{aligned} I_\nu(z) &\sim \left(\tfrac{1}{2}z\right)^\nu / \Gamma(\nu+1), \quad \nu \neq -1, -2, \ldots, \\ K_\nu(z) &\sim \tfrac{1}{2}\Gamma(\nu) \left(\tfrac{1}{2}z\right)^{-\nu}, \quad \Re\nu > 0. \end{aligned} \tag{9.37}$$

These estimates also hold when z is fixed and $\Re\nu \to +\infty$.

9.6. Integral Representations for I– and K–Functions

Apart from the integral representations (9.3) and (9.6) introduced earlier, many more interesting integrals for the Bessel functions exist. It is not possible to present more than a brief overview. The integrals introduced thusfar have the slight drawback of being expressed in terms of complex contour integrals.

The benefit is that they are valid for a large range of the complex parameters z and ν. In this section we restrict ourselves to the representations that are useful in deriving asymptotic expansions, or that can be used in deriving other new relations.

Combining the third line of (9.26) with (9.3) ($j = 1$) and replacing t with $t - \frac{1}{2}\pi i$ (and integrating with respect to real values of the parameter t) we obtain the integral representation

$$K_\nu(z) = \frac{1}{2}\int_{-\infty}^{\infty} e^{-z\cosh t + \nu t}\, dt. \tag{9.38}$$

This is often written in the form

$$\boxed{K_\nu(z) = \int_0^{\infty} e^{-z\cosh t} \cosh(\nu t)\, dt.} \tag{9.39}$$

It holds for $|\arg z| < \frac{1}{2}\pi$ and $\nu \in \mathbb{C}$. For the I–function a contour integral reads:

$$\boxed{I_\nu(z) = \frac{1}{2\pi i}\int_{\infty - i\pi}^{\infty + i\pi} e^{z\cosh t - \nu t}\, dt,} \tag{9.40}$$

which is valid for $|\arg z| < \frac{1}{2}\pi$ and $\nu \in \mathbb{C}$. For other phases of z the contour may be shifted upwards or downwards. See the analog for $J_\nu(z)$ in (9.6), from which (9.40) can be derived.

In the same manner we can verify that the integrals in (9.3) for the Hankel functions can be written as

$$\begin{aligned} H_\nu^{(1)}(z) &= \frac{e^{-\nu\pi i/2}}{\pi i}\int_{-\infty}^{\infty} e^{iz\cosh t - \nu t}\, dt, \quad \Im z > 0, \\ H_\nu^{(2)}(z) &= -\frac{e^{\nu\pi i/2}}{\pi i}\int_{-\infty}^{\infty} e^{-iz\cosh t - \nu t}\, dt, \quad \Im z < 0. \end{aligned}$$

We take in (9.38): $e^{-t} = u$. Then

$$K_\nu(z) = \frac{1}{2}\int_0^{\infty} e^{-z(u+1/u)/2} u^{-\nu-1}\, du. \tag{9.41}$$

By using simple transformations this can also be written as:

$$\boxed{K_\nu\left(2\sqrt{z\zeta}\right) = \frac{1}{2}\left(\frac{\zeta}{z}\right)^{\frac{1}{2}\nu}\int_0^{\infty} e^{-zt - \zeta/t} t^{-\nu-1}\, dt.} \tag{9.42}$$

Substitution of the relation

$$u^{-\nu-\frac{1}{2}} = \frac{1}{\Gamma\left(\nu+\frac{1}{2}\right)} \int_0^\infty e^{-ux} x^{\nu-\frac{1}{2}}\, dx, \quad \Re\nu > -\frac{1}{2},$$

in (9.41) gives

$$K_\nu(z) = \frac{1}{2\Gamma\left(\nu+\frac{1}{2}\right)} \int_0^\infty x^{\nu-\frac{1}{2}} \left[\int_0^\infty e^{-u(x+z/2)-z/(2u)} \frac{du}{\sqrt{u}}\right] dx.$$

The order of integration may be interchanged by absolute convergence of the repeated integrals. The inner integral is a special case of (9.41) with $\nu = -\frac{1}{2}$. Hence it equals (see Exercise 9.3)

$$2\left(\frac{z/2}{x+z/2}\right)^{\frac{1}{4}} K_{\frac{1}{2}}\left[2\sqrt{\frac{1}{2}z(x+\frac{1}{2}z)}\,\right] = \sqrt{\frac{\pi}{x+z/2}}\, \exp\left[-2\sqrt{\frac{1}{2}z(x+\frac{1}{2}z)}\,\right]$$

The substitution $t = \sqrt{x+z/2}\,/\sqrt{z/2}$ in the remaining x−integral then finally gives

$$K_\nu(z) = \frac{\sqrt{\pi}\,(\frac{1}{2}z)^\nu}{\Gamma\left(\nu+\frac{1}{2}\right)} \int_1^\infty e^{-zt}(t^2-1)^{\nu-\frac{1}{2}}\, dt. \tag{9.43}$$

A simple substitution brings this in the form:

$$\boxed{K_\nu(z) = \frac{\sqrt{\pi}\,\left(\frac{2}{z}\right)^\nu e^{-z}}{\Gamma\left(\nu+\frac{1}{2}\right)} \int_0^\infty e^{-2zt} t^{\nu-\frac{1}{2}}(t+1)^{\nu-\frac{1}{2}}\, dt.} \tag{9.44}$$

Hence, from equation (6.12) we find the connection with the confluent hypergeometric function:

$$\boxed{K_\nu(z) = \sqrt{\pi}\,(2z)^\nu e^{-z} U\left(\nu+\frac{1}{2}, 2\nu+1, 2z\right).} \tag{9.45}$$

This result can also be verified directly by using the differential equations (7.4) and (9.25), and by using the behavior of both sides of (9.45) for large values of z. All other Bessel functions can be written in terms of Kummer functions. We mention

$$\boxed{I_\nu(z) = \frac{(z/2)^\nu e^{-z}}{\Gamma(\nu+1)} M\left(\nu+\frac{1}{2}, 2\nu+1, 2z\right).}$$

From this result, which also follows from the differential equations, we derive via (7.8) and (3.4) the integral representations

$$\boxed{\begin{aligned} I_\nu(z) &= \frac{(2z)^\nu e^z}{\sqrt{\pi}\,\Gamma\left(\nu+\frac12\right)}\int_0^1 e^{-2zt}[t(1-t)]^{\nu-\frac12}\,dt \\ &= \frac{(z/2)^\nu}{\sqrt{\pi}\,\Gamma\left(\nu+\frac12\right)}\int_{-1}^1 e^{-zt}\left(1-t^2\right)^{\nu-\frac12}\,dt. \end{aligned}} \qquad (9.46)$$

This gives the analog of the Poisson integral as given in Exercise 9.12 for the ordinary Bessel function:

$$I_\nu(z) = \frac{(z/2)^\nu}{\sqrt{\pi}\,\Gamma(\nu+\frac12)}\int_{-1}^1 (1-t^2)^{\nu-\frac12}\cosh zt\,dt, \quad \Re\nu > -\frac12, \quad z\in\mathbb{C}. \quad (9.47)$$

Expanding $\cosh zt$ in powers of zt gives the series in (9.28); so the detour via the Kummer functions is not really needed for (9.47).

Remark 9.3. We have remarked after (9.45) that all Bessel functions can be written in terms of Kummer functions (that is, of confluent hypergeometric functions, see Chapter 7), and we expressed the I–function in terms of the M–function, which in fact is a ${}_1F_1$–function. A similar result holds for the J–Bessel function:

$$J_\nu(z) = \frac{(z/2)^\nu e^{-iz}}{\Gamma(\nu+1)} M\left(\nu+\frac12, 2\nu+1, 2iz\right).$$

We see that the Bessel functions $e^{iz}J_\nu(z)$ and $e^z I_\nu(z)$ are ${}_1F_1$–functions, whereas $J_\nu(z)$ and $I_\nu(z)$ themselves are ${}_0F_1$–functions (see (9.12) and (9.28) and the definition of generalized hypergeometric functions (5.29)).

Remark 9.4. The representations in (9.43) and (9.46) can be interpreted as Laplace transforms. For the ordinary Bessel functions such integrals exist in the form of Fourier integrals. Assume now that the function $y(z)$ is a solution of (9.1). Then $w(z) = z^{-\nu}y(z)$ satisfies the equation

$$L_z[w] = zw'' + (2\nu+1)w' + zw = 0.$$

We will verify that integrals of the form

$$z^\nu \int_\alpha^\beta e^{izt}\left(t^2-1\right)^{\nu-\frac12}\,dt$$

are solutions of this differential equation. Using the method of §4.4 we take $K(z,t) = \exp(izt)$ and, via the relation $L_z[\exp(izt)] = M_t[\exp(izt)]$, we find the operators M_t, M_t^* in the form

$$M_t[v(t)] = \left[i(2\nu+1)t - i\left(1-t^2\right)\frac{d}{dt}\right]v(t),$$

$$M_t^*[v(t)] = i(2\nu+1)tv(t) + i\frac{d}{dt}\left[\left(1-t^2\right)v(t)\right].$$

A solution of $M_t^*[v(t)] = 0$ is $v(t) = (t^2-1)^{\nu-\frac{1}{2}}$; the function P of formula (4.28) is given by $P(u,v) = i(t^2-1)uv$. Next we can choose α, β and the location of the contour of integration, taking into account the branch cuts of the many-valued function $(t^2-1)^{\nu-\frac{1}{2}}$. So we find representations of the Hankel and the Bessel functions. In WATSON (1944, Ch. VI) a number of integral representations are derived in this way for the Bessel functions with the Fourier kernel $\exp(\pm izt)$. In §9.2 we have applied the method of §4.4 directly to equation (9.1) by using the kernel $K(z,t) = \exp(-z\sinh t)$. In this way the perils with the many-valued functions are circumvented. In Exercise 9.12 several Fourier integrals for the Bessel functions will be presented.

9.7. Asymptotic Expansions

The following notation, *Hankel's symbol*, is frequently used in representing the coefficients of the asymptotic expansions for the Bessel functions:

$$\begin{aligned}(\alpha,n) &= \frac{2^{-2n}}{n!}\left\{(4\alpha^2-1)(4\alpha^2-3^2)\cdots\left[4\alpha^2-(2n-1)^2\right]\right\}\\ &= \frac{\Gamma\left(\frac{1}{2}+\alpha+n\right)}{n!\,\Gamma\left(\frac{1}{2}+\alpha-n\right)}, \qquad n = 0,1,2,\ldots \\ &= \frac{(-1)^n\cos(\pi\alpha)}{\pi n!}\Gamma\left(\frac{1}{2}+\alpha+n\right)\Gamma\left(\frac{1}{2}-\alpha+n\right).\end{aligned} \tag{9.48}$$

We have $(\alpha,0) = 1$ and the recursion

$$(\alpha,n+1) = -\frac{\left(n+\frac{1}{2}\right)^2-\alpha^2}{n+1}(\alpha,n), \quad n = 0,1,2,\ldots.$$

In the asymptotic results to be derived in this section we assume that ν is fixed and that $z\to\infty$ (in a sector that will usually be specified).

Expanding the function $(t+1)^{\nu-1/2}$ of (9.44) into a power series and using Watson's lemma, we obtain the asymptotic expansion:

$$\boxed{K_\nu(z) \sim \sqrt{\frac{\pi}{2z}}\,e^{-z}\sum_{n=0}^{\infty}\frac{(\nu,n)}{(2z)^n}, \quad |\arg z| < \frac{3}{2}\pi.} \tag{9.49}$$

By taking $z = \zeta e^{-\frac{1}{2}\pi i}$, $z = \zeta e^{\frac{1}{2}\pi i}$ we obtain for the Hankel functions via the third and fourth formula in (9.26):

$$H_\nu^{(1)}(\zeta) \sim \sqrt{\frac{2}{\pi\zeta}}\, e^{\zeta i - \frac{1}{2}\nu\pi i - \frac{1}{4}\pi i} \sum_{n=0}^{\infty} \frac{i^n(\nu, n)}{(2\zeta)^n}, \qquad -\pi < \arg\zeta < 2\pi,$$

$$H_\nu^{(2)}(\zeta) \sim \sqrt{\frac{2}{\pi\zeta}}\, e^{-\zeta i + \frac{1}{2}\nu\pi i + \frac{1}{4}\pi i} \sum_{n=0}^{\infty} \frac{(-i)^n(\nu, n)}{(2\zeta)^n}, \qquad -2\pi < \arg\zeta < \pi.$$

The terms with odd indexes in these asymptotic series show the factor i, the even indexed terms do not have this factor explicitly. It appears to be very convenient to split up the real and imaginary terms (without taking into account whether ζ is complex or not), and to introduce P and Q by writing:

$$\begin{aligned} H_\nu^{(1)}(z) &= \sqrt{\frac{2}{\pi z}}\,[P(\nu, z) + iQ(\nu, z)]e^{i\chi}, \qquad -\pi < \arg z < 2\pi, \\ H_\nu^{(2)}(z) &= \sqrt{\frac{2}{\pi z}}\,[P(\nu, z) - iQ(\nu, z)]e^{-i\chi}, \qquad -2\pi < \arg z < \pi, \end{aligned} \tag{9.50}$$

where $\chi = z - (\frac{1}{2}\nu + \frac{1}{4})\pi$. Then the functions P and Q have the following asymptotic expansions:

$$P(\nu, z) \sim \sum_{n=0}^{\infty} (-1)^n \frac{(\nu, 2n)}{(2z)^{2n}}, \qquad Q(\nu, z) \sim \sum_{n=0}^{\infty} (-1)^n \frac{(\nu, 2n+1)}{(2z)^{2n+1}}. \tag{9.51}$$

That is, we have, writing $\mu = 4\nu^2$,

$$P(\nu, z) \sim 1 - \frac{(\mu - 1)(\mu - 9)}{2!\,(8z)^2} + \frac{(\mu - 1)(\mu - 9)(\mu - 25)(\mu - 49)}{4!\,(8z)^4} - \ldots,$$

$$Q(\nu, z) \sim \frac{\mu - 1}{8z} - \frac{(\mu - 1)(\mu - 9)(\mu - 25)}{3!\,(8z)^3} + \ldots .$$

From (9.50) and (9.6) it follows that the Bessel functions of the first and second kinds can be written as

$$\begin{aligned} J_\nu(z) &= \sqrt{\frac{2}{\pi z}}\,[P(\nu, z)\cos\chi - Q(\nu, z)\sin\chi], \\ Y_\nu(z) &= \sqrt{\frac{2}{\pi z}}\,[P(\nu, z)\sin\chi + Q(\nu, z)\cos\chi]. \end{aligned} \tag{9.52}$$

Asymptotic expansions for $J_\nu(z)$, $Y_\nu(z)$ now follow from those for P and Q. For $J_\nu(z)$, $Y_\nu(z)$ the expansions hold in the sector $|\arg z| < \pi$.

Observe that (9.50) and (9.52) are exact relations; (9.50) is the definition of P and Q. We can determine P and Q in terms of $J_\nu(z)$, $Y_\nu(z)$ from (9.52):

$$\boxed{\begin{aligned} P(\nu, z) &= \sqrt{\frac{\pi z}{2}}\,[J_\nu(z)\cos\chi + Y_\nu(z)\sin\chi], \\ Q(\nu, z) &= \sqrt{\frac{\pi z}{2}}\,[-J_\nu(z)\sin\chi + Y_\nu(z)\cos\chi]. \end{aligned}} \tag{9.53}$$

Expansions for the derivatives of the Bessel functions follow from formal differentiation of the above relations. For the I–function we can use the results of the J–function and the first two formulas of (9.26). We have

$$I_\nu(z) \sim \frac{e^z}{\sqrt{2\pi z}} \sum_{n=0}^{\infty} (-1)^n \frac{(\nu, n)}{(2z)^n}, \quad |\arg z| < \frac{1}{2}\pi, \tag{9.54}$$

which also follows from applying Watson's lemma to the first integral of (9.46). This result is certainly not valid outside the indicated limited sector. Just as for the M–function, see Exercise 7.7 of Chapter 7, we need an extra series for describing the asymptotic behavior outside the sector $|\arg z| < \frac{1}{2}\pi$. In the case of the function $I_\nu(z)$ we have

$$\boxed{I_\nu(z) \sim \frac{e^z}{\sqrt{2\pi z}} \sum_{n=0}^{\infty} (-1)^n \frac{(\nu, n)}{(2z)^n} + \frac{e^{-z+(\nu+\frac{1}{2})\pi i}}{\sqrt{2\pi z}} \sum_{n=0}^{\infty} \frac{(\nu, n)}{(2z)^n},}$$

when $-\frac{1}{2}\pi < \arg z < \frac{3}{2}\pi$, and

$$\boxed{I_\nu(z) \sim \frac{e^z}{\sqrt{2\pi z}} \sum_{n=0}^{\infty} (-1)^n \frac{(\nu, n)}{(2z)^n} + \frac{e^{-z-(\nu+\frac{1}{2})\pi i}}{\sqrt{2\pi z}} \sum_{n=0}^{\infty} \frac{(\nu, n)}{(2z)^n},}$$

when $-\frac{3}{2}\pi < \arg z < \frac{1}{2}\pi$.

When we take $z = x > 0$, we have for large values of x

$$\boxed{\begin{aligned} J_\nu(x) &= \sqrt{\frac{2}{\pi x}}\,\left[\cos(x - \frac{1}{2}\nu\pi - \frac{1}{4}\pi) + \mathcal{O}\left(x^{-1}\right)\right], \\ Y_\nu(x) &= \sqrt{\frac{2}{\pi x}}\,\left[\sin(x - \frac{1}{2}\nu\pi - \frac{1}{4}\pi) + \mathcal{O}\left(x^{-1}\right)\right]. \end{aligned}} \tag{9.55}$$

Hence the Bessel functions have an oscillatory character, with amplitude $\sqrt{2/(\pi x)}$, which steadily decreases as $x \to \infty$.

Asymptotic representations for large values of ν with z fixed can be obtained from the series (9.12) and (9.28). The remaining functions can be treated by using (9.9), (9.5) and (9.29). When both parameters ν and z are large the asymptotics becomes much more complicated. A good survey is given in ABRAMOWITZ & STEGUN (1964, Ch. 9). The theory can be found in OLVER (1974), where the results, along with the above expansions, are mainly derived by using differential equations and are supplied with bounds for the remainders.

9.8. Zeros of Bessel Functions

Looking at (9.55) we observe that the Bessel functions $J_\nu(z)$ and $Y_\nu(z)$ will have real zeros when ν is real. A *zero of the Bessel function* $J_\nu(z)$ is a value of z such that $J_\nu(z) = 0$, with ν a given fixed number. Similarly for $Y_\nu(z)$ and the derivatives. In this section we assume that the order ν is real. We prove the following theorems, of which the first one is a special case of a more general result for zeros of solutions of second order differential equations.

Theorem 9.1.

1. *All zeros of a solution of the Bessel differential equation are simple (with a possible exception of the point $z = 0$).*
2. *All zeros of the derivative of a solution of the Bessel equation are simple (with a possible exception of the points $z = 0$ and $z = \pm\nu$).*

Proof. If w is a solution of (9.1), and $z = z_0$ is a regular point with $w(z_0) = w'(z_0) = 0$, then Theorem 4.1 leads to $w = 0$. If, on the other hand, $w'(z_0) = w''(z_0) = 0$, then from (9.1) it follows that $w(z_0) = 0$, when $z_0 \neq \pm\nu$. ∎

The following theorem is due to Lommel (1868).

Theorem 9.2. *$J_\nu(z)$ has no non-real zeros when $\nu \geq -1$; $J_\nu'(z)$ has no non-real zeros when $\nu \geq 0$.*

Proof. From (9.28) it follows that $J_\nu(z)$ has no purely imaginary zeros if $\nu \geq -1$. Assume now that $z = \alpha$ is a non-real zero, with $\alpha \notin i\mathbb{R}$. Then $z = \overline{\alpha}$ is also a zero. Consider the identity

$$(\alpha^2 - \beta^2) \int_0^z t J_\nu(\alpha t) J_\nu(\beta t)\, dt = z \left[J_\nu(\alpha z) \frac{dJ_\nu(\beta z)}{dz} - J_\nu(\beta z) \frac{dJ_\nu(\alpha z)}{dz} \right], \tag{9.56}$$

where $\nu \geq -1$, which can be proved by differentiating with respect to z and by using (9.1). We use this formula with $z = 1$ and $\beta = \overline{\alpha}$. Then,

$$\int_0^1 t J_\nu(\alpha t) J_\nu(\overline{\alpha} t)\, dt = 0.$$

Table 9.1. Positive Zeros $j_{\nu,n}, y_{\nu,n}$ of the Bessel Functions $J_\nu(x), Y_\nu(x),\ \nu = 0, 1$.

n	$j_{0,n}$	$j_{1,n}$	$y_{0,n}$	$y_{1,n}$
1	2.40483	3.83171	0.89358	2.19714
2	5.52008	7.01559	3.95768	5.42968
3	8.65373	10.17347	7.08605	8.59601
4	11.79153	13.32369	10.22235	11.74915
5	14.93092	16.47063	13.36110	14.89744
6	18.07106	19.61586	16.50092	18.04340
7	21.21164	22.76008	19.64131	21.18807

This is impossible, since the integrand is positive. The same proof can be used for $J'_\nu(z)$, but now there is a pair of imaginary zeros when $-1 < \nu < 0$. This follows from the power series of $J'_\nu(z)$. ∎

When $\nu < -1$ the proof is not valid, since in this case the integral in (9.56) is not convergent at $t = 0$. Indeed complex zeros occur when $\nu < -1$, see WATSON (1944, §15.27).

In Table 9.1 we give the first seven positive zeros $j_{\nu,n}, y_{\nu,n}$ of the Bessel functions $J_\nu(x), Y_\nu(x),\ \nu = 0, 1$. A first approximation of the zeros of $J_\nu(z)$ follows from (9.55):

$$\cos\left(z - \frac{1}{2}\nu\pi - \frac{1}{4}\pi\right) + \mathcal{O}\left(z^{-1}\right) = 0.$$

Hence,

$$z = n\pi + \frac{1}{2}\nu\pi - \frac{1}{4}\pi + \mathcal{O}\left(n^{-1}\right),$$

where n is a large positive integer. Sharper approximations follow from the equation (see (9.52))

$$P(\nu, z)\cos\chi = Q(\nu, z)\sin\chi$$

and the expansions in (9.51), that is

$$z - \alpha = -\arctan\frac{Q(\nu,z)}{P(\nu,z)} \sim -\frac{4\nu^2-1}{8z} - \frac{(4\nu^2-1)(4\nu^2-25)}{384z^3} - \cdots$$

where $\alpha = (n + \frac{1}{2}\nu - \frac{1}{4})\pi$. After a few formal manipulations we obtain for the large zeros of $J_\nu(z)$ the asymptotic expansion (McMahon (1895))

$$\boxed{z \sim \alpha - \frac{4\nu^2-1}{8\alpha} - \frac{(4\nu^2-1)(28\nu^2-31)}{384\alpha^3} - \ldots, \quad n \to \infty.} \tag{9.57}$$

If $\nu = \frac{1}{2}$ the first (and later!) terms give the exact result $\pi, 2\pi, 3\pi, \ldots$ for the positive zeros. Using continuity of the zeros with respect to the parameter we infer that (9.57) indeed gives an approximation for the n-th positive zero ($n > 0$). When $n = 1$, $\nu = 0$ one has $\alpha = \frac{3}{4}\pi$. Computing the first zero of $J_0(z)$ using the terms shown in (9.57) yields

$$z \simeq 2.35619 + 0.05305 - 0.00617 = 2.40307,$$

while the 'exact' value is $2.40482\ldots$.

Theorem 9.3. *$J_\nu(x)$ has a countable number of positive zeros. The distance between two consecutive zeros of $J_\nu(x)$ is $\begin{smallmatrix}\le\\=\\>\end{smallmatrix} \pi$ if $|\nu| \begin{smallmatrix}\le\\=\\>\end{smallmatrix} \frac{1}{2}$, respectively.*

Proof. Starting from the Bessel differential equation (9.1) it easily follows that the function $w(x) = \sqrt{x}\, J_\nu(x)$ is a real solution of the differential equation

$$w'' + \left(1 - \frac{\nu^2 - \frac{1}{4}}{x^2}\right) w = 0. \tag{9.58}$$

We now use Sturm's comparison theorem (§4.3). We apply Theorem 4.4 for the two following cases (the case $\nu = \pm\frac{1}{2}$ is trivial, see Exercise 9.3).

(i): $|\nu| < \frac{1}{2}$. Take $g_1(x) = 1$, $g_2(x) = 1 - (\nu^2 - \frac{1}{4})/x^2$ with $x > 0$. We compare (9.58) with the equation $w'' + w = 0$, which has the general solution $w(x) = A\cos x + B\sin x$. For each fixed pair A, B this solution has countably many zeros, with distance π. According to Sturm's theorem, between each pair of consecutive zeros of equation $w'' + w = 0$ there is at least one zero of $J_\nu(x)$. Considering that in a finite interval $J_\nu(x)$ only has a finite number of zeros (this follows from a well-known property of analytic functions), we infer that $J_\nu(x)$ has a countable number of positive zeros. Assume next that $J_\nu(x)$ has consecutive zeros α, β. Take a solution w of the equation $w'' + w = 0$, with zeros α and $\alpha + \pi$. According to Sturm's theorem we conclude that $\alpha < \beta < \alpha + \pi$. Hence $\beta - \alpha < \pi$.

(ii): $|\nu| > \frac{1}{2}$. Now take $g_1(x) = 1/(4\nu^2)$, $g_2(x) = 1 - (\nu^2 - \frac{1}{4})/x^2$ with $x > |\nu|$. The general solution of $w'' + (1/4\nu^2)w = 0$ has countably many zeros. According to Sturm's theorem, then $J_\nu(x)$ has a countable number of zeros. For the remaining part of the theorem we take $g_1(x) = 1 - (\nu^2 - \frac{1}{4})/x^2$, $g_2(x) = 1$ and reason as in the first part of the theorem. ∎

The zeros of $J_\nu(x)$ can be arranged as a sequence

$$0 < j_{\nu,1} < j_{\nu,2} < \cdots < j_{\nu,n} < \cdots, \qquad \lim_{n\to\infty} j_{\nu,n} = \infty. \tag{9.59}$$

We will now show that the zeros of $J_{\nu+1}(x)$ are located between these zeros.

Theorem 9.4. *Between two consecutive positive zeros of $J_\nu(x)$ there is exactly one zero of $J_{\nu+1}(x)$. Conversely, between two consecutive positive zeros of $J_{\nu+1}(x)$ there is exactly one zero of $J_\nu(x)$.*

Proof. We use the relations in (9.17) with $k = 1$ and $C_\nu = J_\nu$, in the form

$$\frac{1}{z}\frac{d}{dz}\left[z^{\nu+1}J_{\nu+1}(z)\right] = z^\nu J_\nu(z), \quad \frac{1}{z}\frac{d}{dz}\left[z^{-\nu}J_\nu(z)\right] = -z^{-\nu-1}J_{\nu+1}(z). \tag{9.60}$$

Take two consecutive positive zeros $j_{\nu,n}$, $j_{\nu,n+1}$ of $J_\nu(x)$. Then $j_{\nu,n}$, $j_{\nu,n+1}$ are zeros of $x^{-\nu}J_\nu(x)$ as well. According to Rolle's theorem, at least one zero of the derivative, hence of $J_{\nu+1}(x)$ lies between $j_{\nu,n}$ and $j_{\nu,n+1}$ (see the second relation in (9.60)). Assume now that $J_{\nu+1}(x)$ has two zeros λ_1, λ_2 such that $j_{\nu,n} < \lambda_1 < \lambda_2 < j_{\nu,n+1}$. Then λ_1, λ_2 are zeros of $x^{\nu+1}J_{\nu+1}(x)$ as well. With the help of the first relation of (9.60) and Rolle's theorem it follows that at least one zero of $J_\nu(x)$ would lie between λ_1 and λ_2. This is in contradiction with the assumption that $j_{\nu,n}$ and $j_{\nu,n+1}$ are two consecutive zeros of $J_\nu(x)$. Hence between $j_{\nu,n}$ and $j_{\nu,n+1}$ there is exactly one zero of $J_{\nu+1}(x)$. The proof of the remaining part of the theorem is similar. ■

When $\nu > -1$ the zeros of $J_\nu(x)$ and $J_{\nu+1}(x)$ can be arranged according to

$$0 < j_{\nu,1} < j_{\nu+1,1} < j_{\nu,2} < j_{\nu+1,2} < \cdots < j_{\nu,n} < j_{\nu+1,n} < \cdots \tag{9.61}$$

Much information is available on the zeros of the Bessel functions (also on complex zeros), for instance in WATSON (1944, Ch. 15) and OLVER (1974). The real zeros are extensively tabulated; see ABRAMOWITZ & STEGUN (1964, p. 409). In Exercise 9.5 you are invited to derive more properties of the zeros of Bessel functions.

When ν is large the functions $J_\nu(x), Y_\nu(x)$ are monotonic in the interval $(0, \nu)$ and start oscillating when $x > \nu$; see Figure 9.9.

9.9. Orthogonality Relations, Fourier-Bessel Series

In Theorem 9.3 we have shown that for real positive values of ν the Bessel function $J_\nu(x)$ has a countable number of zeros (and Theorem 9.2 says that there are no complex zeros when $\nu \ge -1$). Here we denote the positive zeros of $J_\nu(z)$ by j_n, arranging them as in (9.59):

$$0 < j_1 < j_2 < j_3 < \cdots < j_n < \cdots, \quad J_\nu(j_n) = 0.$$

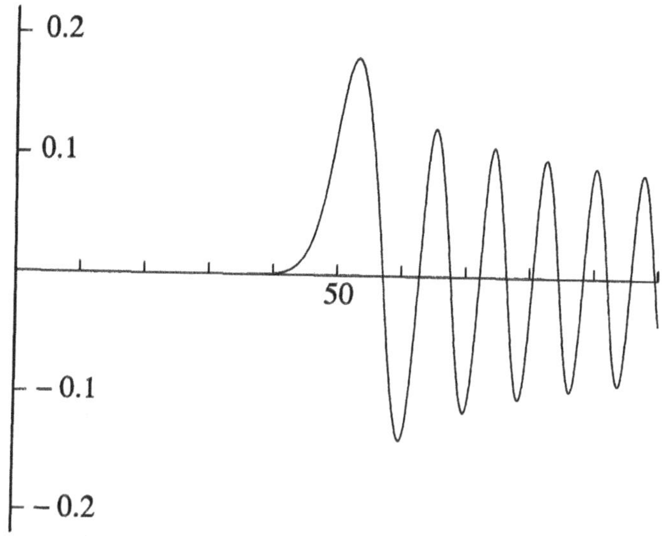

Figure 9.9. The Bessel function $J_{50}(x), 0 \le x \le 100$.

Theorem 9.5. *Let $\nu > -1$. Then, on the interval $[0,1]$, the Bessel functions*

$$J_\nu(j_1 x),\ J_\nu(j_2 x), \ldots,\ J_\nu(j_n x),\ \ldots$$

constitute an orthogonal system with weight function x. Furthermore,

$$\boxed{\int_0^1 x J_\nu(j_m x) J_\nu(j_n x)\, dx = \frac{1}{2} J_{\nu+1}^2(j_m)\delta_{m,n}.} \tag{9.62}$$

Proof. We first consider the case $m \neq n$. Using (9.56) with $\alpha = j_m$, $\beta = j_n$ and $z = 1$, we obtain

$$\int_0^1 x J_\nu(j_m x) J_\nu(j_n x)\, dx = 0.$$

If $m = n$ (9.56) has to be be analyzed in more detail. We bring the factor $\alpha^2 - \beta^2$ to the right-hand side and apply l'Hôpital's rule, as $\alpha \to \beta$. Then we have

$$\int_0^z t J_\nu(\beta t)\, dt = \frac{z}{2\beta}\left\{\beta z \left[J_\nu'(\beta z)\right]^2 - J_\nu(\beta z) J_\nu'(\beta z) - \beta z J_\nu(\beta z) J_\nu''(\beta z)\right\}.$$

Again, take $z = 1$, $\beta = j_n$. Then the terms on the right-hand side containing $J_\nu(j_n)$ vanish. The expression $[J_\nu'(\beta z)]^2$ can be replaced by $[J_{\nu+1}(z)]^2$ upon using one of the recurrence relations in (9.16), namely

$$xJ_\nu'(x) = \nu J_\nu(x) - xJ_{\nu+1}(x). \tag{9.63}$$

It follows that

$$\int_0^1 xJ_\nu^2(j_n x)\,dx = \frac{1}{2}J_{\nu+1}^2(j_n).$$

■

By analogy with the theory of Fourier series we can now investigate series of the form

$$S_\nu(x) = \sum_{m=1}^{\infty} a_m J_\nu(j_m x), \quad 0 < x < 1, \tag{9.64}$$

which are called *Fourier-Bessel series*. Assume that this series converges uniformly on the interval $[0, 1]$. Then the coefficients a_m can be expressed in terms of the function $S_\nu(x)$. For that purpose, multiply (9.64) with $xJ_\nu(j_n x)$ and integrate over the interval $[0, 1]$. Then, on account of the orthogonality of the Bessel functions,

$$\int_0^1 xS_\nu(x)J_\nu(j_n x)\,dx = \sum_{m=1}^{\infty} a_m \int_0^1 xJ_\nu(j_m x)J_\nu(j_n x)\,dx = \frac{1}{2}a_n J_{\nu+1}^2(j_n).$$

We can now solve for a_n:

$$a_n = \frac{2}{J_{\nu+1}^2(j_n)} \int_0^1 xS_\nu(x)J_\nu(j_n x)\,dx. \tag{9.65}$$

This formula is the analog of the formulas for the coefficients in a Fourier series. When $\nu = \pm\frac{1}{2}$ the Fourier-Bessel series (9.64) reduces to a Fourier sine or cosine series.

Next we can consider the problem from a different point of view. Let f be defined on $[0, 1]$. Compute the coefficients a_n according to the rule

$$a_n = \frac{2}{J_{\nu+1}^2(j_n)} \int_0^1 xf(x)J_\nu(j_n x)\,dx \tag{9.66}$$

and form the Fourier-Bessel series $\sum_{n=1}^{\infty} a_n J_\nu(j_n x)$. One can prove, under certain conditions on f, that this Fourier-Bessel series is convergent with sum $f(x)$. One condition is that f is integrable such that $\int_0^1 \sqrt{x}\, f(x)\,dx$ exists, and that this integral (if it is an improper integral) converges absolutely. If $x \in (a, b)$ with $0 < a < b < 1$ and f has bounded variation on (a, b), then, if

$\nu \geq -\frac{1}{2}$ and a_n is given by (9.66), the series $\sum_{n=1}^{\infty} a_n J_\nu(j_n x)$ converges and the sum is equal to $\frac{1}{2}[f(x+0)+f(x-0)]$ (see WATSON (1944, §18.24)).

Example 9.1. We have

$$x^\nu = \sum_{n=1}^{\infty} \frac{2J_\nu(j_n x)}{j_n J_{\nu+1}(j_n)}, \quad 0 \leq x < 1, \quad \nu \geq 0. \tag{9.67}$$

To prove this we derive from (9.17):

$$x^{\nu+1} J_\nu(x) = \frac{d}{dx}[x^{\nu+1} J_{\nu+1}(x)].$$

Hence the result (9.67) follows from integrating by parts in (9.66) with $f(x) = x^\nu$. Observe that the right-hand side of (9.67) vanishes as $x \to 1$, whereas the left-hand side takes the value unity.

9.10. Remarks and Comments for Further Reading

9.1. All asymptotic expansions given in this chapter for the Bessel functions are for large z and fixed order ν. More powerful expansions are available in which ν usually is considered as the large parameter. We have *Debye type expansions* and *Airy type expansions.* An overview of the results can be found in Chapter 9, §9.3, of ABRAMOWITZ & STEGUN (1964). See also OLVER (1974).

9.2. Marcum's function considered in the Introduction of this chapter is discussed in Chapter 11, §11.4, where we discuss asymptotic expansions and numerical aspects.

9.3. The computation of the Bessel functions can be based on the recurrence relations. In Chapter 13 more information is given. In AMOS (1986) a large collection of Fortran programs is described for all kinds of Bessel functions for real order and complex argument. MATVIYENKO (1993) discusses the implementation of several kinds of asymptotic expansions of the Bessel functions. This paper also shows that the Bessel functions $J_\nu(x)$ of integer order ν describe displacements of coupled harmonic oscillators on a line.

9.11. Exercises and Further Examples

9.1. Verify with Abel's identity that the Wronskian $W[u,v](z) = uv' - u'v$ (see (3.13) and (3.14)) for any pair of solutions u, v of (9.1) satisfies

$$W[u,v](z) = \frac{C}{z},$$

where C may depend on ν but not on z. Verify the following cases:

$$u(z)=J_\nu(z),\ v(z)=J_{-\nu}(z),\qquad W[u,v](z)=-\frac{2\sin\nu\pi}{\pi z};$$

$$u(z)=J_\nu(z),\ v(z)=Y_\nu(z),\qquad W[u,v](z)=\frac{2}{\pi z};$$

$$u(z)=H_\nu^{(1)}(z),\ v(z)=H_\nu^{(2)}(z),\qquad W[u,v](z)=-\frac{4i}{\pi z};$$

$$u(z)=I_\nu(z),\ v(z)=I_{-\nu}(z),\qquad W[u,v](z)=-\frac{2\sin\nu\pi}{\pi z};$$

$$u(z)=I_\nu(z),\ v(z)=K_\nu(z),\qquad W[u,v](z)=-\frac{1}{z}.$$

The second, third and fifth relations cannot vanish for any combination of z and ν.

9.2. Verify the following relations, which are related with the relations for the Wronskians in the previous exercise,

$$J_{\nu+1}(z)J_{-\nu}(z)+J_\nu(z)J_{-(\nu+1)}(z)=-\frac{2\sin\nu\pi}{\pi z};$$

$$J_{\nu+1}(z)Y_\nu(z)+J_\nu(z)Y_{\nu+1}(z)=\frac{2}{\pi z};$$

$$H_{\nu+1}^{(1)}(z)H_\nu^{(2)}(z)-H_\nu^{(1)}(z)H_{\nu+1}^{(2)}(z)=-\frac{4i}{\pi z};$$

$$I_\nu(z)I_{-(\nu+1)}(z)-I_{\nu+1}(z)I_{-\nu}(z)=-\frac{2\sin\nu\pi}{\pi z};$$

$$I_\nu(z)K_{\nu+1}(z)+I_{\nu+1}(z)K_\nu(z)=\frac{1}{z},$$

$$P(\nu,z)P(\nu+1,z)+Q(\nu,z)Q(\nu+1,z)=1.$$

9.3. Show that when $\nu=\pm\frac{1}{2}$ the Bessel functions are elementary functions. Verify the following relations:

$$J_{\frac{1}{2}}(z)=Y_{-\frac{1}{2}}(z)=\sqrt{\frac{2}{\pi z}}\,\sin z,$$

$$Y_{\frac{1}{2}}(z)=-J_{-\frac{1}{2}}(z)=-\sqrt{\frac{2}{\pi z}}\,\cos z,$$

$$H_{\frac{1}{2}}^{(1)}(z)=-iH_{-\frac{1}{2}}^{(1)}(z)=-i\sqrt{\frac{2}{\pi z}}\,e^{iz},$$

$$H_{\frac{1}{2}}^{(2)}(z)=iH_{-\frac{1}{2}}^{(2)}(z)=i\sqrt{\frac{2}{\pi z}}\,e^{-iz},$$

$$I_{\frac{1}{2}}(z)=\sqrt{\frac{2}{\pi z}}\,\sinh z,\quad I_{-\frac{1}{2}}(z)=\sqrt{\frac{2}{\pi z}}\,\cosh z,$$

$$K_{\frac{1}{2}}(z)=K_{-\frac{1}{2}}(z)=\sqrt{\frac{\pi}{2z}}\,e^{-z}.$$

With the recurrence relations in (9.16) and (9.34) all Bessel functions of order $n+\frac{1}{2}$ can be expressed in terms of circular functions multiplied by a polynomial in powers of z^{-1}. In the literature on Bessel functions these functions are often called *spherical Bessel functions.* There is a separate notation:

$$j_n(z) = \sqrt{\frac{\pi}{2z}}\, J_{n+\frac{1}{2}}(z), \quad y_n(z) = \sqrt{\frac{\pi}{2z}}\, Y_{n+\frac{1}{2}}(z).$$

Show that

$$j_n(z) = (-z)^n \left[\frac{1}{z}\frac{d}{dz}\right]^n \frac{\sin z}{z}, \quad y_n(z) = -(-z)^n \left[\frac{1}{z}\frac{d}{dz}\right]^n \frac{\cos z}{z}.$$

9.4. Verify, by using (6.42) and the generating function (9.20) for the Bessel coefficients (see also (9.33)), the following Chebyshev expansions for the Bessel functions:

$$\cos xz = 2\sum_{n=0}^{\infty}{}'(-1)^n J_{2n}(z)T_{2n}(x),$$

$$\sin xz = 2\sum_{n=0}^{\infty}(-1)^n J_{2n+1}(z)T_{2n+1}(x),$$

$$e^{xz} = 2\sum_{n=0}^{\infty}{}' I_n(z)T_n(x).$$

The prime in $\sum'$ denotes that the first term of the series is halved:

$$\sum_{n=0}^{\infty}{}' a_n = \frac{1}{2}a_0 + a_1 + a_2 + \cdots .$$

9.5. In addition to the theory in the text we present extra examples on zeros of Bessel functions.

1. Show that between each pair of consecutive positive zeros of $J_\nu(x)$ there is exactly one zero of $Y_\nu(x)$, and conversely. *Hint:* use the Wronskian for both functions, see Exercise 9.1. Draw the graph of the function $J_\nu(x)/Y_\nu(x)$ when $x > 0$.
2. Verify that the first terms of the asymptotic expansion of the positive zeros of $Y_\nu(x)$ follow from (9.57) by replacing α with $\beta = (n+\frac{1}{2}\nu-\frac{3}{4})\pi$. Verify that the positive zeros $j'_{\nu,n}$, $y'_{\nu,n}$ of $J'_\nu(x)$, $Y'_\nu(x)$, respectively, have the following asymptotic expansion (McMahon (1895)):

$$j'_{\nu,n},\, y'_{\nu,n} \sim \alpha - \frac{4\nu^2+3}{8\alpha} - \frac{112\nu^4+328\nu^2-9}{384\alpha^3} - \cdots, \qquad n\to\infty,$$

where $\alpha = (n + \frac{1}{2}\nu - \frac{3}{4})\pi$ for $j'_{\nu,n}$ and $\alpha = (n + \frac{1}{2}\nu - \frac{1}{4})\pi$ for $y'_{\nu,n}$.

3. Introduce the function

$$G(x) = AJ_\nu(x) + xJ'_\nu(x),$$

where $\nu > -1$ and A is a real constant. Prove that, if $A + \nu \geq 0$, all zeros of G are real and simple (with possible exception the point $x = 0$). Show that between each pair of consecutive positive zeros of G there is exactly one zero of $J_\nu(x)$, and conversely.

9.6. Expand the function $\ln x$ in terms of a Fourier-Bessel series with $\nu = 0$.

9.7. Verify the indefinite integral

$$\int J_\mu(x)J_\nu(x)\frac{dx}{x} = \frac{x[J'_\mu(x)J_\nu(x) - J_\mu(x)J'_\nu(x)]}{\mu^2 - \nu^2}, \qquad \mu^2 \neq \nu^2.$$

Investigate the convergence of the integral

$$\int_0^\infty J_\mu(x)J_\nu(x)\frac{dx}{x},$$

and compute this integral.

9.8. Verify the Bessel transforms in connection with the Laguerre polynomials:

$$\int_0^\infty t^{n+\alpha/2} J_\alpha\left(2\sqrt{xt}\right) e^{-t}\, dt = x^{\alpha/2}e^{-x}\; n!\; L_n^\alpha(x), \tag{1}$$

$$\int_0^\infty t^{\alpha/2} J_\alpha\left(\sqrt{xt}\right) e^{-t/2} L_n^\alpha(t)\, dt = 2(-1)^n x^{\alpha/2} e^{-x/2}\; L_n^\alpha(x). \tag{2}$$

(1) is valid when $\Re(n + \alpha) > -1$, (2) is valid when $\alpha > -1$; x may be any complex number. By a slight change of integration variable, both relations may be seen as Laplace transforms of the Bessel function. In fact (1) generalizes the Laplace transform (9.11). Observe that (2) gives an *integral equation for Laguerre polynomials*: let $y_n(x) = x^{\alpha/2}e^{-x/2}L_n(x)$, then $y_n(x)$ solves the integral equation $y_n(x) = \frac{1}{2}(-1)^n \int_0^\infty J_\alpha\left(\sqrt{xt}\right) y_n(t)\, dt$. To prove (1) start with (9.11) in the form

$$e^{-x}x^{n+\alpha} = \int_0^\infty \left(\sqrt{xt}\right)^{n+\alpha} J_{n+\alpha}\left(2\sqrt{xt}\right) e^{-t}\, dt,$$

and assume first that $x > 0$. Next verify that

$$\frac{d}{du}u^{\nu/2}J_\nu\left(2\sqrt{u}\right) = u^{(\nu-1)/2}J_{\nu-1}\left(2\sqrt{u}\right).$$

We find that, for $m = 0, 1, 2, \ldots$,

$$\frac{d^m}{dx^m}\left(e^{-x}x^{m+\alpha}\right) = \int_0^\infty \left(\sqrt{xt}\right)^{n-m+\alpha} J_{n-m+\alpha}\left(2\sqrt{xt}\right) e^{-t}t^m\, dt.$$

Taking $m = n$ and using (6.23), we obtain (1). The result in (1) is valid for all complex values of x. When $n = \nu$ (non-integer) the right-hand side can be replaced by a Kummer M−function; see Exercise 7.10. To verify (2), expand $L_n(t) = \sum c_m^{(n)} t^m$, where $c_m^{(n)}$ follow from (6.40). The resulting integrals can be evaluated by using (1). Finally you need the following identity for Laguerre polynomials:

$$(-1)^n L_n^\alpha(2x) = \sum_{m=0}^{n} m!\, 2^m\, c_m^{(n)}\, L_m^\alpha(x).$$

This follows from Exercise 6.18.

9.9. Verify, by expanding the Bessel function in powers of t, the Laplace integral

$$\int_0^\infty t^{\mu-1} J_\nu(t) e^{-st}\, dt = \frac{\Gamma(\mu+\nu)}{2^\nu s^{\mu+\nu}\Gamma(\nu+1)} F\left(\frac{1}{2}\mu + \frac{1}{2}\nu, \frac{1}{2}\mu + \frac{1}{2}\nu + \frac{1}{2}; \nu+1; -s^{-2}\right), \tag{1}$$

where $\Re s > 0$ and $\Re(\mu+\nu) > 0$. Verify, with the help of (3.4) and (5.2), that the cases $\mu = \nu+1, \mu = \nu+2$ give the elementary results:

$$\int_0^\infty t^\nu J_\nu(t) e^{-st}\, dt = \frac{2^\nu \Gamma\left(\nu + \frac{1}{2}\right)}{\sqrt{\pi}\,(s^2+1)^{\nu+\frac{1}{2}}},$$

$$\int_0^\infty t^{\nu+1} J_\nu(t) e^{-st}\, dt = \frac{2^{\nu+1} \Gamma\left(\nu + \frac{3}{2}\right)}{\sqrt{\pi}\,(s^2+1)^{\nu+\frac{3}{2}}}.$$

Prove, by taking $\mu = 0$ and $\mu = 1$, and using Exercise 5.10, that

$$\int_0^\infty t^{-1} J_\nu(t) e^{-st}\, dt = \frac{\left(\sqrt{s^2+1} - s\right)^\nu}{\nu},$$

$$\int_0^\infty J_\nu(t) e^{-st}\, dt = \frac{\left(\sqrt{s^2+1} - s\right)^\nu}{\sqrt{s^2+1}}.$$

The right-hand side of (1) is a Legendre function, as becomes clear from the right-hand side of (7.57). However, this is not the form that is usually found in the literature. Use the quadratic transformation of (5.28) with $2a = \mu+\nu$, $c = \nu + 1$ and

$$z = \frac{s - \sqrt{s^2+1}}{s + \sqrt{s^2+1}}.$$

The hypergeometric function in (1) then takes the form

$$(1+z)^{\mu+\nu} F\left(\mu+\nu, \mu; \nu+1; z\right).$$

Next use the second transformation in (5.5) and finally (8.75). Thus we obtain for (1):

$$\int_0^\infty t^{\mu-1} J_\nu(t) e^{-st}\, dt = (s^2+1)^{-\frac{1}{2}\mu} \Gamma(\mu+\nu) P_{\mu-1}^{-\nu}\left(\frac{s}{\sqrt{s^2+1}}\right).$$

A further special case is

$$\int_0^\infty J_\nu(t)\, dt = 1, \quad \Re\nu > -1.$$

Prove with the final result and (9.9) that

$$\int_0^\infty Y_\nu(t)\, dt = -\tan\left(\frac{1}{2}\nu\pi\right), \quad |\Re\nu| < 1.$$

9.10. Use (9.43) and an integral for the beta function to prove that

$$\int_0^\infty t^{\mu-1} K_\nu(t)\, dt = 2^{\mu-2} \Gamma\left(\frac{1}{2}\mu + \frac{1}{2}\nu\right) \Gamma\left(\frac{1}{2}\mu - \frac{1}{2}\nu\right), \quad \Re\mu > |\Re\nu|.$$

In fact, this is the *Mellin transform of the $K-$function*. The inversion formula gives (see the Mellin transformation pair in §5.6)

$$K_\nu(t) = \frac{2^{-2}}{2\pi i} \int_{\mathcal{L}} 2^\mu t^{-\mu} \Gamma\left(\frac{1}{2}\mu + \frac{1}{2}\nu\right) \Gamma\left(\frac{1}{2}\mu - \frac{1}{2}\nu\right) d\mu,$$

where $\mathcal{L}$ is a vertical line in the complex $\mu-$plane, on which $\Re\mu > |\Re\nu|$. Shifting the path of integration to the left while picking up the residues of the gamma functions, we obtain a power series that corresponds with (9.29) and (9.28). From Exercise 7.8 and formula (9.45) a different *Mellin-Barnes integral for $K_\nu(z)$* follows.

9.11. Substitute $u = \sinh^2 \frac{1}{2}t$, that is $\cosh t = 2u + 1$, in (9.39) and derive with formula (2) of Exercise 5.4 the integral representation

$$K_\nu(z) = e^{-z} \int_0^\infty e^{-2zu} F\left(\frac{1}{2}+\nu, \frac{1}{2}-\nu; \frac{1}{2}; -u\right) \frac{du}{\sqrt{u}}.$$

Substitution of the power series expansion of the F–function and Watson's lemma again give the asymptotic expansion (9.49). Verify this. Use of the first formula of (5.5) then shows that

$$K_\nu(z) = e^{-z}\int_0^\infty \frac{e^{-2zu}}{(1+u)^{\nu+\frac{1}{2}}} F\left(\frac{1}{2}+\nu, \nu; \frac{1}{2}; \frac{u}{1+u}\right)\frac{du}{\sqrt{u}}.$$

Derive from this result the expansion

$$K_\nu(z) = e^{-z}\sum_{n=0}^{\infty}\frac{\left(\nu+\frac{1}{2}\right)_n(\nu)_n}{\left(\frac{1}{2}\right)_n n!}\int_0^\infty e^{-2zu}\frac{u^{n-\frac{1}{2}}}{(u+1)^{n+\nu+\frac{1}{2}}}\,du,$$

which, by using (7.12), can be written as

$$K_\nu(z) = \sqrt{\pi}\,e^{-z}\sum_{n=0}^{\infty}\frac{\left(\nu+\frac{1}{2}\right)_n(\nu)_n}{n!}U\left(n+\frac{1}{2}, 1-\nu, 2z\right).$$

This expansion is a *convergent* alternative of the asymptotic expansion (9.49) of the K–function. The U–functions in this expansion can be computed by using backward recursion (Miller's algorithm) as described in Chapter 13.

9.12. Verify the analog of (9.47) for the J–function:

$$J_\nu(z) = \frac{(z/2)^\nu}{\sqrt{\pi}\,\Gamma(\nu+\frac{1}{2})}\int_{-1}^{1}\left(1-t^2\right)^{\nu-\frac{1}{2}}\cos zt\,dt, \quad \Re\nu > -\frac{1}{2}, \quad z\in\mathbb{C}.$$

This is the *Poisson integral* and has the form as mentioned in Remark 9.3 at the end of §9.6. It can be proved by expanding the cosine and comparing the result with (9.12). The *Mehler-Sonine integrals* also have the form as mentioned in Remark 9.2 (with $-\nu$ in place of ν):

$$J_\nu(x) = \frac{2\left(\frac{1}{2}x\right)^{-\nu}}{\sqrt{\pi}\,\Gamma\left(\frac{1}{2}-\nu\right)}\int_1^\infty \frac{\sin xt}{\left(t^2-1\right)^{\nu+\frac{1}{2}}}\,dt,$$

$$Y_\nu(x) = -\frac{2\left(\frac{1}{2}x\right)^{-\nu}}{\sqrt{\pi}\,\Gamma\left(\frac{1}{2}-\nu\right)}\int_1^\infty \frac{\cos xt}{\left(t^2-1\right)^{\nu+\frac{1}{2}}}\,dt,$$

where $x>0$ and $|\Re\nu| < \frac{1}{2}$. Prove these formulas by replacing in (9.43) ν with $-\nu$ and z with $-ix$. Next use the third relation of (9.26) in order to have a Hankel function in the result. The remaining step follows from (9.5).

A related integral for the K–function is *Basset's integral*:

$$K_\nu(xz) = \frac{\Gamma\left(\nu+\frac{1}{2}\right)(2z)^\nu}{\sqrt{\pi}\,x^\nu}\int_0^\infty \frac{\cos xt\,dt}{\left(t^2+z^2\right)^{\nu+\frac{1}{2}}}\,dt,$$

where $\Re\nu > -\frac{1}{2}, \quad x > 0, \quad |\arg z| < \frac{1}{2}\pi.$

9.13. The solutions of the equation $y'' - zy = 0$ are called *Airy functions.* They play an important part in several physical problems, for instance in diffraction of light. Airy functions are also used in asymptotics to describe the transition of oscillatory behavior of functions to exponentially small or large behavior. In §4.4 (Example 4.6) three solutions y_i of the Airy differential equation are found, of which $y_1(z) = \mathrm{Ai}(z)$ is given as a contour integral. Verify by expanding e^{-zt} that a representation in terms of modified Bessel functions can be obtained:

$$\begin{aligned}\mathrm{Ai}(z) &= \frac{1}{\pi}\sqrt{\frac{z}{3}}\,K_{\frac{1}{3}}(\zeta) = \frac{1}{3}\sqrt{z}\left[I_{-\frac{1}{3}}(\zeta) - I_{\frac{1}{3}}(\zeta)\right] \\ &= \sum_{n=0}^\infty \frac{z^{3n}}{3^{2n+\frac{2}{3}}n!\,\Gamma\left(n+\frac{2}{3}\right)} - \sum_{n=0}^\infty \frac{z^{3n+1}}{3^{2n+\frac{4}{3}}n!\,\Gamma\left(n+\frac{4}{3}\right)},\end{aligned}$$

where

$$\zeta = \frac{2}{3}z^{\frac{3}{2}}. \tag{1}$$

A second real solution is

$$\begin{aligned}\mathrm{Bi}(z) &= \sqrt{\frac{z}{3}}\left[I_{-\frac{1}{3}}(\zeta) + I_{\frac{1}{3}}(\zeta)\right] \\ &= \sqrt{3}\left[\sum_{n=0}^\infty \frac{z^{3n}}{3^{2n+\frac{2}{3}}n!\,\Gamma\left(n+\frac{2}{3}\right)} + \sum_{n=0}^\infty \frac{z^{3n+1}}{3^{2n+\frac{4}{3}}n!\,\Gamma\left(n+\frac{4}{3}\right)}\right].\end{aligned}$$

Both series representations define entire functions; however, the representations in terms of modified Bessel functions are valid only in the sector $|\arg z| < \frac{2}{3}\pi$. Outside this sector the representations in terms of ordinary Bessel functions read:

$$\begin{aligned}\mathrm{Ai}(-z) &= \frac{1}{3}\sqrt{z}\left[J_{-\frac{1}{3}}(\zeta) + J_{\frac{1}{3}}(\zeta)\right], \\ \mathrm{Bi}(-z) &= \sqrt{\frac{z}{3}}\left[J_{-\frac{1}{3}}(\zeta) - J_{\frac{1}{3}}(\zeta)\right],\end{aligned}$$

again in the sector $|\arg z| < \frac{2}{3}\pi$. The quantity ζ is given in (1). Show that the Wronskian (see Exercise 9.1) for the pair $\mathrm{Ai}(z)$, $\mathrm{Bi}(z)$ is given by

$$W[\mathrm{Ai}, \mathrm{Bi}](z) = \frac{1}{\pi}.$$

Verify that from the asymptotic expansions (see (9.49) and (9.54)) of the modified Bessel functions it follows that:

$$\begin{aligned} \mathrm{Ai}(z) &= \frac{1}{2\sqrt{\pi}} z^{-\frac{1}{4}} e^{-\zeta} \left[1 + \mathcal{O}\left(z^{-\frac{3}{2}}\right)\right], \quad |\arg z| < \pi, \\ \mathrm{Bi}(z) &= \frac{1}{\sqrt{\pi}} z^{-\frac{1}{4}} e^{\zeta} \left[1 + \mathcal{O}\left(z^{-\frac{3}{2}}\right)\right], \quad |\arg z| < \frac{1}{3}\pi, \end{aligned} \qquad z \to \infty.$$

Verify that from the asymptotic expansion (see (9.55)) of the ordinary Bessel function it follows that:

$$\begin{aligned} \mathrm{Ai}(-z) &\sim \frac{1}{\sqrt{\pi}} z^{-\frac{1}{4}} \cos\left(\zeta - \frac{1}{4}\pi\right), \\ \mathrm{Bi}(-z) &\sim -\frac{1}{\sqrt{\pi}} z^{-\frac{1}{4}} \sin\left(\zeta - \frac{1}{4}\pi\right), \end{aligned} \qquad z \to \infty, \quad |\arg z| < \frac{2}{3}\pi,$$

where ζ is defined in (1).

In Figure 4.2 graphs of the Airy functions $\mathrm{Ai}(x), \mathrm{Bi}(x)$ are given.

10

Separating the Wave Equation

As explained in §4.1, the special functions of classical mathematical physics frequently arise when the *potential equation* $\Delta u = 0$, the *diffusion equation* $\Delta u = u_t$, or the *wave equation* $\Delta u = u_{tt}$ are separated with respect to the variables in, say, spherical or cylindrical coordinate systems. The symbol Δ is the *Laplace operator*, which in three dimensional space reads

$$\Delta u = \frac{\partial^2 u}{\partial x^2} + \frac{\partial^2 u}{\partial y^2} + \frac{\partial^2 u}{\partial z^2},$$

and with a similar form in spaces of other dimensions. The special functions treated in this book, in particular the functions from the Chapters 7, 8 and 9, play an important part in the construction of solutions of these equations. The time variable t in the diffusion and the wave equations, the so-called *evolution equations*, is often removed by using Fourier or Laplace transformations, or by introducing special solutions with a time dependent factor e^{ikt}. The result can then be written in terms of the *Helmholtz equation*

$$\left(\Delta + k^2\right) v = 0, \tag{10.1}$$

which is also called the *time independent wave equation*.

When (10.1) has to be solved in domains corresponding with interior or exterior parts of configurations such as spheres or cylinders, it is necessary to write the Helmholtz equation in terms of a different coordinate system (u, v, w), in place of the Cartesian system (x, y, z). Often, the new system is curvilinear and orthogonal. In the case of a sphere, for instance, one introduces *spherical coordinates* $u = r, v = \theta, w = \phi$, which will be described below.

First we give the general transformation formulas that the describe the new forms in terms of the new variables of, for instance, the Laplace operator, the divergence, the gradient, and the rotation of scalar or vector functions in three dimensions. Next we give special cases in which we obtain explicit forms of these quantities and see how the Helmholtz equation can be separated. In a final section we discuss in some detail two boundary value problems from mathematical physics (heat conduction in a cylinder and diffraction of a plane wave to a sphere) that can be solved in terms of special functions by separating the variables, and we give three exercises with further examples.

10.1. General Transformations

Let (u, v, w) be an orthogonal coordinate system related to the original Cartesian system (x, y, z) by the equations

$$x = x(u, v, w), \quad y = y(u, v, w), \quad z = z(u, v, w),$$

where we assume that the surfaces defined by $u = c_1, v = c_2, w = c_3$ are mutually orthogonal. In order to describe equation (10.1) (or a different equation) in terms of the new coordinate system it is useful to know how the *element of arc length* ds and the *element of volume* $d\tau$ are expressed in the new system. We have the following relations

$$(ds)^2 = (dx)^2 + (dy)^2 + (dz)^2 = \frac{1}{U^2}(du)^2 + \frac{1}{V^2}(dv)^2 + \frac{1}{W^2}(dw)^2,$$

$$d\tau = dx\,dy\,dz = \frac{1}{UVW}\,du\,dv\,dw,$$

while the new *elements of surface* read

$$\frac{du}{U}\frac{dv}{V}, \quad \frac{dv}{V}\frac{dw}{W}, \quad \frac{dw}{W}\frac{du}{U}.$$

The functions U, V, W are given by

$$\begin{aligned}
\frac{1}{U^2} &= \left(\frac{\partial x}{\partial u}\right)^2 + \left(\frac{\partial y}{\partial u}\right)^2 + \left(\frac{\partial z}{\partial u}\right)^2,\\
\frac{1}{V^2} &= \left(\frac{\partial x}{\partial v}\right)^2 + \left(\frac{\partial y}{\partial v}\right)^2 + \left(\frac{\partial z}{\partial v}\right)^2,\\
\frac{1}{W^2} &= \left(\frac{\partial x}{\partial w}\right)^2 + \left(\frac{\partial y}{\partial w}\right)^2 + \left(\frac{\partial z}{\partial w}\right)^2.
\end{aligned}$$

Now, let Φ and $\mathbf{F} = (F_x, F_y, F_z)$ be a scalar and a vector function, respectively (the subscripts denote components, not differentiations). Then the quantities

$$\nabla\Phi = \operatorname{grad}\Phi, \quad \Delta\Phi = \nabla^2\Phi = \operatorname{div}(\operatorname{grad}\Phi), \quad \nabla\cdot\mathbf{F} = \operatorname{div}\mathbf{F}, \quad \nabla\times\mathbf{F} = \operatorname{curl}\mathbf{F}$$

are represented, in terms of the new coordinates u, v, w, by

$$
\begin{aligned}
\nabla\Phi &= \left(U\frac{\partial\Phi}{\partial u}, V\frac{\partial\Phi}{\partial v}, W\frac{\partial\Phi}{\partial w}\right),\\
\Delta\Phi &= UVW\left[\frac{\partial}{\partial u}\left(\frac{U}{VW}\frac{\partial\Phi}{\partial u}\right)+\frac{\partial}{\partial v}\left(\frac{V}{UW}\frac{\partial\Phi}{\partial v}\right)+\frac{\partial}{\partial w}\left(\frac{W}{UV}\frac{\partial\Phi}{\partial w}\right)\right],\\
\nabla\cdot\mathbf{F} &= UVW\left[\frac{\partial}{\partial u}\left(\frac{F_u}{VW}\right)+\frac{\partial}{\partial v}\left(\frac{F_v}{UW}\right)+\frac{\partial}{\partial w}\left(\frac{F_w}{UV}\right)\right],\\
\nabla\times\mathbf{F} &= [(\nabla\times\mathbf{F})_u, (\nabla\times\mathbf{F})_v, (\nabla\times\mathbf{F})_w],
\end{aligned}
$$

with

$$
\begin{aligned}
(\nabla\times\mathbf{F})_u &= VW\left[\frac{\partial}{\partial v}\left(\frac{F_w}{W}\right)-\frac{\partial}{\partial w}\left(\frac{F_v}{V}\right)\right],\\
(\nabla\times\mathbf{F})_v &= UW\left[\frac{\partial}{\partial w}\left(\frac{F_u}{U}\right)-\frac{\partial}{\partial u}\left(\frac{F_w}{W}\right)\right],\\
(\nabla\times\mathbf{F})_w &= UV\left[\frac{\partial}{\partial u}\left(\frac{F_v}{V}\right)-\frac{\partial}{\partial v}\left(\frac{F_u}{U}\right)\right],
\end{aligned}
$$

where F_u, F_v, F_w denote the components of $\mathbf{F}$ in the new (u, v, w)system. The following relations express the transformation of the vector components:

$$
\begin{aligned}
F_u &= U\left(F_x\frac{\partial x}{\partial u}+F_y\frac{\partial y}{\partial u}+F_z\frac{\partial z}{\partial u}\right),\\
F_v &= V\left(F_x\frac{\partial x}{\partial v}+F_y\frac{\partial y}{\partial v}+F_z\frac{\partial z}{\partial v}\right),\\
F_w &= W\left(F_x\frac{\partial x}{\partial w}+F_y\frac{\partial y}{\partial w}+F_z\frac{\partial z}{\partial w}\right).
\end{aligned}
$$

10.2. Special Coordinate Systems

We give the transformation rules for a number of commonly used systems of coordinates and the standard functions which arise in the method of separating the variables for the Helmholtz equation.

10.2.1. Cylindrical Coordinates

$$
\begin{array}{lll}
x = r\cos\phi & u = r & U = 1\\
y = r\sin\phi & v = \phi & V = 1/r\\
z = z & w = z & W = 1
\end{array}
$$

This case can also be used in two (instead of three) dimensions, in order to treat the well-known polar coordinates; some of the following vector relations

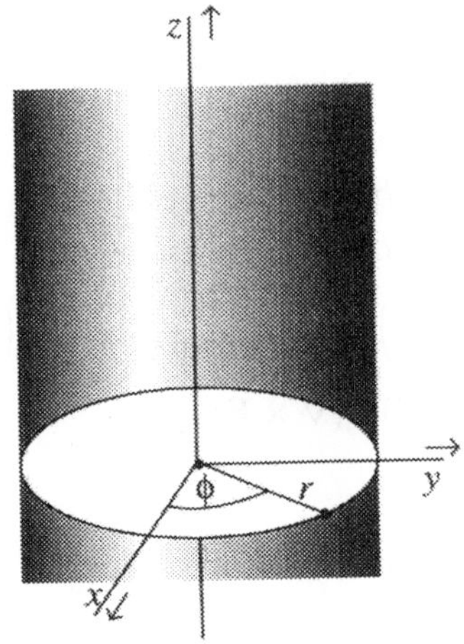

Figure 10.1. Cylindrical coordinates r, ϕ, z.

are not relevant in that case. The surfaces r = constant are right circular cylinders with the $z-$axis as axis of symmetry, and the surfaces $\phi =$ constant are planes through the $z-$axis. In Figure 10.1 we show the cylindrical coordinates.

The transformation rules are:

$$(ds)^2 = (dr)^2 + (rd\phi)^2 + (dz)^2,$$

$$\nabla\Phi = \left(\frac{\partial\Phi}{\partial r}, \frac{1}{r}\frac{\partial\Phi}{\partial\phi}, \frac{\partial\Phi}{\partial z}\right),$$

$$\nabla\cdot\mathbf{F} = \frac{1}{r}\frac{\partial}{\partial r}(rF_r) + \frac{1}{r}\frac{\partial F_\phi}{\partial\phi} + \frac{\partial F_z}{\partial z},$$

$$\Delta\Phi = \frac{\partial^2\Phi}{\partial r^2} + \frac{1}{r}\frac{\partial\Phi}{\partial r} + \frac{1}{r^2}\frac{\partial^2\Phi}{\partial\phi^2} + \frac{\partial^2\Phi}{\partial z^2},$$

$$(\nabla\times\mathbf{F})_r = \frac{1}{r}\left[\frac{\partial F_z}{\partial\phi} - \frac{\partial F_\phi}{\partial z}\right],$$

$$(\nabla\times\mathbf{F})_\phi = \frac{\partial F_r}{\partial z} - \frac{\partial F_z}{\partial r},$$

$$(\nabla\times\mathbf{F})_z = \frac{1}{r}\left[\frac{\partial}{\partial r}(rF_\phi) - \frac{\partial F_r}{\partial\phi}\right],$$

$$F_r = F_x\cos\phi + F_y\sin\phi,$$

$$F_\phi = -F_x\sin\phi + F_y\cos\phi,$$

$$F_z = F_z.$$

We write $\Phi(r, \phi, z) = f_1(r) f_2(\phi) f_3(z)$. Then the Helmholtz equation $\Delta\Phi + k^2\Phi = 0$ separates into the three equations:

$$f_1'' + \frac{1}{r} f_1' + \left(k^2 - \alpha^2 - \frac{\mu^2}{r^2}\right) f_1 = 0, \qquad f_1(r) = C_\mu\left(r\sqrt{k^2 - \alpha^2}\right),$$

$$f_2'' + \mu^2 f_2 = 0, \qquad f_2(\phi) = e^{\pm i\mu\phi},$$

$$f_3'' + \alpha^2 f_3 = 0, \qquad f_3(z) = e^{\pm i\alpha z}.$$

Here α and μ are arbitrary constants, the separation constants; f_1 is a Bessel function. The occurrence in this case of a Bessel function has given this function the name *cylinder function.*

10.2.2. Spherical Coordinates

$$\begin{array}{lll} x = r\sin\theta\cos\phi & u = r & U = 1 \\ y = r\sin\theta\sin\phi & v = \theta & V = 1/r \\ z = r\cos\theta & w = \phi & W = 1/(r\sin\theta) \end{array}$$

The surfaces r = constant are spheres and the surfaces ϕ = constant are half planes with the z-axis as boundary. The surfaces θ = constant are right circular cones with vertex at the origin and the z-axis as the axis of symmetry. In Figure 10.2 we show the spherical coordinates.

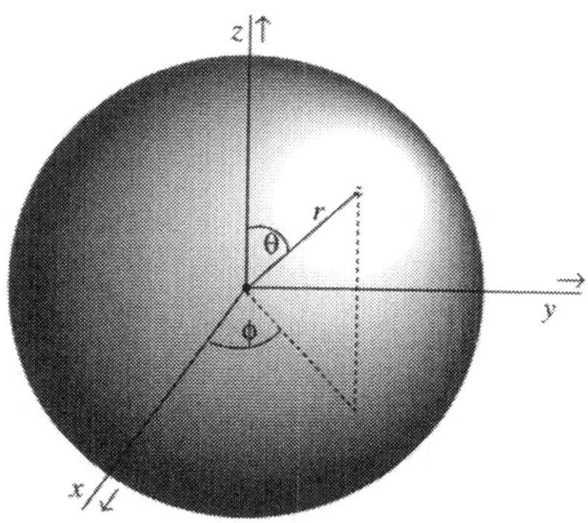

Figure 10.2. Spherical coordinates r, ϕ, θ.

The transformation rules are:

$$(ds)^2 = (dr)^2 + (rd\theta)^2 + (r\sin\theta d\phi)^2,$$

$$\nabla\Phi = \left(\frac{\partial\Phi}{\partial r}, \frac{1}{r}\frac{\partial\Phi}{\partial\theta}, \frac{1}{r\sin\theta}\frac{\partial\Phi}{\partial\phi}\right),$$

$$\nabla\cdot\mathbf{F} = \frac{1}{r^2}\frac{\partial}{\partial r}(r^2F_r) + \frac{1}{r\sin\theta}\frac{\partial}{\partial\theta}(\sin\theta F_\theta) + \frac{1}{r\sin\theta}\frac{\partial F_\phi}{\partial\phi},$$

$$\Delta\Phi = \frac{1}{r^2}\frac{\partial}{\partial r}\left(r^2\frac{\partial\Phi}{\partial r}\right) + \frac{1}{r^2\sin\theta}\frac{\partial}{\partial\theta}\left(\sin\theta\frac{\partial\Phi}{\partial\theta}\right) + \frac{1}{r^2\sin^2\theta}\frac{\partial^2\Phi}{\partial\phi^2},$$

$$(\nabla\times\mathbf{F})_r = \frac{1}{r\sin\theta}\left[\frac{\partial}{\partial\theta}(\sin\theta F_\phi) - \frac{\partial F_\theta}{\partial\phi}\right],$$

$$(\nabla\times\mathbf{F})_\theta = \frac{1}{r\sin\theta}\left[\frac{\partial F_r}{\partial\phi} - \sin\theta\frac{\partial}{\partial r}(rF_\phi)\right],$$

$$(\nabla\times\mathbf{F})_\phi = \frac{1}{r}\left[\frac{\partial}{\partial r}(rF_\theta) - \frac{\partial F_r}{\partial\phi}\right],$$

$$F_r = F_x\sin\theta\cos\phi + F_y\sin\theta\sin\phi + F_z\cos\theta,$$

$$F_\theta = F_x\cos\theta\cos\phi + F_y\cos\theta\sin\phi - F_z\sin\theta,$$

$$F_\phi = -F_x\sin\phi + F_y\cos\phi.$$

We write $\Phi(r,\theta,\phi) = f_1(r)f_2(\theta)f_3(\phi)$. Then the Helmholtz equation $\Delta\Phi + k^2\Phi = 0$ separates into the three equations:

$$f_1'' + \frac{2}{r}f_1' + \left[k^2 - \frac{\nu(\nu+1)}{r^2}\right]f_1 = 0, \qquad f_1(r) = \frac{1}{\sqrt{r}}C_{\nu+\frac{1}{2}}(kr),$$

$$\sin^2\theta f_2'' + \sin\theta\cos\theta f_2' + \left[\nu(\nu+1)\sin^2\theta - \mu^2\right]f_2 = 0,$$
$$f_2(\theta) = P_\nu^\mu(\cos\theta),$$

$$f_3'' + \mu^2 f_3 = 0, \qquad f_3(\phi) = e^{\pm i\mu\phi}.$$

Here ν and μ are the separation constants; f_1 is a *cylinder function*, or *Bessel function* and f_2 is a *Legendre function.* Often ν and μ are integers. In that case f_3 becomes a periodic function of ϕ and the Bessel function in f_1 is of 'half odd integer' order; such Bessel functions are called *spherical Bessel function.* Also, the Legendre functions become simple functions in that case. If $k = 0$ the Helmholtz equation reduces to the *potential equation.* In that case the function f_1 is given by

$$f_1'' + \frac{2}{r}f_1' - \frac{\nu(\nu+1)}{r^2}f_1 = 0,$$

with solutions $f_1(r) = r^\nu$ or $f_1(r) = r^{-\nu-1}$.

10.2.3. Elliptic Cylinder Coordinates

$$\begin{aligned} x &= c\cosh\xi\cos\eta & u &= \xi & U &= 1/(c\tau) \\ y &= c\sinh\xi\sin\eta & v &= \eta & V &= 1/(c\tau) \\ z &= z & w &= z & W &= 1 \end{aligned}$$

where $\tau = \sqrt{\sinh^2\xi + \sin^2\eta}$. The results can also be used in two dimensions for treating elliptic coordinates; some of the following vector relations are then not relevant. The domain of the new parameters is given by:

$$0 \le \xi < \infty, \quad 0 \le \eta \le 2\pi, \quad -\infty < z < \infty.$$

The surfaces $\xi = \xi_0$ (constant) are right elliptic cylinders with generators parallel to the z-axis. The cylinders have semi-axes $c\cosh\xi_0$, $c\sinh\xi_0$ and they are defined by the equation

$$\frac{x^2}{c^2\cosh^2\xi_0} + \frac{y^2}{c^2\sinh^2\xi_0} = 1.$$

The surfaces $\eta = \eta_0$ (constant) are hyperbolic cylinders with generators parallel to the z-axis and defined by the equation

$$\frac{x^2}{c^2\cos^2\eta_0} - \frac{y^2}{c^2\sin^2\eta_0} = 1.$$

In Figure 10.3 we show the elliptic cylinder coordinates in the $z = 0$ plane.

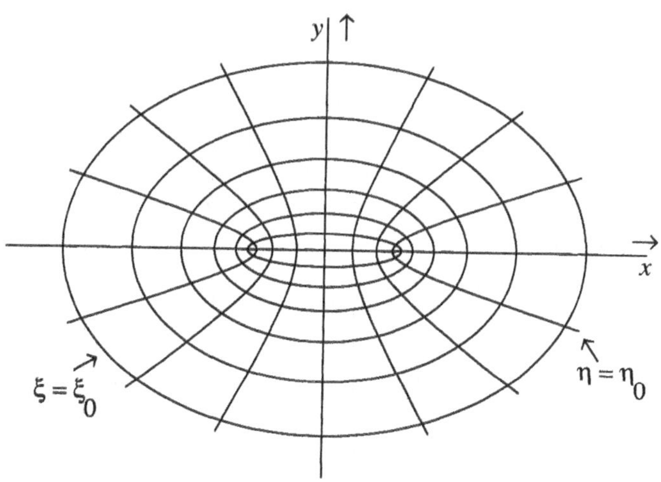

Figure 10.3. Elliptic cylinder coordinates ($z = 0$).

The transformation rules are:

$$(ds)^2 = c^2\tau^2[(d\xi)^2 + (d\eta)^2] + (dz)^2,$$

$$\nabla\Phi = \left(\frac{1}{c\tau}\frac{\partial\Phi}{\partial\xi}, \frac{1}{c\tau}\frac{\partial\Phi}{\partial\eta}, \frac{\partial\Phi}{\partial z}\right),$$

$$\nabla\cdot\mathbf{F} = \frac{1}{c\tau^2}\left[\frac{\partial}{\partial\xi}(\tau F_\xi) + \frac{\partial}{\partial\eta}(\tau F_\eta) + c\tau^2\frac{\partial F_z}{\partial z}\right],$$

$$\Delta\Phi = \frac{1}{c^2\tau^2}\left(\frac{\partial^2\Phi}{\partial\xi^2} + \frac{\partial^2\Phi}{\partial\eta^2}\right) + \frac{\partial^2\Phi}{\partial z^2},$$

$$(\nabla\times\mathbf{F})_\xi = \frac{1}{c\tau}\left(\frac{\partial F_z}{\partial\eta} - c\tau\frac{\partial F_\eta}{\partial z}\right),$$

$$(\nabla\times\mathbf{F})_\eta = \frac{1}{c\tau}\left(c\tau\frac{\partial F_\xi}{\partial z} - \frac{\partial F_z}{\partial\xi}\right),$$

$$(\nabla\times\mathbf{F})_z = \frac{1}{c\tau}\left(\frac{\partial\tau F_\eta}{\partial\xi} - \frac{\partial\tau F_\xi}{\partial\eta}\right),$$

$$F_\xi = \frac{1}{\tau}[F_x\sinh\xi\cos\eta + F_y\cosh\xi\sin\eta],$$

$$F_\eta = \frac{1}{\tau}[-F_x\cosh\xi\sin\eta + F_y\sinh\xi\cos\eta],$$

$$F_z = F_z.$$

We write $\Phi(\xi,\eta,z) = f_1(\xi)f_2(\eta)f_3(z)$. Then the Helmholtz equation $\Delta\Phi + k^2\Phi = 0$ separates into the three equations:

$$f_1'' + \left[-\lambda + \tfrac{1}{2}c^2\left(k^2-\alpha^2\right)\cosh 2\xi\right]f_1 = 0,$$

$$f_2'' + \left[\lambda - \tfrac{1}{2}c^2\left(k^2-\alpha^2\right)\cos 2\eta\right]f_2 = 0,$$

$$f_3'' + \alpha^2 f_3 = 0, \quad f_3(z) = e^{\pm i\alpha z}.$$

Here α and λ are the separation constants. The differential equations for f_1 and f_2 are called *Mathieu equations* and the solutions *Mathieu functions*. The typical form is that for f_2 because of the periodic function $\cos 2\eta$; the equation for f_1 follows from that for f_2 by writing $i\xi$ in place of η. The standard form of a Mathieu equation is

$$\frac{d^2f}{dx^2} + (a - 2q\cos 2x)f = 0.$$

10.2.4. Parabolic Cylinder Coordinates

$x = \frac{1}{2}(\xi^2 - \eta^2)$	$u = \xi$	$U = 1/\rho$
$y = \xi\eta$	$v = \eta$	$V = 1/\rho$
$z = z$	$w = z$	$W = 1$

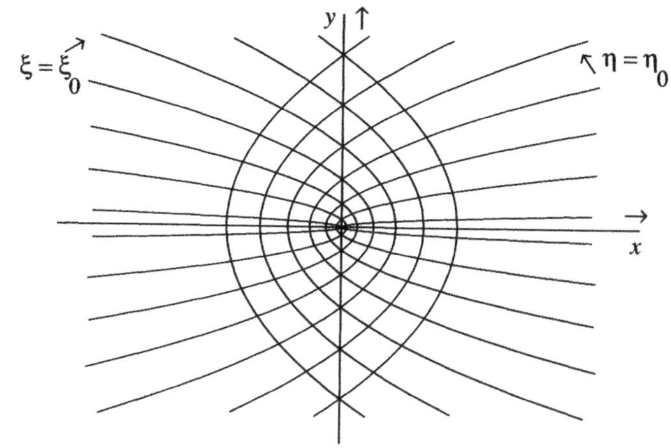

Figure 10.4. Parabolic cylinder coordinates ($z = 0$).

where $\rho = \sqrt{\xi^2 + \eta^2}$. This case is described in the (x, y)–plane by the conformal mapping $x + iy = \frac{1}{2}(\xi + i\eta)^2$. The lines $\xi = \xi_0$ and $\eta = \eta_0$ correspond with mutual orthogonal parabolas, of which the x–axis is the axis of symmetry and the origin is the focal point. Taking into account the z–variable, we see parabolic cylinders with generators parallel to the z–axis. In Figure 10.4 we show the parabolic cylinder coordinates in the $z = 0$ plane.

The transformation rules are:

$$(ds)^2 = \rho^2[(d\xi)^2 + (d\eta)^2] + (dz)^2,$$

$$\nabla\Phi = \left(\frac{1}{\rho}\frac{\partial\Phi}{\partial\xi}, \frac{1}{\rho}\frac{\partial\Phi}{\partial\eta}, \frac{\partial\Phi}{\partial z}\right),$$

$$\nabla \cdot \mathbf{F} = \frac{1}{\rho^2}\left[\frac{\partial}{\partial\xi}(\rho F_\xi) + \frac{\partial}{\partial\eta}(\rho F_\eta) + \rho^2\frac{\partial F_z}{\partial z}\right],$$

$$\Delta\Phi = \frac{1}{\rho^2}\left(\frac{\partial^2\Phi}{\partial\xi^2} + \frac{\partial^2\Phi}{\partial\eta^2}\right) + \frac{\partial^2\Phi}{\partial z^2},$$

$$(\nabla \times \mathbf{F})_\xi = \frac{1}{\rho}\left(\frac{\partial F_z}{\partial\eta} - \rho\frac{\partial F_\eta}{\partial z}\right),$$

$$(\nabla \times \mathbf{F})_\eta = \frac{1}{\rho}\left(\rho\frac{\partial F_\xi}{\partial z} - \frac{\partial F_z}{\partial\xi}\right),$$

$$(\nabla \times \mathbf{F})_z = \frac{1}{\rho^2}\left[\frac{\partial}{\partial\xi}(\rho F_\eta) - \frac{\partial}{\partial\eta}(\rho F_\xi)\right],$$

$$F_\xi = \frac{1}{\rho}(\xi F_x + \eta F_y), \quad F_\eta = \frac{1}{\rho}(\xi F_y - \eta F_x), \quad F_z = F_z.$$

We write $\Phi(\xi, \eta, z) = f_1(\xi) f_2(\eta) f_3(z)$. Then the Helmholtz equation $\Delta\Phi + k^2\Phi = 0$ separates into the three equations:

$$f_1'' + \left(\lambda + l^2\xi^2\right) f_1 = 0, \qquad l^2 = k^2 - \alpha^2,$$

$$f_2'' + \left(-\lambda + l^2\eta^2\right) f_2 = 0, \quad l^2 = k^2 - \alpha^2,$$

$$f_3'' + \alpha^2 f_3 = 0, \qquad\qquad f_3(z) = e^{\pm i\alpha z}.$$

Here α and λ are the separation constants. The functions f_1 and f_2 can be expressed in terms of *Weber functions*, or *parabolic cylinder functions*. Such functions are special cases of the *confluent hypergeometric functions*, see (7.21). One can take the following solutions:

$$f_1(\xi) = U\left[-\frac{1}{2} + (l + i\lambda)/(2l), \pm\xi\sqrt{l}\,(1+i)\right],$$
$$f_2(\eta) = U\left[-\frac{1}{2} + (l - i\lambda)/(2l), \pm\eta\sqrt{l}\,(1+i)\right].$$

10.2.5. Oblate Spheroidal Coordinates

$$\begin{array}{lll} x = c\tau\cos\phi & u = \xi & U = \gamma/(c\rho) \\ y = c\tau\sin\phi & v = \eta & V = \delta/(c\rho) \\ z = c\xi\eta & w = \phi & W = 1/(c\gamma\delta) \end{array}$$

where

$$\rho = \sqrt{\xi^2 + \eta^2}, \quad \gamma = \sqrt{1 + \xi^2}, \quad \delta = \sqrt{1 - \eta^2}, \quad \tau = \gamma\,\delta.$$

The domain of the parameters is

$$0 \le \xi < \infty, \quad -1 \le \eta \le 1, \quad 0 \le \phi \le 2\pi.$$

The surfaces $\xi = \xi_0$ (constant) are ellipsoids of revolution around the $z-$axis and are given by

$$\frac{x^2 + y^2}{c^2\left(1 + \xi_0^2\right)} + \frac{z^2}{c^2\xi_0^2} = 1.$$

The surfaces $\eta = \eta_0$ (constant) are hyperboloids and are given by

$$\frac{x^2 + y^2}{c^2\left(1 - \eta_0^2\right)} - \frac{z^2}{c^2\eta_0^2} = 1.$$

The surfaces $\phi = \phi_0$ (constant) are half planes through the $z-$axis and are defined by $y = x\tan\phi_0$.

In Figure 10.5 we show an oblate ellipsoid of revolution ($\xi = \xi_0$) that intersects a hyperboloid of revolution ($\eta = \eta_0$).

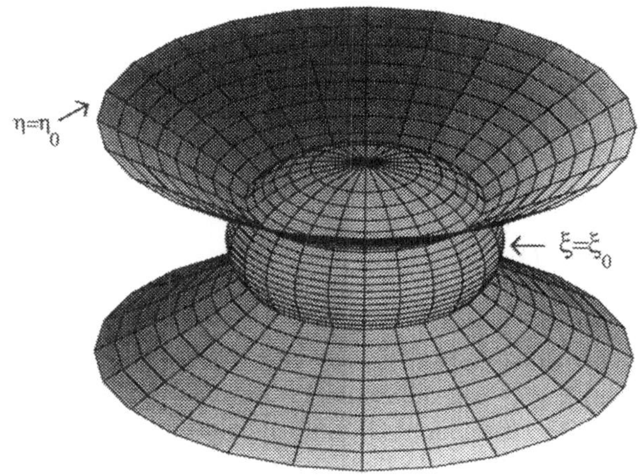

Figure 10.5. An oblate ellipsoid of revolution ($\xi = \xi_0$) intersecting a hyperboloid of revolution ($\eta = \eta_0$).

The transformation rules are:

$$(ds)^2 = c^2\rho^2\left[\frac{1}{\gamma}(d\xi)^2 + \frac{1}{\delta}(d\eta)^2\right] + c^2\tau^2(d\phi)^2,$$

$$\nabla\Phi = \left(\frac{\gamma}{c\rho}\frac{\partial\Phi}{\partial\xi}, \frac{\delta}{c\rho}\frac{\partial\Phi}{\partial\eta}, \frac{1}{c\tau}\frac{\partial\Phi}{\partial\phi}\right),$$

$$\nabla\cdot\mathbf{F} = \frac{1}{c\rho^2}\left[\frac{\partial}{\partial\xi}(\gamma\rho F_\xi) + \frac{\partial}{\partial\eta}(\delta\rho F_\eta) + \frac{\xi^2-\eta^2}{\tau}\frac{\partial F_\phi}{\partial\phi}\right],$$

$$\Delta\Phi = \frac{1}{c^2\rho^2}\left[\frac{\partial}{\partial\xi}\left(\gamma^2\frac{\partial\Phi}{\partial\xi}\right) + \frac{\partial}{\partial\eta}\left(\delta^2\frac{\partial\Phi}{\partial\eta}\right) + \frac{\rho^2}{\tau^2}\frac{\partial^2\Phi}{\partial\phi^2}\right],$$

$$(\nabla\times\mathbf{F})_\xi = \frac{1}{c\rho\gamma}\left[\frac{\partial}{\partial\eta}(\tau F_\phi) - \frac{\rho}{\delta}\frac{\partial F_\eta}{\partial\phi}\right],$$

$$(\nabla\times\mathbf{F})_\eta = \frac{1}{c\delta\rho}\left[\frac{\rho}{\gamma}\frac{\partial F_\xi}{\partial\phi} - \frac{\partial}{\partial\xi}(\tau F_\phi)\right],$$

$$(\nabla\times\mathbf{F})_\phi = \frac{\tau}{c\rho^2}\left[\frac{1}{\delta}\frac{\partial}{\partial\xi}(\rho F_\eta) - \frac{1}{\gamma}\frac{\partial}{\partial\eta}(\rho F_\xi)\right],$$

$$F_\xi = \frac{1}{\rho}[\xi\delta(F_x\cos\phi + F_y\sin\phi) + \eta\gamma F_z],$$

$$F_\eta = \frac{1}{\rho}[-\eta\gamma(F_x\cos\phi + F_y\sin\phi) + \xi\delta F_z],$$

$$F_\phi = -F_x\sin\phi + F_y\cos\phi.$$

We write $\Phi(\xi,\eta,\phi) = f_1(\xi)f_2(\eta)f_3(\phi)$. Then the Helmholtz equation $\Delta\Phi + k^2\Phi = 0$ separates into the three equations:

$$(1+\xi^2)f_1'' + 2\xi f_1' + \left[-\lambda + k^2c^2\xi^2 + \frac{\mu^2}{1+\xi^2}\right] f_1 = 0,$$

$$(1-\eta^2)f_2'' - 2\eta f_2' + \left[+\lambda + k^2c^2\eta^2 - \frac{\mu^2}{1-\eta^2}\right] f_2 = 0,$$

$$f_3'' + \mu^2 f_3 = 0, \quad f_3(z) = e^{\pm i\mu\phi}.$$

Here λ and μ are the separation constants. For the Laplace equation, that is, $k = 0$, the functions f_1 and f_2 can be expressed in terms of the *Legendre functions*. When $k \neq 0$, the differential equations for f_1 and f_2 are called the Lamé differential equations and the solutions are called *Lamé functions*. See BATEMAN PROJECT (1953, Vol. 3).

10.3. Boundary Value Problems

By means of examples of boundary value problems from mathematical physics we demonstrate in this section the method of separating the variables. Many properties of the resulting special functions are given in the previous chapters.

10.3.1. Heat Conduction in a Cylinder

We consider a cylinder described by the cylindrical coordinates r, ϕ, z with domain $0 \le r \le 1$, $0 \le \phi < 2\pi$, $-\infty < z < \infty$. Let $u(r,\phi,z,t)$ be the temperature at (r,ϕ,z) at time t. Then u satisfies the *heat conduction equation*

$$\Delta u = \frac{\partial u}{\partial t}. \tag{10.2}$$

The boundary of the cylinder is cooled and kept at constant temperature $u = 0$. Furthermore, the initial temperature is taken as $u(r,\phi,z,0) = \Psi(r)$. From these data it follows that the solution will show symmetry: u will not depend on ϕ and z. Therefore we write $u(r,\phi,z,t) = u(r,t)$ and we obtain the following *boundary value problem* for $u(r,t)$ (see §10.2.1):

$$\frac{\partial^2 u}{\partial r^2} + \frac{1}{r}\frac{\partial u}{\partial r} = \frac{\partial u}{\partial t}, \quad 0 < r < 1, \quad t > 0, \tag{10.3}$$

$$u(1,t) = 0, \quad t > 0, \quad \text{boundary condition,} \tag{10.4}$$

$$u(r,0) = \Psi(r), \quad 0 \le r \le 1, \quad \text{initial condition.} \tag{10.5}$$

We observe that the differential equation (10.3) and the boundary condition (10.4) are homogeneous and linear. This means that, if the functions u_1 and

u_2 satisfy the differential equation and the boundary condition, any linear combination $Au_1 + Bu_2$ also satisfies the equation and the condition (the principle of *linear superposition*). The initial condition (10.5) does not have this property; this condition is not homogeneous. In the methods of solving equations like (10.2) homogeneous and inhomogeneous conditions may play an essentially different part.

By writing $u(r,t) = f(r)g(t)$, separation of the variables gives the relation

$$\frac{f''}{f} + \frac{1}{r}\frac{f'}{f} = \frac{g'}{g}.$$

The left-hand side is a function of r while the right-hand side is a function of z. This is only possible when both sides are equal to a constant λ, say. Hence we arrive at the equations

$$f'' + \frac{1}{r}f' - \lambda f = 0, \quad g' - \lambda g = 0. \tag{10.6}$$

We assume that $f(0)$ is finite. Furthermore, it follows from (10.4) that it is convenient to prescribe $f(1) = 0$. Taking $\lambda = 0$ we obtain $f(r) = A + B\ln r$. But with this solution we cannot have $f(1) = 0$ and a finite value $f(0)$, unless $f = 0$. Next we try $\lambda = -p^2$. Then we find the following Bessel function solution of the first equation in (10.6):

$$f(r) = AJ_0(pr) + BY_0(pr).$$

Since $f(0)$ is finite we must take $B = 0$. The condition $f(1) = 0$ now reads $J_0(p) = 0$; hence p is one of the zeros of the Bessel function $J_0(x)$: $p = j_n$, the n–th positive zero (see §9.8). For the function g in (10.6) we find the solution $g(t) = e^{\lambda t}$. As the materials for building the final solution we thus find the functions

$$u_n(r,t) = J_0(j_n r)e^{-j_n^2 t}, \quad n = 1, 2, 3, \ldots$$

and we combine these functions in the form of a trial solution (linear superposition)

$$u(r,t) = \sum_{n=1}^{\infty} c_n J_0(j_n r)e^{-j_n^2 t}, \tag{10.7}$$

where the coefficients c_n are still to be determined. This series formally satisfies the differential equation in (10.3) and the boundary condition in (10.4). Next we determine the coefficients c_n such that (10.7) also satisfies (10.5):

$$u(r,0) = \sum_{n=1}^{\infty} c_n J_0(j_n r) = \Psi(r), \quad 0 \le r < 1. \tag{10.8}$$

Hence, the series in (10.8) is a the Fourier-Bessel expansion of the function Ψ. By using (9.66) the coefficients c_n can be represented as follows:

$$c_n = \frac{2}{J_1^2(j_n)} \int_0^1 r\Psi(r) J_0(j_n r)\, dr. \tag{10.9}$$

The solution $u(r, t)$ of the boundary value problem (10.3)-(10.5) has been found in the form (10.7) with coefficients given by (10.9). After this we have to verify that

(*i*) the series indeed converges and can be differentiated termwise with respect to r and t;

(*ii*) the function $\Psi(r)$ equals the sum of the Fourier-Bessel series; see the conditions formulated after (9.66).

10.3.2. Diffraction of a Plane Wave Due to a Sphere

We consider spherical coordinates (r, θ, ϕ), as introduced in §10.2.2, and the surface S of a sphere given by

$$r = 1, \quad 0 \le \theta \le \pi, \quad 0 \le \phi \le 2\pi.$$

The sphere is hit by a plane scalar wave

$$u_0 = e^{ikz - i\omega t} = e^{ikr\cos\theta} e^{-i\omega t}, \tag{10.10}$$

which propagates in the direction of the positive $z-$axis. In (10.10) we write $k = \omega/c$, where ω and c represent the frequency and the velocity of the wave. The incoming wave u_0 will be diffracted by the sphere. We call the diffracted wave $u = v(r, \theta)e^{-i\omega t}$; observe that the problem has some symmetry and that, hence, u will not depend on ϕ. We now can formulate the following boundary value problem for the function $v(r, \theta)$:

$$\begin{aligned} \Delta v + k^2 v = \frac{\partial^2 v}{\partial r^2} + \frac{2}{r}\frac{\partial v}{\partial r} + \frac{1}{r^2}\frac{\partial^2 v}{\partial \theta^2} + \frac{\cos\theta}{r^2 \sin\theta}\frac{\partial v}{\partial \theta} + k^2 v = 0,& \\ r > 1, \quad 0 \le \theta \le \pi,& \\ v(1, \theta) = -e^{ik\cos\theta}, \quad 0 \le \theta \le \pi,& \\ v(r, \theta) \sim A(\theta)\frac{e^{ikr}}{r}, \quad r \to \infty, \quad 0 \le \theta \le \pi.& \end{aligned} \tag{10.11}$$

Explanation: the boundary condition for $v(1, \theta)$ tells us that at the surface S the total wave satisfies $u + u_0 = 0$. The remaining condition is called the *Sommerfeld radiation condition*; this says that the diffracted wave should behave like an outgoing wave (emitted from S), as $r \to \infty$.

We write $v(r,\theta) = f(r)g(\theta)$. Then, as mentioned in §10.2.2, we obtain for f and g the differential equations of Bessel and Legendre, respectively. On account of the desired regularity of the function g at $\theta = 0$ and $\theta = \pi$ we choose $g(\theta) = P_n(\cos\theta)$, $n = 0, 1, 2, \ldots$. For f we choose Hankel functions:

$$f(r) = \frac{A}{\sqrt{r}} H^{(1)}_{n+\frac{1}{2}}(kr) + \frac{B}{\sqrt{r}} H^{(2)}_{n+\frac{1}{2}}(kr).$$

On account of the radiation condition in (10.11) and the asymptotic behavior of the Hankel functions we must take $B = 0$. In this way we arrive at the following series representation:

$$v(r,\theta) = \frac{1}{\sqrt{r}} \sum_{n=0}^{\infty} A_n H^{(1)}_{n+\frac{1}{2}}(kr) P_n(\cos\theta). \tag{10.12}$$

This solution formally satisfies the differential equation in (10.11) and also the radiation condition. The coefficients A_n are determined such that this solution also satisfies the boundary condition at the surface S, that is, for $r = 1$. This gives the relation

$$v(1,\theta) = \sum_{n=0}^{\infty} A_n H^{(1)}_{n+\frac{1}{2}}(k) P_n(\cos\theta) = -e^{ik\cos\theta}, \quad 0 \le \theta \le \pi.$$

Comparing this with the series in (6.64), we conclude that

$$A_n = -\sqrt{\frac{\pi}{2k}}\, (2n+1)\, i^n\, \frac{J_{n+\frac{1}{2}}(k)}{H^{(1)}_{n+\frac{1}{2}}(k)}.$$

On account of the asymptotic behavior of the Bessel functions given in (9.13) we infer that, with the present coefficients A_n, the series (10.12) converges in a suitable way, and that, hence, the right-hand side of (10.12) is a solution of the diffraction problem.

10.4. Remarks and Comments for Further Reading

10.1. Extensive use of the method of separation of variables can be found in any book on classical mathematical physics. See for instance the classics Morse & Feshbach (1953) and Carslaw & Jaeger (1959). Collections of exercises can be found in Budak et al. (1964) and Lebedev et al. (1965).

10.5. Exercises and Further Examples

10.1. Consider the Dirichlet problem for the interior of a sphere:

$$\begin{aligned} \Delta u &= 0, \quad r < 1, \quad 0 \le \theta \le \pi, \quad 0 \le \phi \le 2\pi, \\ u &= f, \quad r = 1, \quad 0 \le \theta \le \pi, \quad 0 \le \phi \le 2\pi, \end{aligned}$$

where $f = f(\theta, \phi)$ is a given function on the unit sphere. Verify that a formal solution can be written in the form

$$u(r, \theta, \phi) = \sum_{n=0}^{\infty} \sum_{m=0}^{n} r^n P_n^m(\cos\theta)[A_{m,n}\cos(m\phi) + B_{m,n}\sin(m\phi)],$$

where the associated Legendre functions of integer values of order and degree (and $m \le n$) are chosen on account of regularity. Verify that, by using Exercise 8.6 and the well-known relations for Fourier coefficients, that the values of $A_{m,n}$ and $B_{m,n}$ follow from

$$A_{m,n} = \frac{n+\frac{1}{2}}{\eta_m \pi} \frac{(n-m)!}{(n+m)!} \int_0^{\pi} \int_0^{2\pi} f(\theta, \phi) P_n^m(\cos\theta)\cos(m\phi)\sin\theta \, d\theta \, d\phi,$$

$$B_{m,n} = \frac{n+\frac{1}{2}}{\pi} \frac{(n-m)!}{(n+m)!} \int_0^{\pi} \int_0^{2\pi} f(\theta, \phi) P_n^m(\cos\theta)\sin(m\phi)\sin\theta \, d\theta \, d\phi,$$

where $n = 0, 1, 2, \ldots; m = 0, 1, 2, \ldots, n$ and $\eta_0 = 2$, $\eta_m = 1$ when $m \ge 1$.

10.2. At the point P (with spherical coordinates $r = a < 1, \theta = 0$) an electrical dipole is located with dipole moment 1 directed parallel to the positive x-axis. The dipole is situated inside the grounded surface of a sphere S given by $r = 1$. Determine the potential $V(r, \theta, \phi)$ of the electrical field inside S. The potential of the field due to the dipole is given by

$$V_d(r, \theta, \phi) = -\frac{1}{4\pi}\frac{\partial}{\partial x}\left(\frac{1}{R}\right) = -\frac{1}{4\pi}\frac{\partial}{\partial x}\frac{1}{\sqrt{r^2 + a^2 - 2ar\cos\theta}},$$

where R is the distance between the point (r, θ, ϕ) and P. Show that on the surface S:

$$V_d(1, \theta, \phi) = -\frac{1}{4\pi}\sum_{n=0}^{\infty} a^n P_{n+1}^1(\cos\theta)\cos\phi.$$

The required potential can be written as

$$V(r, \theta, \phi) = V_d(r, \theta, \phi) + u(r, \theta, \phi),$$

where the function $u(r, \theta, \phi)$ will be a solution of the boundary value problem

$$\begin{aligned} \Delta u &= 0, \quad 0 < r < 1, \quad 0 \le \theta \le \pi, \quad 0 \le \phi < 2\pi, \\ u(1, \theta, \phi) &= -V_d(1, \theta, \phi), \quad 0 \le \theta \le \pi, \quad 0 \le \phi < 2\pi. \end{aligned}$$

10.3. A circular disc with radius 1 is given by $x^2 + y^2 \le 1$, $z = 0$. The disc is an electrical conductor with potential V maintained at the value $V = 1$. The potential in space is the solution of the boundary value problem:

$$\begin{aligned} &\Delta V = 0 && \text{outside the disc,} \\ &V = 1 && \text{on the disc,} \\ &V(x,y,z) \to 0 \quad \text{as} && x^2 + y^2 + z^2 \to \infty. \end{aligned}$$

Determine V in space and the current density on the disc (that is, $-\partial V/\partial z$ evaluated at $z = 0$).

Hint. Use oblate spheroidal coordinates introduced in §10.2.5. That is, introduce ξ, η, ϕ, which correspond to x, y, z in the following way:

$$x = \sqrt{(\xi^2+1)(1-\eta^2)}\,\cos\phi, \quad y = \sqrt{(\xi^2+1)(1-\eta^2)}\,\sin\phi, \quad z = \xi\eta, \quad (1)$$

where $\xi \ge 0$, $-1 \le \eta \le 1$, $0 \le \phi < 2\pi$. Investigate the surfaces in x, y, z–space corresponding to the coordinate planes in ξ, η, ϕ–space given by $\xi =$ constant, $\eta =$ constant, $\phi =$ constant. Verify that the disc is described by $\xi = 0$. Use the method of separation of variables to obtain the solution. Explain the following choice of the solution, after separating the variables: $V = c_0 Q_0(i\xi)$, with $c_0 = 1/Q_0(+i0)$, see (8.22). Verify that this solution reduces to $V = 1 - \frac{2}{\pi}\arctan(\xi) = \frac{2}{\pi}\arcsin(1/\sqrt{1+\xi^2}\,)$, and that it satisfies all conditions. Translate this answer to x, y, z coordinates; use the relations in (1) to show that $u = 1+\xi^2$ is one of the solutions of the quadratic equation $u^2 - (1 + r^2 + z^2)u + r^2 = 0$ where $r^2 = x^2 + y^2$. Finally, when $z = 0$, $r \ge 1$, we have $V = \frac{2}{\pi}\arcsin(1/r)$, and the current density on the disc is given by

$$j(r) = \frac{2}{\pi}\,\frac{1}{\sqrt{1-r^2}}.$$

11

Special Statistical Distribution Functions

In this chapter we consider several statistical distribution functions that have relations with special functions mentioned in earlier chapters. In particular, we consider error functions, which are related to the normal (Gaussian) distribution, the incomplete gamma functions, which are related to the gamma distribution (or χ^2–distribution), the incomplete beta function, which is related to the beta distribution, with as special case Student's distribution, and the non-central χ^2–distribution.

11.1. Error Functions

The *error functions* are defined by

$$\boxed{\operatorname{erf} z = \frac{2}{\sqrt{\pi}} \int_0^z e^{-t^2}\, dt, \quad \operatorname{erfc} z = 1 - \operatorname{erf} z = \frac{2}{\sqrt{\pi}} \int_z^\infty e^{-t^2}\, dt.}$$

These functions are used in statistics and probability theory as the *normal distribution functions*, with somewhat different notation. For instance, let

$$\boxed{Z(x) = \frac{1}{\sqrt{2\pi}} e^{-x^2/2}, \quad P(x) = \int_{-\infty}^x Z(t)\, dt, \quad Q(x) = \int_x^\infty Z(t)\, dt.}$$

Then it is an easy exercise to show that

$$\boxed{P(x) = \tfrac{1}{2}\operatorname{erfc}\left(-x/\sqrt{2}\right), \quad Q(x) = \tfrac{1}{2}\operatorname{erfc}\left(x/\sqrt{2}\right).}$$

In physics the *plasma dispersion function* is used. The definition is

$$\boxed{w(z) = \frac{1}{i\pi} \int_{-\infty}^\infty \frac{e^{-t^2}}{t - z}\, dt, \quad \Im z > 0.} \qquad (11.1)$$

For $\Im z \le 0$ the function $w(z)$ is defined by analytic continuation, or by lowering the path of integration in (11.1). To be more precise, let c be a real number less than $\Im z$. Then, in particular when $\Im z \le 0$, (11.1) can be written as

$$w(z) = \frac{1}{i\pi}\int_{-\infty+ic}^{+\infty+ic} \frac{e^{-t^2}}{t-z}\,dt, \quad c < \Im z.$$

When $\Im z < 0$ we can again integrate along the real axis, by shifting the line of integration upwards, across the pole and picking up the residue:

$$w(z) = 2e^{-z^2} + \frac{1}{i\pi}\int_{-\infty}^{+\infty} \frac{e^{-t^2}}{t-z}\,dt, \quad \Im z < 0.$$

Just as the error functions, w is an entire function. It is not difficult to verify that

$$\boxed{w(z) = e^{-z^2}\operatorname{erfc}(-iz).}$$

By using analytic continuation it follows that this relation holds without restriction on $z \in \mathbb{C}$. We have the symmetry relations

$$\boxed{\operatorname{erf}(-z) = -\operatorname{erf} z, \quad \operatorname{erfc}(-z) = 2 - \operatorname{erfc} z,}$$

from which we have $w(-z) = 2e^{-z^2} - w(z)$. This also follows from the above manipulations with the integrals.

Another member of this family is *Dawson's integral:*

$$\boxed{F(z) = e^{-z^2}\int_0^z e^{t^2}\,dt.}$$

It is not difficult to verify that

$$\boxed{w(z) = e^{-z^2} + \frac{2i}{\sqrt{\pi}}F(z).}$$

11.1.1. The Error Function and Asymptotic Expansions

The error functions play an important part in uniform asymptotic expansions of integrals. Some examples are given in later subsections, see §§11.2.4, 11.3, 11.4.2 and 11.4.3. In all these cases the functions to be approximated can be interpreted as probability density functions. New applications of the error function arose recently, starting with the paper BERRY (1989), in which the so-called *Stokes phenomenon* has been given a new interpretation. This phenomenon is related with the different asymptotic expansions a function may have in certain sectors in the complex plane and with the changing of constants multiplying asymptotic series when the complex variable crosses certain lines (also called *Stokes lines*). Berry explained that the constants are in fact rapidly changing smooth functions, which can be approximated in terms of the error function. His approach was followed by a series of papers by himself and other writers. At the same time interest arose in earlier work by Stieltjes, Airey and Dingle to re-expand remainders in asymptotic expansions and to improve the accuracy obtainable from asymptotic expansions by considering exponentially small terms. An introductory paper on the Stokes phenomenon and exponential asymptotics is PARIS & WOOD (1995). In OLVER (1991a, 1991b) Berry's approach is rigorously treated for integrals representing the confluent hypergeometric functions $U(a, c, z)$ of Chapter 7. Other rigorous work is done for solutions of differential equations; see OLVER (1993) and OLDE DAALHUIS & OLVER (1994). Many other references can be found in the paper by Paris and Wood.

11.2. Incomplete Gamma Functions

The *incomplete gamma functions* are generalizations of the exponential integrals defined in (7.24) by taking n to be an arbitrary complex number. However, the name 'incomplete gamma function' comes from splitting up the interval of integration of the Euler gamma function. The definitions are

$$\boxed{\gamma(a,z) = \int_0^z t^{a-1}e^{-t}\,dt, \quad \Gamma(a,z) = \int_z^\infty t^{a-1}e^{-t}\,dt.} \tag{11.2}$$

For $\gamma(a, z)$ we assume the condition $\Re a > 0$; with respect to z we assume $|\arg z| < \pi$. It is useful to have the normalizations

$$\boxed{P(a,z) = \frac{\gamma(a,z)}{\Gamma(a)}, \quad Q(a,z) = \frac{\Gamma(a,z)}{\Gamma(a)}.} \tag{11.3}$$

Then $P(a, z) + Q(a, z) = 1$.

In statistics and probability theory one is more familiar with the *chi-square probability functions*, which are defined by

$$\boxed{P(\chi^2|\nu) = P(a,x), \quad Q(\chi^2|\nu) = Q(a,x), \quad \nu = 2a, \quad \chi^2 = 2x.}$$

In other words,

$$P\left(\chi^2|\nu\right) = \frac{1}{2^{\nu/2}\Gamma\left(\frac{1}{2}\nu\right)}\int_0^{\chi^2} t^{\nu/2-1}e^{-t/2}\,dt,$$

$$Q\left(\chi^2|\nu\right) = \frac{1}{2^{\nu/2}\Gamma\left(\frac{1}{2}\nu\right)}\int_{\chi^2}^{\infty} t^{\nu/2-1}e^{-t/2}\,dt.$$

When ν is even we have the *Poisson distribution*, which reads

$$Q\left(\chi^2|\nu\right) = \sum_{j=0}^{c-1} e^{-m}\frac{m^j}{j!}, \quad c = \frac{1}{2}\nu, \quad m = \frac{1}{2}\chi^2.$$

The point $z = 0$ is a singularity for the incomplete gamma function, except when $a = 0, 1, 2, 3, \ldots$. The singularity at $z = 0$ becomes apparent in the representation

$$\boxed{\gamma^*(a, z) = z^{-a}P(a, z) = \frac{z^{-a}}{\Gamma(a)}\gamma(a, z);} \tag{11.4}$$

$\gamma^*(a, z)$ is an entire function of z and a.

The relations with the Kummer functions (see Chapter 7) are as follows:

$$\begin{aligned}
\gamma(a, z) &= a^{-1}z^a e^{-z}M(1, a+1, z) \\
&= a^{-1}z^a M(a, a+1, -z), \\
\gamma^*(a, z) &= \frac{e^{-z}}{\Gamma(a+1)}M(1, a+1, z) \\
&= \frac{1}{\Gamma(a+1)}M(a, a+1, -z), \\
\Gamma(a, z) &= z^a e^{-z}U(1, a+1, z) \\
&= e^{-z}U(1-a, 1-a, z).
\end{aligned} \tag{11.5}$$

When a is a non-negative integer the incomplete gamma functions are very simple:

$$\begin{aligned}
\gamma(n+1, z) &= n!\left[1 - e^{-z}e_n(z)\right] \\
\Gamma(n+1, z) &= n!\,e^{-z}e_n(z).
\end{aligned} \qquad n = 0, 1, 2, \ldots, \tag{11.6}$$

in which $e_n(z)$ is the first part of the Taylor series of the exponential function:

$$e_n(z) = \sum_{m=0}^{n}\frac{z^m}{m!}, \quad n = 0, 1, 2, \ldots .$$

The following recurrence relations are easily derived from the integral representations:

$$\boxed{\gamma(a+1,z) = a\gamma(a,z) - z^a e^{-z}, \quad \Gamma(a+1,z) = a\Gamma(a,z) + z^a e^{-z}.} \tag{11.7}$$

The relations for the normalized functions are

$$P(a+1,z) = P(a,z) - \frac{z^a e^{-z}}{\Gamma(a+1)}, \quad Q(a+1,z) = Q(a,z) + \frac{z^a e^{-z}}{\Gamma(a+1)}. \tag{11.8}$$

These relations are important for numerical computations. However, as will be explained in Chapter 13, the relations for $\gamma(a,z)$ and $P(a,z)$ are unstable when applied repeatedly (when the parameters a and z are positive). The relations for $\Gamma(a,z)$ and $Q(a,z)$ are stable.

11.2.1. Series Expansions

Important series expansions are

$$\boxed{\gamma(a,z) = e^{-z} \sum_{n=0}^{\infty} \frac{z^{a+n}}{(a)_{n+1}} = \sum_{n=0}^{\infty} \frac{(-1)^n}{n!} \frac{z^{a+n}}{a+n}.} \tag{11.9}$$

The first one is very suitable for numerical calculations, in particular when $a > z$. The series are convergent for each complex a and z, with the exception of $a = 0, -1, -2, \ldots$. Speed of convergence of the first series depends on the ratio $|a/z|$; when this is smaller than unity convergence is quite fast.

The second series is obtained by expanding the function $\exp(-t)$ in the first integral of (11.2). The first series arises after a transformation $t = z(1-u)$ in the integral, which gives,

$$\gamma(a,z) = z^a e^{-z} \int_0^1 (1-u)^{a-1} e^{uz}\, du.$$

Expanding the exponential function again, and using the beta integral (3.2), we obtain the first series in (11.9). Series expansions for $\Gamma(a,z)$ follow from the relation $\Gamma(a,z) = \Gamma(a) - \gamma(a,z)$.

In this way many complementary results become available. It is not sufficient to concentrate on just one function $\gamma(a,z)$ or $\Gamma(a,z)$, however. Especially, from a numerical point of view, one needs relations for both functions. See §11.2.5 for more information on this point.

When $z > a$ we concentrate on $\Gamma(a,z)$. When $z \gg a$, we can use an asymptotic expansion of $\Gamma(a,z)$. This expansion follows from the representation

$$\boxed{\Gamma(a,z) = z^a e^{-z} \int_0^\infty (t+1)^{a-1} e^{-zt}\, dt,} \tag{11.10}$$

and integrating by parts. We obtain for $N = 0, 1, 2, \ldots$

$$\Gamma(a,z) = z^{a-1}e^{-z}\left[\sum_{n=0}^{N-1}\frac{(-1)^n\,(1-a)_n}{z^n} + \theta_N\frac{(-1)^N(1-a)_N}{z^N}\right], \tag{11.11}$$

(when $N = 0$ the sum is empty). The quantity θ_N is the remainder:

$$\theta_N = z\int_0^\infty (t+1)^{a-N-1}e^{-zt}\,dt.$$

For positive values of the parameters we can obtain an interesting estimate for θ_N. We need the condition $z > a - N$. Then we write

$$\theta_N = z^{N+1-a}e^z\int_z^\infty t^{a-N-1}e^{-t}\,dt$$

and next we integrate with respect to $u = t - (a-N)\ln t$. We obtain

$$\theta_N = z^{N+1-a}e^z\int_{z_0}^\infty \frac{e^{-u}}{t-(a-N)}\,du < \frac{z}{z-(a-N)},$$

in which $z_0 = z - (a-N)\ln z$.

We can also apply Watson's lemma (Theorem 2.5) to (11.10). This gives the expansion

$$\boxed{\Gamma(a,z) \sim z^{a-1}e^{-z}\sum_{n=0}^{\infty}\frac{(-1)^n\,(1-a)_n}{z^n}, \quad |\arg z| < \tfrac{3}{2}\pi.} \tag{11.12}$$

11.2.2. Continued Fraction for $\Gamma(a,z)$

When z is not large enough to apply the asymptotic series (11.12) we can use a continued fraction for $\Gamma(a,z)$. Several continued fractions are available for this function and the next example is due to Legendre. The starting point is the integral representation (see (7.12) and (11.5)):

$$\boxed{\Gamma(a,z) = \frac{e^{-z}}{\Gamma(1-a)}\int_0^\infty \frac{e^{-zt}t^{-a}}{t+1}\,dt, \quad \Re a < 0, \quad \Re z > 0.} \tag{11.13}$$

Now let

$$U^{\nu,\rho} = \int_0^\infty e^{-zt}t^\nu(1+t)^\rho\,dt, \quad \Re\nu > -1, \quad \Re z > 0. \tag{11.14}$$

Then according to (11.13) we have

$$U^{-a,-1} = \Gamma(1-a)e^{z}\Gamma(a,z). \tag{11.15}$$

Integrating by parts in (11.14), writing $t^{\nu}dt = \frac{1}{\nu+1}dt^{\nu+1}$, we obtain

$$zU^{\nu+1,\rho} = (\nu+1)U^{\nu,\rho} + \rho U^{\nu+1,\rho-1},$$

with as a variant

$$\frac{U^{\nu+1,\rho}}{U^{\nu,\rho}} = \frac{\nu+1}{z - \rho\dfrac{U^{\nu+1,\rho-1}}{U^{\nu+1,\rho}}}. \tag{11.16}$$

On the other hand we can write in (11.14)

$$t^{\nu+1}(1+t)^{\rho} = t^{\nu+1}(1+t)^{\rho-1} + t^{\nu+2}(1+t)^{\rho-1},$$

so that (11.16) can be written as

$$\frac{U^{\nu+1,\rho}}{U^{\nu,\rho}} = \cfrac{\nu+1}{z - \cfrac{\rho}{1 + \cfrac{U^{\nu+2,\rho-1}}{U^{\nu+1,\rho-1}}}}. \tag{11.17}$$

Applying this formula repeatedly we obtain a formal expansion in the form of a continued fraction. From (11.15) it follows that

$$\frac{U^{1-a,-1}}{U^{-a,-1}} = \frac{\Gamma(2-a)\Gamma(a-1,z)}{\Gamma(1-a)\Gamma(a,z)} = \frac{(1-a)\Gamma(a-1,z)}{\Gamma(a,z)}.$$

From this and the second relation in (11.7) we get

$$\frac{U^{1-a,-1}}{U^{-a,-1}} = -1 + \frac{e^{-z}z^{a-1}}{\Gamma(a,z)}.$$

Thus, (11.17) finally gives with $\nu = -a, \rho = -1$

$$\Gamma(a,z) = \cfrac{e^{-z}z^{a}}{z + \cfrac{1-a}{1 + \cfrac{1}{z + \cfrac{2-a}{1 + \cfrac{2}{z + \cfrac{3-a}{1+ \ddots}}}}}} \tag{11.18}$$

This expansion converges for all $z \neq 0$, $|\arg z| < \pi$ and any complex value of a. It is a splendid addition to the asymptotic expansion (11.11). The continued fraction converges better as the ratio z/a increases. A more compact notation of the continued fraction is

$$\Gamma(a,z) = e^{-z} z^a \left(\frac{1}{z+} \, \frac{1-a}{1+} \, \frac{1}{z+} \, \frac{2-a}{1+} \, \frac{2}{z+} \, \frac{3-a}{1+} \, \cdots \right). \tag{11.19}$$

More information on the analytic theory of continued fractions can be found in PERRON (1950), WALL (1948) and JONES & THRON (1980), where also convergence aspects are discussed. A recent book is LORENTZEN AND WAADELAND (1992). In §13.5 we point out how to compute a continued fraction.

11.2.3. Contour Integral for the Incomplete Gamma Functions

We verify the contour integral

$$P(a,z) = \frac{1}{2\pi i} \int_{c-i\infty}^{c+i\infty} e^{zs} \frac{ds}{s(s+1)^a}, \quad c > 0. \tag{11.20}$$

This representation holds for a wide range of the parameters. We assume $\Re a > 0$ and $|\arg z| < \frac{1}{2}\pi$, $z \neq 0$. The contour of integration can be deformed into the Hankel contour (see Figure 3.3), which is used for the reciprocal gamma function in (3.6). We assume that the branch cut of $(s+1)^{-a}$ runs from -1 to $-\infty$. The phase of $s+1$ is zero when $s > -1$. By turning the branch cut and the loop integral around it, we can extend the z–domain to $|\arg z| < \pi$, $z \neq 0$.

The above representation can be proved by using some elementary properties of Laplace transforms. We know the Laplace transform

$$\frac{1}{(s+1)^a} = \int_0^\infty e^{-st} \frac{d\,P(a,t)}{dt} \, dt.$$

Since $P(a,0) = 0$, we obtain by integrating by parts:

$$\int_0^\infty e^{-st} P(a,t) \, dt = \frac{1}{s(s+1)^a}.$$

On inverting this Laplace transform we obtain (11.20).

The contour in (11.20) can be shifted to the left of the origin, but then we have to take into account the residue at this point. It follows that

$$P(a,z) = 1 + \frac{1}{2\pi i} \int_{d-i\infty}^{d+i\infty} e^{zs} \frac{ds}{s(s+1)^a}, \quad -1 < d < 0,$$

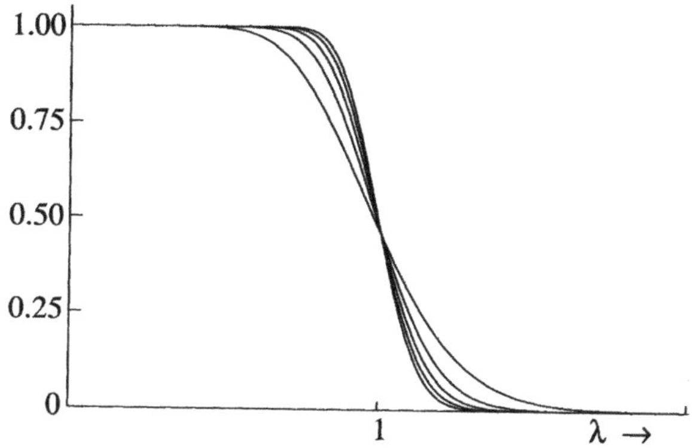

Figure 11.1. Incomplete gamma function $Q(a, \lambda a)$ as function of λ with $a = 20, 40, \ldots, 100$. As a increases the graphs become steeper when λ passes the transition point $\lambda = 1$.

from which we conclude that

$$Q(a, z) = \frac{-1}{2\pi i} \int_{d-i\infty}^{d+i\infty} e^{zs} \frac{ds}{s(s+1)^a}, \quad -1 < d < 0.$$

The above representations have a striking relation with the Hankel contour integral representation of the reciprocal gamma function (3.6) and with the integrals used in §3.6.3 for deriving the asymptotic expansion of $1/\Gamma(z)$. To describe this relationship in more detail we write

$$Q(a, z) = \frac{e^{-a\phi(\lambda)}}{2\pi i} \int_{c-i\infty}^{c+i\infty} e^{a\phi(t)} \frac{dt}{\lambda - t},$$

where

$$\phi(t) = t - 1 - \ln t, \quad \lambda = \frac{z}{a}.$$

Again we can integrate along the path of steepest descent, defined in (3.34). More details on this will be given in the next subsection.

11.2.4. Uniform Asymptotic Expansions

The asymptotic expansion (11.11) becomes useless when $a = \mathcal{O}(z)$. Also, the first series in (11.9), which has an asymptotic property when $|a/z| \gg 1$, converges slowly when $z = \mathcal{O}(a)$. When a is large, the functions $P(a, \lambda a), Q(a, \lambda a)$ (see (11.3)) change rapidly when λ crosses the value 1; see Figure 11.1. This change in behavior can be described by using the error function.

In this section we derive asymptotic expansions for $P(a,z), Q(a,z)$ in which a is a large positive parameter. The expansions hold uniformly with respect to $z \in [0,\infty)$, in particular in the neighborhood of $z = a$.

From the previous subsection it follows that

$$Q(a,z) = \frac{e^{-a\phi(\lambda)}}{2\pi i}\int_{\mathcal{L}} e^{a\phi(t)}\frac{dt}{\lambda - t},$$

where

$$\phi(t) = t - 1 - \ln t, \quad \lambda = \frac{z}{a}.$$

$\mathcal{L}$ is the path in the complex t−plane defined by (3.34). When we take temporarily $z > a$, the pole at $t = \lambda$ is located to the right of the saddle point at $t = 1$. On the path $\mathcal{L}$ the function $\phi(t)$ is non-negative, and we transform

$$\frac{1}{2}u^2 = -\phi(t), \tag{11.21}$$

with the condition that $t \in \mathcal{L}$ corresponds with $u \in \mathbb{R}$, and $\text{sign}(u) = \text{sign}(\Im t)$. We have

$$u = -i(t-1) + \mathcal{O}\left[(t-1)^2\right], \quad t \to 1.$$

The result of transformation (11.21) is

$$Q(a,z) = \frac{e^{-a\phi(\lambda)}}{2\pi i}\int_{-\infty}^{\infty} e^{-\frac{1}{2}au^2}\frac{dt}{du}\frac{du}{\lambda - t}, \quad \frac{dt}{du} = \frac{ut}{1-t}. \tag{11.22}$$

In the u−plane the point $u_1 = \sqrt{-2\phi(\lambda)}$ is a pole of the integrand. The sign of the square root follows from the conditions imposed on the mapping $t \mapsto u(t)$. In fact we have

$$u_1 = -i(\lambda - 1)\sqrt{2\frac{\lambda - 1 - \ln\lambda}{(\lambda-1)^2}},$$

where the square root is positive for positive values of the argument ($\lambda > 0$). The pole is at the negative imaginary axis; when integrating from $-\infty$ to ∞, the pole is at the right of the path of integration, just as in the u−plane. The conformal mapping $t \mapsto u(t)$ preserves this orientation.

When $\lambda \sim 1$ the pole is near the saddle point at the origin. We can remove the pole by writing

$$\frac{dt}{du}\frac{du}{\lambda - t} = \left[\frac{dt}{du}\frac{du}{\lambda - t} + \frac{1}{u - u_1}\right] - \frac{1}{u - u_1}.$$

The part between [ ] is analytic at $u = u_1$. To verify this, use l'Hôpital's rule:

$$\lim_{u \to u_1} \frac{dt}{du} \frac{u - u_1}{\lambda - t} = -1.$$

By using (11.22) and the above splitting of the integrand we obtain the representation

$$\begin{aligned}
Q(a,z) &= \frac{1}{2}\,\mathrm{erfc}(\eta\sqrt{a/2}\,) + R_a(\eta),\\
P(a,z) &= \frac{1}{2}\,\mathrm{erfc}(-\eta\sqrt{a/2}\,) - R_a(\eta),\\
\eta &= (\lambda - 1)\sqrt{2\frac{\lambda - 1 - \ln\lambda}{(\lambda-1)^2}}, \quad \lambda = \frac{z}{a}, \qquad (11.23)\\
R_a(\eta) &= \frac{e^{-\frac{1}{2}a\eta^2}}{2\pi i}\int_{-\infty}^{\infty} e^{-\frac{1}{2}au^2} g(u)\,du,\\
g(u) &= \frac{dt}{du}\frac{1}{\lambda - t} + \frac{1}{u + i\eta}.
\end{aligned}$$

where we have used the relations for the error functions in §11.1. Note that the symmetry relation $P(a,z) + Q(a,z) = 1$ is preserved in the above representations. The condition $\lambda > 1$ can now be dropped, since the error function and $R_a(\eta)$ are analytic with respect to λ, in particular at $\lambda = 1$.

By expanding $g(u) = \sum_{n=0}^{\infty} g_n(\eta)u^n$, we obtain the asymptotic expansion

$$R_a(\eta) \sim \frac{e^{-\frac{1}{2}a\eta^2}}{\sqrt{2\pi a}} \sum_{n=0}^{\infty} \frac{c_n(\eta)}{a^n}, \quad a \to \infty, \qquad (11.24)$$

where

$$c_n(\eta) = -i2^n \frac{\Gamma\left(n + \frac{1}{2}\right)}{\Gamma\left(\frac{1}{2}\right)}\, g_{2n}(\eta), \quad n = 0, 1, 2, \ldots .$$

The function g is analytic at the origin, the nearest singularities being located at the points $2\sqrt{\pi}\,\exp(\pm 3\pi i/4)$. This follows from a further study of the mapping defined in (11.21). The point $t = 0$ is mapped to $u = -\infty$, and is not of interest. When we consider (11.21) as a conformal mapping from the t–plane to the u–plane, the finite singularities can be found by examining du/dt given in (11.22). Especially, t–values at which du/dt or dt/du vanish are of interest. At $t = 1$, $u = 0$ the mapping is conformal and analytic; we have $u(t) = (t-1) + \mathcal{O}\left[(t-1)^2\right]$ as $t \to 1$. However, to give a proper description of the mapping we need more than the principal Riemann sheet defined by $|\arg t| < \pi$. Outside this principal sheet we have points $t = 1$ with phases equal to $2\pi i k$ where $k = \pm 1, \pm 2, \ldots$. The many-valued logarithmic function in $\phi(t) = t - 1 - \ln(t)$ treats these values of unity differently, according to their phases. At such points the derivative du/dt vanishes. The $u-$values corresponding to $t_k = \exp(2\pi ik)$ follow from (11.21) and are given by

$$\frac{1}{2}u_k^2 = -2\pi ik \quad \Longrightarrow \quad u_k = 2\sqrt{\pi k}\; e^{\pm 3\pi i/4}, \quad k = 1, 2, 3, \ldots .$$

These points t_k are the only finite singular points for the mapping in (11.21). The corresponding points u_k are the only finite singularities of the function $g(u)$ defined in (11.23). It follows that $g(u)$ is analytic in the strip $|\Im u| < \sqrt{2\pi}$, uniformly with respect to $\eta \in \mathbb{R}$. Observe that the relation between η and λ is the same as that between u and t. In fact, g is also a (bounded) analytic function of η in the strip $|\Im\eta| < \sqrt{2\pi}$. It follows that the coefficients $g_n(\eta)$, and hence $c_n(\eta)$ in (11.24) are analytic functions of η.

We can obtain a bound for $c_n(\eta)$. Let ρ be any number less than $2\sqrt{\pi}$ and let

$$M_\rho := \sup_{|u|=\rho} |g(u)|.$$

Then we have, using Cauchy's inequality, $|g_n(\eta)| \le M_\rho \rho^{-n}$, which gives

$$|c_n(\eta)| \le \left(\frac{2}{\rho^2}\right)^n \frac{\Gamma\left(n+\frac{1}{2}\right)}{\Gamma\left(\frac{1}{2}\right)} M_\rho.$$

Since M_ρ is bounded (qua function of η), it follows that all $c_n(\eta)$ are bounded functions of η. When we define remainders for the expansion (11.24) by writing

$$R_a(\eta) = \frac{e^{-\frac{1}{2}a\eta^2}}{\sqrt{2\pi a}} \left[\sum_{n=0}^{N-1} \frac{c_n(\eta)}{a^n} + a^{-N} T_N(a,\eta)\right], \quad N = 0, 1, 2, \ldots,$$

we can obtain estimates of $|T_N|$ in a similar way. This gives the result that (11.24) holds uniformly with respect to $\eta \in \mathbb{R}$, that is with respect to $\lambda \in [0, \infty)$. Extension of this result to complex variables is also possible. A few relations for the coefficients $c_n(\eta)$ are given in Exercise 11.4.

When $a \sim z$, the parameter η in (11.23) is small. When η is small enough to make $\eta\sqrt{a}$ small as well, we have $R_a(\eta) = \mathcal{O}(1/\sqrt{a})$, $a \to \infty$ and both P and Q approach $\frac{1}{2}$.

11.2.5. Numerical Aspects

In the computational problem we concentrate on positive values of a and $z = x$, and the normalized functions $P(a, x)$ and $Q(a, x)$. We compute the function that is less than $\frac{1}{2}$. The relation $P(a, x) + Q(a, x) = 1$ gives the other value. With slight corrections for small values of x and a we have the following rule:

$0 \le a \le x$: first compute Q, then $P = 1 - Q$;
$0 \le x < a$: first compute P, then $Q = 1 - P$.

This follows from the asymptotic relation

$$P(a,a), Q(a,a) \sim \frac{1}{2}, \quad \text{as} \quad a \to \infty,$$

which becomes clear from the previous subsection. The method for Q is usually based on Legendre's continued fraction (11.19), say when $x > 1$. When $x \gg a$ the asymptotic expansion (11.11) may be used for Q. The computation of P may be based on the first series in (11.9), that is,

$$P(a,x) = x^a e^{-x} \sum_{n=0}^{\infty} \frac{x^n}{\Gamma(a+n+1)}. \tag{11.25}$$

For large values of the parameters x and a, in particular when the parameters are nearly equal, both methods for computing P and Q need much effort. For example, GAUTSCHI (1979) reports for 8-digit accuracy and with $x = 10236$ and $a = x \cdot (1 + 0.001)$ for the power series the need of 536 terms, and for $a = x \cdot (1 - 0.001)$ of 124 iterations for the continued fraction (11.19) (with the remark that the continued fraction is $2 - 2\frac{1}{2}$ times as expensive, per iteration, as the Taylor series). This motivates the construction of other algorithms, in particular for the case $x \sim a$, both large.

We conclude with a discussion of how to compute the dominant part

$$D(a,x) := \frac{x^a e^{-x}}{\Gamma(a+1)}, \tag{11.26}$$

that usually turns up in algorithms for P and Q. Especially when a and x are large, the computation of this part needs some care. When a is large we write (11.26) in the form

$$D(a,x) = \frac{e^{-a[\mu - \ln(1+\mu)]}}{\sqrt{2\pi a}\,\Gamma^*(a)}, \quad \mu = \frac{x-a}{a}, \tag{11.27}$$

where

$$\Gamma^*(a) = \frac{\Gamma(a)}{\sqrt{2\pi}\, a^{a-\frac{1}{2}} e^{-a}}, \quad a > 0. \tag{11.28}$$

Γ^* is introduced in (3.29), with a recommendation to have this function available in a computing environment. When $|\mu|$ is small, one needs a special routine to compute the function

$$\mu - \ln(1+\mu) = \frac{1}{2}\mu^2 - \frac{1}{3}\mu^3 + \ldots$$

(say indeed by using a Taylor series), since otherwise precision is lost in the subtraction $\mu - \ln(1+\mu)$. (Observe that already the computation of $\ln(1+\mu)$ needs some care when $|\mu|$ is small). An error in μ itself (from the evaluation of $x - a$) is the remaining problem in (11.27). When the parameters a and x are both large this may still cause a serious error in the numerical evaluation of P and Q, and this error may dominate the errors originating from the evaluation of the continued fraction or the Taylor series.

11.3. Incomplete Beta Functions

The *incomplete beta function* is defined by

$$B_x(p,q) = \int_0^x t^{p-1}(1-t)^{q-1}\,dt, \quad \Re p > 0, \quad \Re q > 0,$$

with as normalized version

$$\boxed{I_x(p,q) = \frac{1}{B(p,q)} \int_0^x t^{p-1}(1-t)^{q-1}\,dt,} \tag{11.29}$$

where $B(p,q)$ is the beta integral introduced in (3.2). We have the complementary relation

$$I_{1-x}(p,q) = 1 - I_x(q,p). \tag{11.30}$$

Usually x belongs to the interval $(0,1)$ and the parameters p, q are positive, in particular when the function is considered in statistical problems. However, the range of x, $p\,q$ may be extended to complex values.

By expanding $(1-t)^{q-1}$ in powers of t one obtains (see Exercise 3.6)

$$I_x(p,q) = \frac{x^p}{B(p,q)} \sum_{n=0}^{\infty} \frac{(1-q)_n}{p+n} \frac{x^n}{n!} \quad |x| < 1. \tag{11.31}$$

With (11.30) one can obtain an expansion with powers of $1-x$. The substitution $t = (1-s)x$ in (11.29) gives (with (3.2))

$$I_x(p,q) = \frac{x^p(1-x)^{q-1}}{pB(p,q)} \sum_{n=0}^{\infty} \frac{(1-q)_n}{(1+p)_n} \left(\frac{x}{x-1}\right)^n, \quad \left|\frac{x}{x-1}\right| < 1. \tag{11.32}$$

Note that for real x the condition in this series reads:$-\infty < x < \frac{1}{2}$.

Another expansion is

$$I_x(p,q) = \frac{x^p(1-x)^q}{pB(p,q)} \sum_{n=0}^{\infty} \frac{(p+q)_n}{(p+1)_n} x^n, \tag{11.33}$$

again for $|x| < 1$. This expansion easily follows by taking derivatives with respect to x on both sides and dividing both sides through $x^{p-1}(1-x)^{q-1}$. Comparing coefficients of equal powers of x yields a simple recursion in terms of ratios of gamma functions. The above formulas follow also from the relations between the incomplete beta function and the Gauss hypergeometric function, which is introduced in Chapter 5. We have (see Exercise 5.3)

$$\begin{aligned} I_x(p,q) &= \frac{x^p}{pB(p,q)} F(p, 1-q; p+1; x) \\ &= \frac{x^p(1-x)^{q-1}}{pB(p,q)} F\left(1, 1-q; p+1; \frac{x}{x-1}\right) \\ &= \frac{x^p(1-x)^q}{pB(p,q)} F(p+q, 1; p+1; x). \end{aligned} \tag{11.34}$$

We also have the continued fraction

$$\boxed{I_x(p,q) = \frac{x^p(1-x)^q}{pB(p,q)} \left(\frac{1}{1+}\,\frac{d_1}{1+}\,\frac{d_2}{1+}\,\frac{d_3}{1+}\cdots\right),} \tag{11.35}$$

where

$$d_{2m+1} = -\frac{(p+m)(p+q+m)}{(p+2m)(p+2m+1)}\, x, \quad d_{2m} = \frac{m(q-m)}{(p+2m-1)(p+2m)}\, x.$$

A proof of this expansion is given in Exercise 11.5. A few remarks on the numerical evaluation of the incomplete beta function follow in §11.3.6.

11.3.1. Recurrence Relations

We derive several recurrence relation from the integral (11.29). Some results appear in pairs; each result in the pair then follows from the other one via (11.30). Integrating by parts in (11.29), that is, writing $t^{p-1}dt = \frac{1}{p}\, d(t^p)$, we obtain

$$\begin{aligned} I_x(p,q) &= I_x(p+1, q-1) + \frac{x^p(1-x)^{q-1}}{pB(p,q)}, \\ I_x(p,q) &= I_x(p-1, q+1) - \frac{x^{p-1}(1-x)^q}{qB(p,q)}. \end{aligned} \tag{11.36}$$

Integrating by parts with respect to the term $(1-t)^{q-1}$ gives the pair

$$\begin{aligned} I_x(p,q) &= I_x(p+1, q) + \frac{x^p(1-x)^q}{pB(p,q)}, \\ I_x(p,q) &= I_x(p, q+1) - \frac{x^p(1-x)^q}{qB(p,q)}. \end{aligned} \tag{11.37}$$

This pair is very useful in numerical computations. Note that the first of this pair is stable in backward p–direction, that is for the evaluation of $I_x(p,q)$ from $I_x(p+1,b)$, while the second one is stable in the forward q–direction.

Writing in (11.29) $t^{a-1} = t^{a-2}(t-1+1)$ we obtain

$$(p+q)I_x(p,q) = pI_x(p+1,b) + qI_x(p,q+1), \tag{11.38}$$

which is invariant under (11.30). Combining the above results we obtain

$$I_x(p,q) = xI_x(p-1,q) + (1-x)I_x(p,q-1), \tag{11.39}$$

$$\begin{aligned} [p+qx]I_x(p,q) &= qxI_x(p-1,q+1) + pI_x(p+1,q), \\ [q+p\bar{x}]I_x(p,q) &= p\bar{x}I_x(p+1,q-1) + qI_x(p,q+1), \end{aligned} \tag{11.40}$$

where $\bar{x} = 1-x$. Finally we mention the pair with recursion only in the p and q direction:

$$\begin{aligned} pI_x(p+1,q) &= [p+(p+q-1)x]I_x(p,q) - (p+q-1)xI_x(p-1,q), \\ qI_x(p,q+1) &= [q+(p+q-1)\bar{x}]I_x(p,q) - (p+q-1)\bar{x}I_x(p,q-1). \end{aligned} \tag{11.41}$$

11.3.2. Contour Integral for the Incomplete Beta Function

We verify the representation

$$I_x(p,q) = \frac{x^p(1-x)^q}{2\pi i}\int_{c-i\infty}^{c+i\infty} t^{-p}(1-t)^{-q}\,\frac{dt}{t-x}, \quad 0 < c < 1, \tag{11.42}$$

with $p+q > 0$ and $0 < x < c$. When $t \in (0,1)$ the phases of t and $1-t$ and of the multi-valued function are assumed to be zero. A proof follows by expanding

$$\frac{1}{t-x} = \frac{1}{t\,(1-x/t)} = \frac{1}{t}\sum_{n=0}^{\infty}\left(\frac{x}{t}\right)^n.$$

We substitute this in (11.42) (observe that $|x/t| < 1$ on the path of integration), and obtain

$$\int_{c-i\infty}^{c+i\infty} t^{-p}(1-t)^{-q}\,\frac{dt}{t-x} = \sum_{n=0}^{\infty} x^n \int_{c-i\infty}^{c+i\infty} t^{-p-k-1}(1-t)^{-q}\,dt.$$

By comparing the contour integrals in this result with one of the contour integrals in Exercise 3.13, it follows that the series expansion of (11.33) is obtained.

Observe that in (11.42) the normalizing factor $B(p, q)$ is not present. We further remark that property (11.30) is contained in (11.42): when shifting the contour to the right, across the pole at $t = x$, we pick up a residue equal to 1; the remaining integral has a c–value satisfying $0 < c < x < 1$, and can be transformed with $t \to 1 - t$ into $-I_{1-x}(q, p)$.

11.3.3. Asymptotic Expansions

We consider asymptotic behavior of the incomplete beta function as p and or q are large with $x \in (0, 1)$. Elementary expansions can be given when just one of the parameters p, q is large and x is fixed.

When p is large, with q and x fixed, a simple expansion follows from (11.32). Denoting the terms of the series by c_n we observe that, when n is fixed,

$$\frac{c_{n+1}}{c_n} = \frac{1-q+n}{1+p+n}\,\frac{x}{x-1} \sim \frac{x}{p(x-1)}, \qquad p \to \infty.$$

It follows that the series has an asymptotic character when $p \gg x/(1-x)$; recall that when $x < 1/2$ the series is convergent as well.

When q is large, with p and x fixed, we can use the same method through formula (11.30), that is,

$$I_x(p, q) = 1 - \frac{x^{p-1}(1-x)^q}{qB(p, q)} \sum_{n=0}^{\infty} \frac{(1-p)_n}{(1+q)_n} \left(\frac{x-1}{x}\right)^n,$$

which converges when $x > 1/2$; when $q \gg (1-x)/x$ the series has an asymptotic character.

In the following subsections we give asymptotic expansions in which the incomplete gamma function and the error function are used as main approximant in the expansions.

11.3.3.1. Uniform Expansion: q Fixed

We derive an asymptotic expansion that holds as $p \to \infty$, q fixed, and with $x \in [0, 1]$. Recall that (11.32) breaks down when $x \to 1$; this case is allowed now. Substituting in (11.29) $t = \exp(-u)$, we obtain

$$I_x(p, q) = \frac{1}{B(p, q)} \int_{\xi}^{\infty} u^{q-1} e^{-pu} f(u)\, du, \tag{11.43}$$

where

$$\xi = -\ln x, \quad f(u) = \left(\frac{1-e^{-u}}{u}\right)^{q-1}.$$

Expanding $f(u) = \sum_{n=0}^{\infty} c_n (u-\xi)^n$, and substituting this in (11.43) we obtain the formal expansion

$$I_x(p, q) \sim \frac{\Gamma(p+q)}{\Gamma(p)} \sum_{n=0}^{\infty} c_n F_n(p, q, \xi), \tag{11.44}$$

with

$$F_n(p,q,\xi) = \frac{1}{\Gamma(q)} \int_\xi^\infty u^{q-1} e^{-pu} (u-\xi)^n \, du.$$

F_0 is an incomplete gamma function, and the remaining F_n can be expressed in terms of this function. We have

$$F_0(p,q,\xi) = p^{-q} Q(q, p\xi), \quad F_1(p,q,\xi) = \frac{q-p\xi}{p} F_0(p,q,\xi) + \frac{\xi^q e^{-p\xi}}{p\Gamma(q)}.$$

By integrating by parts in the integral defining F_n we obtain the recurrence relation

$$pF_{n+1} = (n+q-p\xi)F_n + n\xi F_{n-1}, \quad n = 1,2,3,\ldots.$$

Although it is rather easy to compute the functions F_n, expansion (11.44) is quite complicated. The fact is that the coefficients are quite difficult to obtain. The first few are

$$c_0 = \left(\frac{x-1}{\ln x}\right)^{q-1}, \quad c_1 = (q-1)\left[\frac{x\ln x + 1 - x}{(1-x)\ln x}\right] c_0.$$

Remark 11.1. We have expanded the function f in (11.43) at the lower limit of integration. This may not be optimal. Observe that the function $u^q e^{-pu}$ assumes its maximal value at $u = u_0 := q/p$. When $\xi < u_0$ it seems better to expand the function at the point u_0. In particular the functions F_n become simpler then. For instance we have in that case $F_1 = \xi^q e^{-p\xi}/[p\Gamma(q)]$.

11.3.3.2. Uniform Expansion: The Symmetric Case

Let $r := p + q$ and assume that $r \to \infty$. In this subsection we derive an asymptotic expansion of the incomplete beta function that holds uniformly with respect to $x \in (0,1)$ and with respect to $p/r, q/r \in [\delta, 1-\delta]$, where δ is a fixed small positive number. The uniform expansion contains an error function that properly describes the change in behavior of $I_x(p,q)$ when x crosses the critical value $x_0 := p/(p+q)$.

Consider the integral in (11.42). The dominant part $t^{-p}(1-t)^{-q}$ has a saddle point at x_0. We have assumed that this point is bounded away from 0 and 1, and we observe that when $x \sim t_0$ the saddle point and pole are close together. As we did in §11.2.4 for the incomplete gamma function, we transform the integral and split off the pole.

First we derive a contour for (11.42) on which the phase of $t^{-p}(1-t)^{-q}$ is zero. This is the path of steepest descent, and it goes through the saddle point at x_0. The path of steepest descent is defined by

$$\mathcal{L} = \left\{ t = \tau e^{i\sigma} \;\middle|\; \tau = \frac{\sin(\rho\sigma)}{\sin[(1+\rho)\sigma]}, \quad |\sigma| < \frac{\pi}{1+\rho} \right\}, \tag{11.45}$$

where $\rho = p/q$. We deform the path in (11.45) into $\mathcal{L}$, assuming temporarily that $x < x_0$. Along $\mathcal{L}$ the function $-p\ln t - q\ln(1-t) + p\ln x_0 + q\ln(1-x_0)$ is non-positive and we transform

$$\frac{1}{2}u^2 = x_0 \ln\frac{t}{x_0} + (1-x_0)\ln\frac{1-t}{1-x_0} \tag{11.46}$$

with the condition $t \in \mathcal{L} \Longleftrightarrow u \in \mathbb{R}$, $\text{sign}(\Im t) = \text{sign}(u)$. Using this transformation we write (11.46) in the form

$$I_x(p,q) = \left(\frac{x}{x_0}\right)^p \left(\frac{1-x}{1-x_0}\right)^q \frac{1}{2\pi i}\int_{-\infty}^{\infty} e^{-\frac{1}{2}ru^2}\,\frac{dt}{du}\frac{du}{t-x}.$$

The mapping $t \mapsto u(t)$ transforms the pole at $t = x$ in the t–plane into a pole $u_1 = u(x)$ in the u–plane. When $x < x_0$ (as assumed) the point u_1 lies on the positive imaginary axis. This follows from the conditions imposed on the mapping in (11.46). We split off the pole by writing

$$\frac{dt}{du}\frac{1}{t-x} = \left[\frac{dt}{du}\frac{1}{t-x} - \frac{1}{u-u_1}\right] + \frac{1}{u-u_1}.$$

Thus we obtain, using the relations for the error functions given in §11.1,

$$I_x(p,q) = \frac{1}{2}\text{erfc}\left(-\eta\sqrt{q/2}\right) + R_r(\eta), \quad r = p+q,$$

$$R_r(\eta) = \left(\frac{x}{x_0}\right)^p \left(\frac{1-x}{1-x_0}\right)^q \frac{1}{2\pi i}\int_{-\infty}^{\infty} e^{-\frac{1}{2}ru^2} g(u)\,du,$$

$$x_0 = \frac{p}{p+q}, \quad g(u) = \frac{dt}{du}\frac{1}{t-x} - \frac{1}{u+i\eta},$$

$$-\frac{1}{2}\eta^2 = x_0 \ \ln\frac{x}{x_0} + (1-x_0)\ \ln\frac{1-x}{1-x_0}, \quad \text{sign}(\eta) = \text{sign}(x-x_0).$$

In this representation the condition $x < x_0$ can be dropped, and we assume now that $x \in (0,1)$. The function $g(u)$ is analytic at the origin, and we can expand $g(u) = \sum_{n=0}^{\infty} g_n u^n$. This gives the expansion

$$R_r(\eta) = \frac{1}{\sqrt{2\pi r}}\left(\frac{x}{x_0}\right)^p \left(\frac{1-x}{1-x_0}\right)^q S_r(\eta), \quad S_r(\eta) \sim \sum_{n=0}^{\infty}\frac{c_n(\eta)}{r^n}, \quad r \to \infty,$$

with

$$c_n(\eta) = -i2^n\frac{\Gamma\left(n+\frac{1}{2}\right)}{\Gamma\left(\frac{1}{2}\right)}\, g_{2n}, \quad n = 0, 1, 2, \ldots.$$

The first coefficient is

$$c_0(\eta) = \frac{1}{\eta} - \frac{\sqrt{x_0(1-x_0)}}{x - x_0}.$$

To derive this we need dt/du at $u = 0$. From (11.46) it follows that

$$\frac{1}{2}u^2 = -\frac{(t-x_0)^2}{2x_0(1-x_0)} + \mathcal{O}\left[(t-x_0)^3\right], \quad t \to x_0.$$

Hence

$$u = -i(t-x_0)/\sqrt{x_0(1-x_0)} + \mathcal{O}\left[(t-x_0)^2\right], \quad \left.\frac{du}{dt}\right|_{u=0} = -\frac{i}{\sqrt{x_0(1-x_0)}}.$$

This gives the required value of $g_0 = g(0)$.

The singularities of g occur at points u where the mapping (11.46) is not conformal. We have

$$\frac{dt}{du} = \frac{ut(1-t)}{x_0 - t}.$$

The point $t = x_0$, that is $u = 0$, is a regular point. The finite singularities of the mapping occur at the points $t_n, \tilde{t}_n$ defined by

$$t_n = x_0 e^{2\pi i n}, \quad (1 - \tilde{t}_m) = (1 - x_0)e^{2\pi i m}, \quad n, m = \pm 1, \pm 2, \ldots.$$

The many-valued logarithmic functions in (11.46) give corresponding points in the u–plane defined by

$$\frac{1}{2}u_n^2 = 2\pi i n x_0, \quad \frac{1}{2}\tilde{u}_m^2 = 2\pi i m(1 - x_0), \quad n, m = \pm 1, \pm 2, \ldots.$$

The first group approaches the origin when $x_0 \to 0$, the second group when $x_0 \to 1$. Because of this, the expansion of $S_r(\eta)$ becomes invalid when the ratio p/r approaches zero or unity.

11.3.3.3. Uniform Expansion: General Case

The expansion in §11.3.3.1 holds for fixed q (a very "skew" beta distribution), and the expansion in the previous case for $q = \mathcal{O}(p)$ (a rather symmetric, or Gaussian shaped, beta distribution); in both cases p is large and x may range through the full interval $(0, 1)$. The asymptotic expansions obtained for both cases do not have an overlapping q–domain of validity. In the present subsection we derive an expansion in which q may run through $(0, \infty)$, without restrictions on its size with respect to p. We assume that $p + q$ is large. In the previous case the error function (a function of one argument) is used. It can be expected that the incomplete beta function, a function of three variables, cannot be approximated in an optimal way by a function of one variable. In §11.3.3.1 the main approximant is the function F_0, that is, the incomplete gamma function Q, a function of two arguments. However, one argument is not used in an optimal way, since q is fixed. We again use the incomplete gamma function $Q(a, z)$ as main approximant for the general case, but now both parameters may range through the interval $(0, \infty)$.

We use the following transformation in (11.42):

$$-\ln t - \mu \ln(1-t) = s - \mu \ln s + A(\mu), \quad \mu = \frac{q}{p}, \tag{11.47}$$

where $A(\mu)$ follows from conditions on the mapping. First we consider real values of t and s. By drawing graphs of the $t-$ and $s-$ functions it is clear that we can define a one-to-one correspondence between t and s when we prescribe that the point $t = x_0 = p/(p+q)$ corresponds with the point $s = \mu$; in both points the functions have saddle points. The function $A(\mu)$ follows from prescribing that indeed these points should correspond, which gives

$$A(\mu) = -\ln x_0 - \mu \ln(1-x_0) - \mu + \mu \ln \mu = -\mu + (\mu+1)\ln(\mu+1). \tag{11.48}$$

We further note that $s(t)$ has the following properties:

$$s(0) = +\infty, \quad s(x_0) = \mu, \quad s(1) = 0, \quad \operatorname{sign}(x_0 - x) = \operatorname{sign}(s - \mu)$$

and that, hence, the derivative ds/dt is negative for $t \in (0,1)$. The derivative is given by

$$\frac{ds}{dt} = \frac{[(\mu+1)t - 1]s}{t(1-t)(s-\mu)}. \tag{11.49}$$

At $t = x_0, s = \mu$ the functions $s(t), t(s)$ are analytic. We compute the derivative at $t = x_0$. We have (note that $x_0 = 1/(\mu+1)$)

$$\lim_{t \to x_0} \frac{ds}{dt} = \frac{\mu(\mu+1)}{x_0(1-x_0)} \lim_{t \to x_0} \frac{t - x_0}{s - \mu} = \frac{\mu(\mu+1)}{x_0(1-x_0)} \left.\frac{dt}{ds}\right|_{t=x_0},$$

where we have used l'Hôpital's rule. It follows that

$$\left.\frac{ds}{dt}\right|_{t=x_0} = -\sqrt{\frac{\mu(\mu+1)}{x_0(1-x_0)}} = -(\mu+1)^{3/2}. \tag{11.50}$$

The mapping (11.47) transforms the path of steepest descent in the $t-$plane (see (11.45)) through x_0 into the path of steepest descent in the $s-$plane through μ. The latter is given by

$$\gamma = \left\{ s = \rho e^{i\theta} \,\middle|\, \rho = \mu \frac{\theta}{\sin \theta}, \quad |\theta| < \pi \right\},$$

a similar path as we used in (3.34) for the reciprocal gamma function. The transformation (11.47) brings (11.42), when integrated along $\mathcal{L}$ defined in (11.45), into

$$I_x(p,q) = -\frac{x^p(1-x)^q}{2\pi i} e^{pA(\mu)} \int_{-\infty}^{(0^+)} e^{ps} s^{-q} \frac{dt}{ds} \frac{ds}{t-x}, \tag{11.51}$$

where dt/ds follows from (11.49). The integration runs over γ. Initially the integration starts in the upper half s−plane. The minus-sign in front of the integral is used to change this. The pole in $t = x$ is mapped into a point $s = \eta$ defined by (11.47), that is, by

$$-\ln x - \mu \ln(1-x) = \eta - \mu \ln \eta + A(\mu), \quad \operatorname{sign}(x_0 - x) = \operatorname{sign}(\eta - \mu). \tag{11.52}$$

Known corresponding points are

$$x = 0 \Longleftrightarrow \eta = +\infty, \quad x = x_0 \Longleftrightarrow \eta = \mu, \quad x = 1 \Longleftrightarrow \eta = 0.$$

Note that at the moment $\eta > \mu$, since we have assumed that $x < x_0$. We split off the pole by writing

$$\frac{dt}{ds}\frac{1}{t-x} = \frac{dt}{ds}\frac{s-\eta}{t-x}\frac{1}{s-\eta} = \left[\frac{dt}{ds}\frac{1}{t-x} - \frac{1}{s-\eta}\right] + \frac{1}{s-\eta}.$$

The part representing the pole gives in (11.24) an incomplete gamma function, see §11.2.3. The result is

$$\begin{aligned} I_x(p,q) &= Q(q,\eta p) - R_x(p,q), \\ R_x(p,q) &= \frac{x^p(1-x)^q}{2\pi i} e^{pA(\mu)} \int_{-\infty}^{(0^+)} e^{ps} s^{-q-1}\, h(s)\, ds, \\ h(s) &= s\left[\frac{dt}{ds}\frac{1}{t-x} - \frac{1}{s-\eta}\right]. \end{aligned} \tag{11.53}$$

In this representation of $I_x(p,q)$ we can drop the condition $x < x_0$, and consider $x \in (0,1)$. The function $h(s)$ is analytic at the saddle point μ and we can obtain an expansion for the function $R_x(p,q)$ by expanding: $h(s) = \sum_{n=0}^{\infty} h_n(\eta,\mu)(s-\mu)^n$. So we obtain

$$R_x(p,q) \sim \frac{x^p(1-x)^q e^{pA(\mu)} p^q}{\Gamma(q+1)} \sum_{n=0}^{\infty} \frac{h_n(\eta,\mu)}{p^n} \Phi_n, \quad p \to \infty, \tag{11.54}$$

where

$$\begin{aligned} \Phi_n &= \frac{p^{n-q}\Gamma(q+1)}{2\pi i} \int_{-\infty}^{(0^+)} e^{ps} s^{-q-1}\, (s-\mu)^n\, ds \\ &= \frac{\Gamma(q+1)}{2\pi i} \int_{-\infty}^{(0^+)} e^t t^{-q-1}\, (t-q)^n\, dt. \end{aligned} \tag{11.55}$$

The functions Φ_n are polynomials of q, $\Phi_0 = 1, \Phi_1 = 0, \Phi_2 = 2q$, with the recurrence relation

$$\Phi_{n+1} = -n(\Phi_n + q\Phi_{n-1}), \quad n = 1, 2, 3, \ldots .$$

The expansion in (11.54) holds uniformly with respect to $x \in (0,1), q > 0$. To verify this one needs information on the singularities of the function $h(s)$, and in particular the singularities of the mapping defined in (11.47), with dt/ds given in (11.49). This problem will not be discussed here.

We consider in more detail the first coefficient of (11.54):

$$h_0(\eta, \mu) = \mu \left[\frac{1}{\eta - \mu} - \frac{(\mu+1)^{-3/2}}{x_0 - x} \right]. \tag{11.56}$$

This coefficient is regular at $\eta = \mu$, that is at $x = x_0$. To evaluate $h_0(\eta, \mu)$ at $\eta = \mu$ we have to investigate the relation between x and η in more detail. From (11.52) it follows that

$$\eta \left[(\mu+1)x - 1 \right] \frac{dx}{d\eta} = x(1-x)(\eta - \mu). \tag{11.57}$$

Substituting the expansion

$$x = x_0 + x_1(\eta - \mu) + x_2(\eta - \mu)^2 + \ldots, \qquad x_0 = \frac{1}{\mu+1}, \tag{11.58}$$

we obtain

$$x_1 = -(\mu+1)^{-3/2}, \quad x_1 x_2 = \frac{1}{3\mu(\mu+1)^3} \left[\frac{1-\mu}{\sqrt{\mu+1}} - 1 \right].$$

The value of x_1 also follows from (11.50). Using the expansion of x in (11.58) we obtain

$$h_0(\eta, \mu) = \mu x_2 / x_1 + \mathcal{O}(\mu - \eta), \quad \eta \to \mu.$$

Hence

$$h_0(\mu, \mu) = \frac{1}{3} \left[\frac{1-\mu}{\sqrt{\mu+1}} - 1 \right].$$

11.3.4. Numerical Aspects

It is sufficient to concentrate on the computation of the incomplete beta function when $x \le x_0$, with $x_0 = p/(p+q)$. The fact is that, in particular when $p+q$ is large, $I_x(p,q) \sim \frac{1}{2}$ when $x \sim x_0$. When $x > x_0$ one can use (11.30).

When $p+q$ is not large efficient algorithms can be based on the power series (11.31), (11.32) and (11.33). When $p+q$ is large, some care is needed with respect to convergence, however. When q is large with respect to p, (11.33) converges slowly. The series in (11.31) and (11.32) terminate when q is a positive integer. The series in (11.32) can be used when $p \gg x/(1-x)$, whether or not the series converges; when q is also large we need $p \gg qx/(1-x)$.

The continued fraction (11.35) can be used for a wide range of the parameters p and q, also when $p+q$ is large. The convergents have the following interesting property: the $4n$ and $4n+1$ convergents are less than the fraction's limit, and the $4n+2$ and $4n+3$ convergents are greater than the limit (see §13.5 for some details on how to compute a continued fraction). This provides excellent numerical checks for terminating the computations, since the limit is approached from above and below.

When the parameters p and q are large, and $|x-x_0|$ is small, it is important to have an algorithm for the expression

$$D_x(p,q) = \frac{x^p(1-x)^q}{B(p,q)}. \tag{11.59}$$

See also (11.26) for a similar discussion. We can write

$$\begin{aligned} D_x(p,q) &= \sqrt{\frac{pq}{\pi(p+q)}}\,\frac{\Gamma^*(p+q)}{\Gamma^*(p)\Gamma^*(q)}\,e^{p[\ln(1+\sigma)-\sigma]+q[\ln(1+\tau)-\tau]} \\ &= \sqrt{\frac{pq}{\pi(p+q)}}\,\frac{\Gamma^*(p+q)}{\Gamma^*(p)\Gamma^*(q)}\,e^{q[\ln(1+\chi)-\chi]}, \end{aligned} \tag{11.60}$$

where the function $\Gamma^*(z)$ is defined in (11.28), and

$$\sigma = \frac{x-x_0}{x_0}, \quad \tau = \frac{x_0-x}{1-x_0}, \quad \chi = \frac{\eta-\mu}{\mu}, \quad x_0 = \frac{p}{p+q},$$

with η defined in (11.52). A small error in σ, τ or η may have a great influence in the numerical evaluation of $D_x(p,q)$, and this may result into a serious error in the evaluation of the incomplete beta function.

11.4. Non-Central Chi-Squared Distribution

The *non-central* χ^2*-distribution* is defined by the series

$$\boxed{Q_\mu(x,y) = e^{-x}\sum_{n=0}^{\infty}\frac{x^n}{n!}Q(\mu+n,y),} \tag{11.61}$$

where $Q(a, z)$ is the incomplete gamma function defined in (11.2), (11.3). Another starting point is

$$\boxed{P_\mu(x, y) = e^{-x} \sum_{n=0}^{\infty} \frac{x^n}{n!} P(\mu + n, y),} \tag{11.62}$$

with $P(a, z) = 1 - Q(a, z)$. It follows that $P_\mu(x, y) + Q_\mu(x, y) = 1$. In problems on radar communications the function $Q_\mu(x, y)$ is known as the *generalized Marcum Q−function*, which for $\mu = 1$ reduces to the ordinary Marcum function. In this field μ is the number of independent samples of the output of a square-law detector. In our analysis μ is a not necessarily an integer number. We assume that $\mu > 0$; the parameters x, y are assumed to be non-negative.

In statistics and probability theory one is more familiar with the definition through the χ^2 *probability functions*, which are defined by

$$P\left(\chi^2 | \nu\right) = P(a, x), \quad Q\left(\chi^2 | \nu\right) = Q(a, x), \quad \nu = 2a, \quad \chi^2 = 2x.$$

The non-central χ^2−distribution functions are defined then by

$$P\left(\chi^2 | \nu, \lambda\right) = \sum_{n=0}^{\infty} e^{-\frac{1}{2}\lambda} \frac{(\frac{1}{2}\lambda)^n}{n!} P\left(\chi^2 | \nu + 2n\right),$$

$$Q\left(\chi^2 | \nu, \lambda\right) = \sum_{n=0}^{\infty} e^{-\frac{1}{2}\lambda} \frac{(\frac{1}{2}\lambda)^n}{n!} Q\left(\chi^2 | \nu + 2n\right),$$

where $\lambda \geq 0$ is called the non-centrality parameter.

The functions $P_\mu(x, y), Q_\mu(x, y)$ can be written as Bessel function integrals:

$$\boxed{\begin{aligned} P_\mu(x, y) &= \int_0^y \left(\frac{z}{x}\right)^{\frac{1}{2}(\mu-1)} e^{-z-x} I_{\mu-1}\left(2\sqrt{xz}\right) dz, \\ Q_\mu(x, y) &= \int_y^\infty \left(\frac{z}{x}\right)^{\frac{1}{2}(\mu-1)} e^{-z-x} I_{\mu-1}\left(2\sqrt{xz}\right) dz, \end{aligned}} \tag{11.63}$$

where $I_\mu(z)$ is the modified Bessel function introduced in §9.5. A proof of the relation with the Bessel function integral follows from substitution of the series expansion (9.28) in (11.63).

We derive a recurrence relations for $Q_\mu(x, y)$ with respect to μ. Using (11.8) and (9.28) we obtain

$$Q_{\mu+1}(x, y) = Q_\mu(x, y) + \left(\frac{y}{x}\right)^{\mu/2} e^{-x} I_\mu\left(2\sqrt{xy}\right). \tag{11.64}$$

We can eliminate the Bessel function in (11.64) using (see the first relation in (9.34)) $I_{\mu-1}(z) = I_{\mu+1}(z) + (2\mu/z)\,I_\mu(z)$. This gives the homogeneous third order recurrence relation:

$$xQ_{\mu+2}(x,y) = (x-\mu)Q_{\mu+1}(x,y) + (y+\mu)Q_\mu(x,y) - yQ_{\mu-1}(x,y). \quad (11.65)$$

In the following subsections we derive asymptotic expansions of the function defined in (11.61) and (11.62). When x and y are large, and $|x-y|$ is small compared to x and y, the integrals in (11.63) have a peculiar behavior. To see this, consider the integral for $Q_\mu(x,y)$ and the asymptotic behavior of $I_{\mu-1}(2\sqrt{xz}\,)$ which follows from (9.54). We see that the term $\exp(-z-x)I_{\mu-1}(2\sqrt{xz}\,)$ (having dominant exponential part $-(\sqrt{x}-\sqrt{z}\,)^2$) is exponentially small, except when $x \geq y$ and $z \sim x$. It follows that, when x and y are large, the behavior of $Q_\mu(x,y)$ significantly changes when y crosses the value x. It will appear that when μ is large too, this change in behavior occurs when y crosses the value $x+\mu$. In both cases, the asymptotic behavior can be described by using the error function introduced in §11.1.

11.4.1. A Few More Integral Representations

First we show that the Bessel function integrals in (11.63) essentially reduce to sums of two simpler functions. Moreover we obtain symmetrical representations for the cases $x < y$ and $x > y$. The auxiliary function is defined by

$$F_\mu(\xi,\sigma) := \int_\xi^\infty e^{-(\sigma+1)t} I_\mu(t)\,dt, \quad \sigma > 0. \quad (11.66)$$

To show that $Q_\mu(x,y)$ can be expressed in terms of this function, we use (see (9.10) and the first relation in (9.26))

$$\left(\frac{z}{x}\right)^{\frac{1}{2}(\mu-1)} I_{\mu-1}(2\sqrt{xz}\,) = \frac{1}{2\pi i}\int e^{sz+x/s}s^{-\mu}\,ds, \quad \mu > 0, \quad (11.67)$$

where the path of integration may be any vertical line in the half plane $\Re s > 0$. The path may be deformed into a Hankel contour $\mathcal{L}$ shown in Figure 9.7. Substituting the loop integral into the second integral in (11.63), we obtain

$$Q_\mu(x,y) = e^{-x}\int_y^\infty \frac{1}{2\pi i}\int_{\mathcal{L}} e^{z(s-1)+x/s}\frac{ds}{s^\mu}\,dz.$$

Take $\mathcal{L}$ such that $\Re s < 1$ for any $s \in \mathcal{L}$. Then, by absolute convergence of the repeated integrals, we may interchange the order of integration. Deforming $\mathcal{L}$ back into a vertical line with $0 < \Re s < 1$ we obtain

$$Q_\mu(x,y) = \frac{e^{-x-y}}{2\pi i}\int_{c-i\infty}^{c+i\infty} \frac{e^{x/s+ys}}{1-s}\,\frac{ds}{s^\mu}, \quad 0 < c < 1. \quad (11.68)$$

When we move the vertical line to the right, across the pole at $s = 1$ and taking into account the residue, we obtain

$$P_\mu(x,y) = \frac{e^{-x-y}}{2\pi i}\int_{c-i\infty}^{c+i\infty} \frac{e^{x/s+ys}}{s-1}\,\frac{ds}{s^\mu}, \quad c > 1. \tag{11.69}$$

In (11.68) we substitute $s = t/\rho$ with $\rho = \sqrt{y/x}$. It follows that

$$Q_\mu(x,y) = e^{-x-y-2z\lambda}\rho^\mu \Phi(z),$$

$$\Phi(z) = \frac{1}{2\pi i}\int_{c-i\infty}^{c+i\infty} \frac{e^{z(t+1/t+2\lambda)}}{\rho - t}\,\frac{dt}{t^\mu}, \quad 0 < c < \rho, \tag{11.70}$$

where $z = \sqrt{xy}$. We now assume, for the time being, that $\rho > 1$. Taking $2\lambda = -(\rho + 1/\rho)$, and assuming (again, for the time being) that ρ does not depend on x, y, z, we obtain

$$\frac{d\Phi(z)}{dz} = -\frac{e^{2\lambda z}}{2\pi i}\int_{c-i\infty}^{c+i\infty} e^{z(t+1/t)}\left(t - \frac{1}{\rho}\right)\frac{dt}{t^{\mu+1}}.$$

Invoking (11.67), we derive

$$\frac{d\Phi(z)}{dz} = -e^{2\lambda z}\left[I_{\mu-1}(2z) - \frac{1}{\rho}I_\mu(2z)\right].$$

To integrate this we use $\Phi(\infty) = 0$. This follows from standard techniques from asymptotics applied on (11.70), for instance the saddle point method. Observe that the exponential function of the integrand in (11.70) has a saddle point at $s = 1$. We obtain

$$Q_\mu(x,y) = \rho^\mu\Phi(z) = \tfrac{1}{2}\rho^\mu\left[F_{\mu-1}(\xi,\sigma) - \frac{1}{\rho}F_\mu(\xi,\sigma)\right], \quad y > x; \tag{11.71}$$

ρ has regained its original meaning and the F-function is defined in (11.66). Furthermore

$$\xi = 2\sqrt{xy}, \quad \sigma = -\lambda - 1 = \frac{(\sqrt{y}-\sqrt{x})^2}{\xi}, \quad \rho = \sqrt{\frac{y}{x}}. \tag{11.72}$$

Now let $\rho < 1$. Repeating the analysis that leads to (11.71), but now with starting point the integral in (11.69), we obtain for this case

$$P_\mu(x,y) = \frac{1}{2}\rho^\mu\left[\frac{1}{\rho}F_\mu(\xi,\sigma) - F_{\mu-1}(\xi,\sigma)\right], \quad y < x, \tag{11.73}$$

where the parameters are as in (11.72).

In the following subsections the large ξ–behavior of $Q_\mu(x,y)$ is discussed. We have, as $\xi \to \infty$ and ρ fixed,

$$Q_\mu(x,y) \sim \begin{cases} 1, & \text{if } \rho < 1; \\ \frac{1}{2}, & \text{if } \rho = 1; \\ 0, & \text{if } \rho > 1. \end{cases} \tag{11.74}$$

It will be shown that a smooth transition can be described in terms of the error function (the normal distribution function).

11.4.2. Asymptotic Expansion; μ Fixed, ξ Large

We concentrate on the function $F_\mu(\xi,\sigma)$ given in (11.66). We point out that this function with ξ and σ as in (11.72) is symmetric in x and y, and occurs in both (11.71) and (11.73). Hence, it is sufficient to assume $x < y$. The case $x = y$ follows from the asymptotic results when we let $x \to y$.

The asymptotic feature is that ξ is large, whereas σ tends to zero when $x \to y$. We give an asymptotic expansion that holds uniformly with respect to $\sigma \in [0,\infty)$. Note that the integral defining $F_\mu(\xi,\sigma)$ becomes undetermined when $\sigma = 0$. However, since we use a combination of two F–functions in (11.71), and ρ tends to unity as $x \to y$, the function $Q_\mu(x,y)$ is well defined in this limit.

We substitute in (11.66) the expansion in (9.54) written in the form

$$e^{-t}I_\mu(t) \sim \frac{1}{\sqrt{2\pi t}}\sum_{n=0}^{\infty}(-1)^n\frac{A_n(\mu)}{t^n},$$

where $A_n(\mu) = 2^{-n}(\mu,n)$, with recursion

$$A_{n+1}(\mu) = -\frac{(2n+1)^2 - 4\mu^2}{8(n+1)}A_n(\mu), \quad n \ge 0, \quad A_0(\mu) = 1.$$

This gives the formal expansion

$$F_\mu(\xi,\sigma) \sim \frac{1}{\sqrt{2\pi}}\sum_{n=0}^{\infty}(-1)^n A_n(\mu)\phi_n, \tag{11.75}$$

where ϕ_n is an incomplete gamma function:

$$\phi_n = \int_\xi^\infty e^{-\sigma t}t^{-n-\frac{1}{2}}\,dt = \sigma^{n-1/2}\Gamma\left(\tfrac{1}{2} - n, \sigma\xi\right). \tag{11.76}$$

The function ϕ_0 is an error function (see §11.1):

$$\phi_0 = \sqrt{\pi/\sigma}\,\mathrm{erfc}\,\sqrt{\sigma\xi} = \sqrt{\pi/\sigma}\,\mathrm{erfc}\left(\sqrt{y} - \sqrt{x}\right). \tag{11.77}$$

Further terms can be obtained from the recursion

$$(n - \tfrac{1}{2})\phi_n = -\sigma\phi_{n-1} + e^{-\sigma\xi}\xi^{-n+\frac{1}{2}}, \quad n = 1, 2, 3, \ldots, \tag{11.78}$$

which follows from (11.7) or from integrating by parts in (11.76).

Using (11.71) and (11.75) we obtain

$$Q_\mu(x, y) \sim \sum_{n=0}^{\infty} \psi_n, \quad \psi_n = \frac{\rho^\mu}{2\sqrt{2\pi}}(-1)^n \left[A_n(\mu - 1) - \frac{1}{\rho}A_n(\mu)\right]\phi_n. \tag{11.79}$$

Expansion (11.75) holds for large values of ξ, uniformly with respect to $\sigma \in [0, \infty)$.

Because $(\rho - 1)/\sqrt{2\sigma} = \sqrt{\rho}$, the first term approximation of the series in (11.79) reads,

$$Q_\mu(x, y) \sim \psi_0 = \frac{1}{2}\rho^{\mu-\frac{1}{2}}\mathrm{erfc}\left(\sqrt{y} - \sqrt{x}\right). \tag{11.80}$$

We remark that the right-hand side reduces to $\frac{1}{2}$ when $x \uparrow y$.

Remark 11.2. When $x > y$, that is $\rho < 1$, the expression $(\rho - 1)/\sqrt{2\sigma}$ should be interpreted as $-\sqrt{\rho}$, and (11.73) gives

$$Q_\mu(x, y) \sim 1 - \psi_0 = 1 - \frac{1}{2}\rho^{\mu-\frac{1}{2}}\mathrm{erfc}\left(\sqrt{x} - \sqrt{y}\right).$$

Again, the right-hand side reduces to $\frac{1}{2}$ when $x \downarrow y$.

11.4.3. Asymptotic Expansion; ξ Large, μ Arbitrary

In this case we consider (11.68). We write

$$Q_\mu(x, y) = \rho^\mu e^{-x-y}\frac{1}{2\pi i}\int_{\mathcal{L}} e^{\xi\alpha(t)}\frac{dt}{\rho - t},$$

where ρ and ξ are as in (11.72), and

$$\alpha(t) = \frac{1}{2}(t + 1/t) - \beta\ln t, \quad \beta = \frac{\mu}{\xi}.$$

The path of integration $\mathcal{L}$ is a vertical line $\Re t = c$, with $0 < c < \rho$. However, $\mathcal{L}$ may be deformed into a different contour, for instance into the path of steepest descent through a saddle point. The saddle points are solutions of the equation $\alpha'(t) = 0$. We select the positive saddle point

$$t_0 = \beta + \sqrt{\beta^2 + 1}.$$

It is convenient to write

$$\beta = \sinh, \quad \rho = e^{\theta}. \tag{11.81}$$

Then we have

$$t_0 = e^{\gamma}, \quad \alpha(t_0) = \cosh\gamma - \gamma\sinh\gamma.$$

Observe that when $\gamma \sim \theta$ the saddle point and the pole are close together. As in earlier sections we handle this case by using an error function. The case $\gamma = \theta$ corresponds to $y = x + \mu$. When ξ is large and y crosses the value $x + \mu$, the function $Q_\mu(x, y)$ suddenly changes. We have (cf. (11.74))

$$Q_\mu(x,y) \sim \begin{cases} 1, & \text{if } x + \mu > y; \\ \frac{1}{2}, & \text{if } x + \mu = y; \\ 0, & \text{if } x + \mu < y. \end{cases}$$

In terms of μ and γ these cases read $\theta < \gamma$, $\theta = \gamma$, $\theta > \gamma$, respectively. When $t_0 < \rho$, that is, when $\theta > \gamma$ or $y > x + \mu$, we can shift $\mathcal{L}$ through the saddle point, without passing the pole at $t = \rho$. We temporarily assume that $t_0 < \rho$.

The path of steepest descent $\mathcal{L}$ through t_0 follows from the equation $\Im\alpha(t) = 0$. Let $t = re^{i\phi}$. Then we can describe $\mathcal{L}$ by

$$r = \sinh\gamma\frac{\phi}{\sin\phi} + \sqrt{1 + \sinh^2\gamma\frac{\phi^2}{\sin^2\phi}}, \quad -\pi < \phi < \pi.$$

We define a mapping $t \mapsto u(t)$ that maps $\mathcal{L}$ to $\mathbb{R}$ by writing

$$\frac{1}{2}u^2 = \alpha(t_0) - \alpha(t).$$

When t follows $\mathcal{L}$ we take $u \in \mathbb{R}$, with $\text{sign}(u) = \text{sign}(\Im t)$. The pole at $t = \rho$ is mapped to the point iu_0, where u_0 is defined by

$$\frac{1}{2}u_0^2 = \cosh\theta - \cosh\gamma + (\gamma - \theta)\sinh\gamma,$$

where θ and γ are introduced in (11.81). The sign of u_0 follows from the definition of the mapping $t \mapsto u(t)$: we have $\text{sign}(u_0) = \text{sign}(\gamma - \theta) = \text{sign}(x + \mu - y)$. Integrating with respect to u, and splitting off the pole at $u = iu_0$, we obtain

$$Q_\mu(x,y) = \frac{1}{2}\text{erfc}\left(-u_0\sqrt{\xi/2}\right) + \frac{e^{-\frac{1}{2}\xi u_0^2}}{2\pi i}\int_{-\infty}^{\infty} e^{-\frac{1}{2}\xi u^2} f(u)\, du, \tag{11.82}$$

where

$$f(u) = \frac{dt}{du}\frac{1}{\rho - t} + \frac{1}{u - iu_0}.$$

In deriving the term with the error function we have used the representations of the error functions in §11.1. The asymptotic expansion of $Q_\mu(x,y)$ now follows by expanding

$$f(u) = i\sum_{n=0}^{\infty} c_n u^n$$

and by substituting this in the above integral. This gives

$$Q_\mu(x,y) \sim \frac{1}{2}\operatorname{erfc}\left(-u_0\sqrt{\xi/2}\right) + \frac{e^{-\frac{1}{2}\xi u_0^2}}{\sqrt{2\pi\xi}}\sum_{n=0}^{\infty} c_{2n}\frac{\Gamma\left(n+\frac{1}{2}\right)}{\Gamma\left(\frac{1}{2}\right)}\left(\frac{2}{\xi}\right)^n, \quad (11.83)$$

as $\xi \to \infty$. This expansion holds uniformly with respect to $\mu \in [0,\infty)$. The first coefficients are

$$c_0 = \frac{1}{\sqrt{\cosh\gamma}}\frac{1}{e^{\theta-\gamma}-1} + \frac{1}{u_0},$$

$$c_2 = -\frac{e^{2\gamma}+e^{-2\gamma}-8+e^{\theta-\gamma}A+e^{2\theta-2\gamma}B}{48\cosh^{7/2}\gamma\,\left(e^{\theta-\gamma}-1\right)^3} - \frac{1}{u_0^3},$$

where

$$A = 10e^{2\gamma} - 2e^{-2\gamma} + 28, \quad B = e^{2\gamma} + 13e^{-2\gamma} + 4.$$

Remark 11.3. We have temporarily assumed $t_0 < \rho$, that is $y > x + \mu$. In (11.82) this condition can be dropped. The expansion in (11.83) also holds for $y \le x + \mu$. Note that a single error function describes the transition from $y > x + \mu$ to $y < x + \mu$, and we do not need different representations for $Q_\mu(x,y)$ as in the previous subsection; confer (11.71) and (11.73). The method of this subsection can also be used when μ is fixed. However, the method of the previous subsection gives very simple coefficients in expansion (11.71).

11.4.4. Numerical Aspects

In applications it is of interest to have available algorithms for $Q_\mu(x,y)$ when $0 < Q_\mu(x,y) \le \frac{1}{2}$ and for $P_\mu(x,y)$ otherwise. The inequalities apply when (roughly speaking) $y \ge x + \mu$ (this follows from the asymptotic expansion of the previous section).

Recurrence relation (11.64) is very useful for computing $Q_\mu(x,y)$. It is numerically stable in forward direction, since the right-hand side of (11.64) has positive terms. An algorithm for the modified Bessel function is needed. A point of warning: the recurrence relation for the modified Bessel function should not be used in forward direction; see §13.4. Observe that the function $P_\mu(x,y)$ satisfies the recursion

$$P_\mu(x,y) = P_{\mu+1}(x,y) + \left(\frac{y}{x}\right)^{\mu/2} e^{-x} I_\mu\left(2\sqrt{xy}\right),$$

which is stable in backward direction. In the homogeneous recurrence relation (11.65) Bessel functions do not occur. It is attractive to use this equation in order to avoid the forward recursion of the Bessel functions. However, one needs to investigate the stability of (11.65) in more detail, and for several combinations of the parameters, which is not a trivial problem. Observe that any constant function (with respect to μ) solves (11.65), and that, hence, $P_\mu(x,y)$ satisfies the same recurrence relation.

For small and moderate values of x, y, μ the expansions (11.61) and (11.62) can be used. Both series have positive terms and both series require the evaluation of one incomplete gamma function. The series in (11.61) requires the value $Q(\mu, y)$, and the remaining terms follow from the stable recursion

$$Q(\mu+n+1,y) = Q(\mu+n,y) + \frac{y^{n+\mu}e^{-y}}{\Gamma(\mu+n+1)}, \qquad n = 0,1,2,\ldots .$$

The series in (11.62) requires an initial $P-$value. The corresponding recursion should be used in the backward direction:

$$P(\mu+n,y) = P(\mu+n+1,y) + \frac{y^{n+\mu}e^{-y}}{\Gamma(\mu+n+1)},$$

because the forward form is not stable. Let n_0 be the (smallest) number such that

$$P_\mu(x,y) \simeq e^{-x}\sum_{n=0}^{n_0}\frac{x^n}{n!}P(\mu+n,y), \tag{11.84}$$

within the required relative accuracy. Then as starting value we need to compute $P(\mu+n_0,y)$, and the remaining values follow from the above recursion. To estimate n_0 we may use

$$P(\mu+n,y) \sim \frac{y^{n+\mu}e^{-y}}{\Gamma(\mu+n+1)}, \qquad \text{as} \quad \mu+n\to\infty.$$

For obtaining *relative* accuracy, we need an estimate of $P_\mu(x,y)$. One can use the value of the integrand of the first integral of (11.63) at $z = y$, that is,

$$P_\mu(x,y) \sim \rho^\mu e^{-x-y} I_\mu(\xi).$$

Table 11.1. n_0 Is the Number of Terms Used in the Series (11.61) or (11.62); $\mu = 8192,\ y = 1.05\mu$; the Relative Accuracy is 10^{-10}

x/μ	n_0	$Q_\mu(x,y)$	$P_\mu(x,y)$
0.01	150	1.984527803e−4	9.998015472e−1
0.03	355	4.000364970e−2	9.599963503e−1
0.05	543	4.985354536e−1	5.014645464e−1
0.07	727	9.556573418e−1	4.434265825e−2
0.09	894	9.996249724e−1	3.750276164e−4
0.11	1054	9.999997188e−1	2.811864384e−7
0.13	1207	1.000000000e+0	1.999694515e−11

Table 11.1 shows the number of terms n_0 used in the series of (11.61), for several values of x. In all cases $\mu = 8192,\ y = 1.05\mu$.

For large values of the parameters the computation can be based on the uniform expansion (11.83). Special care has to be taken when $y \sim x + \mu$, that is, $\theta \sim \gamma$. First it is convenient to have an expansion of u_0. We have

$$u_0 = (\gamma - \theta)\sqrt{2[\cosh\theta - \cosh\gamma - (\theta-\gamma)\sinh]/(\theta-\gamma)^2}\,,$$

where the square root should be taken positive. The expression inside the square root can easily be expanded in powers of $\theta - \gamma$.

The coefficient c_0 has the expansion

$$c_0 = \frac{1}{6\cosh^{3/2}\gamma}\left[a_0 + a_1\zeta + a_2\zeta^2 + \mathcal{O}\left(\zeta^3\right)\right],$$

as $\zeta \to 0$, where

$$a_0 = \sinh\gamma - 3\cosh\gamma,\ a_1 = \left(\frac{1}{2}\sinh^2\gamma + \frac{3}{4}\right),\ a_2 = -\frac{\sinh\gamma(2\sinh^2\gamma + 27)}{360},$$

and $\zeta = (\theta-\gamma)/\cosh\gamma$. For c_2 we have

$$\begin{aligned} c_2 = &-\frac{e^{-3\gamma}(e^{6\gamma} + 6e^{4\gamma} + 309e^{2\gamma} - 46)}{4320\cosh^{9/2}\gamma} \\ &-\frac{e^{-4\gamma}(1 - 8e^{2\gamma} + e^{4\gamma})(1 + 16e^{2\gamma} + e^{4\gamma})}{4608\cosh^{9/2}\gamma}\zeta + \mathcal{O}\left(\zeta^2\right). \end{aligned}$$

When these approximations are used if $|\theta - \gamma| < 10^{-4}$ for the coefficient c_0, if $|\theta - \gamma| < 0.8 \times 10^{-3}$ for c_2, and (11.83) is used with these two coefficients under the condition $\sqrt{xy} + \mu > 1600$, then the relative accuracy is about ten digits, unless $Q_\mu(x, y)$ or $P_\mu(x, y)$ is quite small, say smaller than 10^{-20}, in which case some digits may be lost.

11.5. An Incomplete Bessel Function

From the Fourier expansion in (9.33) it follows that the modified Bessel function can be written as

$$I_0(z) = \frac{1}{2\pi} \int_{-\pi}^{\pi} e^{z \cos t}\, dt. \tag{11.85}$$

An incomplete version of this integral plays a role as a cumulative distribution function. We define

$$I_0(\theta, \kappa) = \frac{1}{2\pi\, I_0(\kappa)} \int_{-\pi}^{\theta} e^{\kappa \cos t}\, dt, \tag{11.86}$$

which is formally equivalent to the cumulative distribution applied by VON MISES (1918) to study deviations of atomic weights from integer values, representable as points on the circumference of a circle or as circular directions. The parameter θ is the angular deviation and κ is the concentration parameter. This distribution of points on a circle is analogous to the normal or Gaussian distribution of points on a line and has applications to the study of quantal or periodic data, directions of sedimentary bedding, surface fault lines, wildlife movements, etc.

To evaluate $I_0(\theta, \kappa)$ one can substitute the Fourier series of (9.33), and it follows that

$$I_0(\theta, \kappa) = \frac{1}{2} + \frac{\theta}{2\pi} + \frac{1}{\pi I_0(\kappa)} \sum_{n=1}^{\infty} \frac{\sin n\theta}{n} I_n(\kappa). \tag{11.87}$$

The modified Bessel functions can be evaluated by using their recurrence relation given in (9.34); see Example 13.1 at the end of §13.4 for more details.

When κ is large (11.87) cannot be used for computing $I_0(\theta, \kappa)$. For instance, when θ is negative, $I_0(\theta, \kappa)$ is very small, and it is difficult to obtain high relative accuracy when summing the series numerically. As in the previous sections an error function can be used to describe the transition from negative to positive values of θ. We derive an expansion that can be used for large values of κ, and the expansion holds uniformly with respect to $\theta \in [-\pi, \pi]$.

We write

$$I_0(\theta,\kappa) = \frac{1}{2} + \frac{1}{2\pi\, I_0(\kappa)} \int_0^\theta e^{\kappa \cos t}\, dt$$

and substitute $\cos t = 1 + 2\sin^2 t/2 = 1 + 2x^2$, or $x = \sin t/2$. This gives

$$I_0(\theta,\kappa) = \frac{1}{2} + \frac{e^\kappa}{\pi\, I_0(\kappa)} \int_0^{\sin\frac{1}{2}\theta} e^{-2\kappa x^2} \frac{dx}{\sqrt{1-x^2}}.$$

Expanding the square root, we obtain the (convergent) asymptotic expansion

$$I_0(\theta,\kappa) = \frac{1}{2} + \frac{e^\kappa}{\pi\, I_0(\kappa)} \sum_{n=0}^{\infty} (-1)^n \binom{-\frac{1}{2}}{n} \phi_n(\theta,\kappa), \tag{11.88}$$

where ϕ_n can be expressed in terms of incomplete gamma functions (see (11.2)):

$$\begin{aligned} \phi_n(\theta,\kappa) &= \int_0^{\sin\frac{1}{2}\theta} e^{-2\kappa x^2} x^{2n}\, dx \\ &= \frac{1}{2}\frac{1}{(2\kappa)^{n+\frac{1}{2}}} \gamma\left(n+\frac{1}{2}, 2\kappa \sin^2\frac{1}{2}\theta\right). \end{aligned} \tag{11.89}$$

To obtain a first order approximation for large values of κ we observe that the first term is an error function, since $\gamma(\frac{1}{2}, z^2) = \sqrt{\pi}\,\mathrm{erf}\, z$, and we use $I_0(\kappa) \sim \frac{1}{\sqrt{2\pi\kappa}} e^\kappa$, see (9.54). This gives

$$I_0(\theta,\kappa) \sim \frac{1}{2} + \frac{1}{2}\mathrm{erf}\left(\sqrt{2\kappa}\, \sin\frac{1}{2}\theta\right) = \frac{1}{2}\mathrm{erfc}\left(-\sqrt{2\kappa}\, \sin\frac{1}{2}\theta\right).$$

For the computation of the incomplete gamma functions we can use the recurrence relation given in (11.7). We have remarked in §11.4.4 (see also §13.1) that this recursion relation is not stable, and that it should be used in backward direction. Another point is that the second line in (11.89) suggests that ϕ_n is an even function of θ, which is not true. It is better to write (see (11.4))

$$\phi_n(\theta,\kappa) = \frac{1}{2}\Gamma\left(n+\frac{1}{2}\right) \sin^{2n+1}\frac{1}{2}\theta\, \gamma^*\left(n+\frac{1}{2}, 2\kappa \sin^2\frac{1}{2}\theta\right).$$

11.6. Remarks and Comments for Further Reading

11.1. Methods of uniform asymptotic expansions for integrals are given in OLVER (1974) and WONG (1989). For the functions considered in this chapter, see also TEMME (1975), (1976), (1979), (1982), (1987A), (1993).

11.2. Rather complete discussions of the computational problem for the incomplete gamma functions P and Q are given by GAUTSCHI (1979) and

DIDONATO & MORRIS (1986). In the latter an algorithm based on the uniform asymptotic expansion of §11.2.4 is used, of which a more efficient version is given in TEMME (1987b), (1994).

11.3. The computational aspects of the incomplete beta function are treated quite well in DIDONATO & MORRIS (1992).

11.4. The analysis of §11.4 is based on TEMME (1993). Information on the asymptotic nature and error bounds of expansion (11.75) can be found in TEMME (1986), where also numerical aspects of recursion (11.78) are discussed. Table 11.1 has the same x, y, μ values as Table I in ROBERTSON (1969).

11.5. Part of the treatment of the Von Mises distribution of §11.5 is based on HILL (1977), where also a Fortran algorithm is given.

11.7. Exercises and Further Examples

11.1. Show that

$$\operatorname{erfc} z = \frac{e^{-z^2}}{z\sqrt{\pi}} \int_0^\infty e^{-t} \frac{dt}{\sqrt{1+t/z^2}}, \qquad |\arg z| < \frac{1}{2}\pi,$$

and obtain the asymptotic representation

$$\operatorname{erfc} z = \frac{e^{-z^2}}{z\sqrt{\pi}} \left[\sum_{m=0}^{n-1} (-1)^m \left(\frac{1}{2}\right)_m z^{-2m} + (-1)^n \left(\frac{1}{2}\right)_n z^{-2n}\theta_n(z)\right],$$

where

$$\theta_n(z) = \int_0^\infty e^{-t} \left(1 + t/z^2\right)^{-n-\frac{1}{2}} dt.$$

Let $z \in \mathbb{C}$, such that

$$|1 + t/z^2| \geq 1, \quad \text{for all} \quad t \geq 0. \tag{1}$$

Then $|\theta_n(z)| \leq 1$ for this z. Verify that (1) holds when $|\arg z^2| \leq \frac{1}{2}\pi$, $z \neq 0$. To verify this observe that the equation $|1 + \zeta| = 1$ in the ζ−plane is satisfied by the points on the circle $(u+1)^2 + v^2 = 1$, where $\zeta = u + iv$.

11.2. Show by using Exercise 3.7 and the first relation in (11.5) that

$$\gamma(a, z) = \frac{z^a \Gamma(a)\Gamma(1-a)}{2\pi i} \int_{-1}^{(0^+)} t^{a-1} e^{zt}\, dt.$$

The integral is defined for all complex values of z and a. The singularities with respect to a become visible now through those of $\Gamma(a)$. The poles of

$\Gamma(1-a)$ at $a = 1, 2, 3, \ldots$ are removable singularities in this representation, because the integral vanishes for these integer values of a. A proper limiting process should yield the first relation of (11.6). Verify that

$$\gamma^*(-n, z) = z^n, \quad n = 0, 1, 2, \ldots,$$

which also easily follows from the first representation of $\gamma^*(a, z)$ in (11.5).

11.3. Show that the analytic continuation of the incomplete gamma function $\Gamma(a, z)$ is described by

$$\Gamma\left(a, ze^{\pm\pi i}\right) = \Gamma(a) - \Gamma(a)\left(ze^{\pm\pi i}\right)^a \gamma^*(a, -z)$$

where $\gamma^*(a, z)$ is the entire function defined in (11.4). By using the relation with the M–function, given in (11.5), and by using (7.16), show that

$$\Gamma\left(a, ze^{\pm\pi i}\right) = \left(1 - e^{\pm 2\pi i a}\right)\Gamma(a) + e^{\pm 2\pi i a}\Gamma\left(a, ze^{\mp\pi i}\right).$$

This relation is important, for instance for describing the asymptotic behavior of $\Gamma(a, z)$ (see (11.11)) outside the range $-\frac{3}{2}\pi < \arg z < \frac{3}{2}\pi$. Verify what happens when $a = \frac{1}{2}$ (the error function case, as mentioned in Exercise 6.4).

11.4. Consider (11.23) and (11.24). Introduce a function $S_a(\eta)$ by writing:

$$R_a(\eta) = \frac{e^{-\frac{1}{2}a\eta^2}}{\sqrt{2\pi a}} S_a(\eta).$$

Show by differentiating the first line of (11.23) with respect to η and using

$$\frac{dz}{d\eta} = a\frac{d\lambda}{d\eta} = a\frac{\lambda\eta}{\lambda - 1}$$

that $S_a(\eta)$ satisfies the differential equation

$$\frac{1}{a}\frac{d}{d\eta}S_a(\eta) = \eta S_a(\eta) + 1 - \frac{f(\eta)}{\Gamma^*(a)}$$

where $f(\eta) = \eta/(\lambda - 1)$ and

$$\Gamma^*(a) = \frac{\Gamma(a)}{\sqrt{2\pi}\, a^{a-\frac{1}{2}} e^{-a}} \sim \sum_{n=0}^{\infty} \frac{\gamma_n}{a^n}, \quad a \to \infty.$$

See (3.25) and (3.37). Substitute $S_a(\eta) \sim \sum_{n=0}^{\infty} c_n(\eta) a^{-n}$ and the expansion of $1/\Gamma^*(a)$ in the differential equation for $S_a(\eta)$ and compare equal powers of a to derive the following set of relations for $c_n(\eta)$:

$$\eta c_n(\eta) = \gamma_n f(\eta) + \frac{d}{d\eta}c_{n-1}(\eta), \quad n = 1, 2, \ldots,$$

with as first value

$$c_0(\eta) = \frac{f(\eta) - 1}{\eta} = \frac{1}{\lambda - 1} - \frac{1}{\eta}.$$

This also follows from $c_0(\eta) = -ig(0)$, see (11.21) and (11.23); observe that $dt/du = i$ at $u = 0$. Show that the next coefficient is given by

$$c_1(\eta) = \frac{1}{\eta^3} - \frac{1}{(\lambda-1)^3} - \frac{1}{(\lambda-1)^2} - \frac{1}{12(\lambda-1)}.$$

11.5. To derive the continued fraction in (11.35) show first that the hypergeometric functions satisfy the relation

$$F(a, b+1; c+1; x) - F(a, b; c; x) = \frac{a(c-b)}{c(c+1)}\, x\, F(a+1, b+1; c+2; x),$$

form which follows

$$G(a, b, c, x) = \frac{1}{1 - \frac{a(c-b)}{c(c+1)}\, x\, G(b+1, a, c+1, x)}, \tag{1}$$

where

$$G(a, b, c, x) = \frac{F(a, b+1; c+1; x)}{F(a, b; c; x)}.$$

Interchange in (1) the parameters a and b, and next change b to $b+1$, c to $c+1$. This gives

$$G(b+1, a, c+1, x) = \frac{1}{1 - \frac{(b+1)(c+1-a)}{(c+1)(c+2)}\, x\, G(a+1, b+1, c+2, x)}.$$

When we substitute this in the right-hand side of (1), we observe that we have a relation between $G(a, b, c, x)$ and $G(a+1, b+1, c+2, x)$, which is the start of a continued fraction expansion of $G(a, b, c, x)$. Taking $b = 0, a = p + q, c = p$ and using the third line in (11.34), we can derive (11.35).

11.6. Consider the representation of $I_x(p, q)$ given in the first line of (11.53). Introduce a function $S_x(p, q)$ by writing

$$R_x(p, q) = \frac{x^p\,(1-x)^q\, e^{pA(\mu)} p^q}{\Gamma(q+1)} S_x(p, q),$$

where $\mu, A(\mu), \eta$ are defined in (11.47), (11.48), (11.52). Observe that from (11.54) it follows that $S_x(p, q) \sim \sum_{n=0}^{\infty} h_n(\eta, \mu)\Phi_n p^{-n}$. Show, by using (11.57) and (11.59), (11.60), that $S_x(p, q)$ satisfies the differential equation

$$\frac{\eta}{p}\frac{d}{d\eta} S_x(p, q) + (\mu - \eta) S_x(p, q) = \mu\left[\phi(\eta)\Phi(p, q) - 1\right],$$

where

$$\phi(\eta) = \frac{\eta - \mu}{1 - (1+\mu)x} \frac{1}{\sqrt{1+\mu}}, \qquad \Phi(p,q) = \frac{\Gamma^*(p+q)}{\Gamma^*(p)}.$$

Substitute the expansion

$$\Phi(p,q) \sim \sum_{n=0}^{\infty} \frac{c_n(\mu)}{p^n}, \qquad \mu = \frac{q}{p},$$

where $c_0(\mu) = 1$ and

$$c_1(\mu) = -\frac{\mu}{12(1+\mu)}, \; c_2(\mu) = \frac{\mu^2}{288(1+\mu)^2}, \; c_3(\mu) = \frac{\mu(432 + 432\mu + 139\mu^2)}{51840(1+\mu)^3},$$

to determine the coefficients of the expansion

$$S_x(p,q) \sim \sum_{n=0}^{\infty} \frac{d_n(\eta,\mu)}{p^n}.$$

Verify that $d_0(\eta,\mu) = \mu[\phi(\eta) - 1]/(\mu - \eta)$ and that the higher coefficients satisfy the recurrence relation

$$(\mu - \eta) d_n(\eta,\mu) + \eta \frac{d}{d\eta} d_{n-1}(\eta,\mu) = \mu\phi(\eta) c_n(\mu), \qquad n = 1, 2, 3, \ldots .$$

Also, verify that $d_0(\eta,\mu) = h_0(\eta,\mu)$ given in (11.56).

12

Elliptic Integrals and Elliptic Functions

Any integral of the type $\int R(x,y)\,dx$, where $R(x,y)$ is a rational function of x and y, with

$$y^2 = a_0x^4 + a_1x^3 + a_2x^2 + a_3x + a_4, \qquad |a_0| + |a_1| > 0,$$

a polynomial of the third or fourth degree in x, is called an *elliptic integral.* Elliptic integrals cannot, in general, be expressed in terms of elementary functions. Exceptions to this are

- when $R(x,y)$ contains no odd powers of y;
- when the polynomial y^2 has a repeated factor.

One can show using suitable transformations that all elliptic integrals can be expressed in terms of three standard integrals, which are called *Legendre's normal elliptic integrals* of the first, second and third kind. The *elliptic functions* considered here can be expressed as inverse functions of an elliptic integral. We also consider *theta functions*, an important class of functions closely related to elliptic functions.

12.1. Complete Integrals of the First and Second Kind

The basic integrals in this field are the *complete elliptic integrals of the first kind*

$$\boxed{K(k) = \int_0^{\pi/2} \frac{d\theta}{\sqrt{1-k^2\sin^2\theta}} = \int_0^1 \frac{dt}{\sqrt{(1-t^2)(1-k^2t^2)}}} \qquad (12.1)$$

and the *complete integral of the second kind*

$$\boxed{E(k) = \int_0^{\pi/2} \sqrt{1-k^2\sin^2\theta}\,d\theta = \int_0^1 \frac{\sqrt{1-k^2t^2}}{\sqrt{1-t^2}}\,dt.} \qquad (12.2)$$

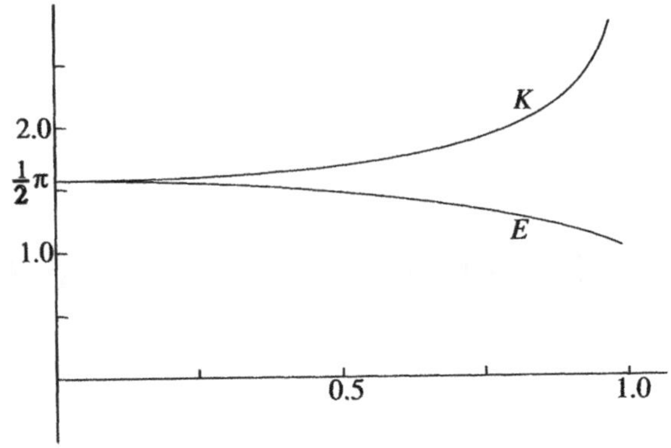

Figure 12.1. The complete elliptic integrals $K(k)$ and $E(k), 0 \le k \le 1$.

In Figure 12.1 we show the graphs of the complete elliptic integrals $K(k)$ and $E(k)$ for $0 \le k \le 1$.

In Exercise 5.2 the relation with hypergeometric functions is given. The *complementary elliptic integrals* E' and K' are the integrals with the complementary variable $k' = \sqrt{1-k^2}$. That is,

$$K'(k) = K\left(\sqrt{1-k^2}\right) = K(k'), \tag{12.3}$$

$$E'(k) = E\left(\sqrt{1-k^2}\right) = E(k'). \tag{12.4}$$

It is common to use the prime when the complementary variable is meant: $f'(k) = f(k')$. Differentiation will be denoted differently. The variable k is called the *modulus* and k' the *complementary modulus*.

The integral in (12.2) is related to the perimeter of an ellipse. In an ellipse with semi-axes a and b the perimeter A equals

$$A = 4\int_0^{\pi/2} \sqrt{a^2\cos^2\theta + b^2\sin^2\theta}\; d\theta = 4aE'\left(\frac{b}{a}\right).$$

Elliptic integrals arise in many physical problems. The integral in (12.1) has an interesting physical interpretation.

12.1.1. The Simple Pendulum

The simple pendulum is shown in Figure 12.1. If p is the period of a simple pendulum with maximal amplitude α and length L, then

$$p = 4\sqrt{L/g}\,K\left(\sin\frac{1}{2}\alpha\right),$$

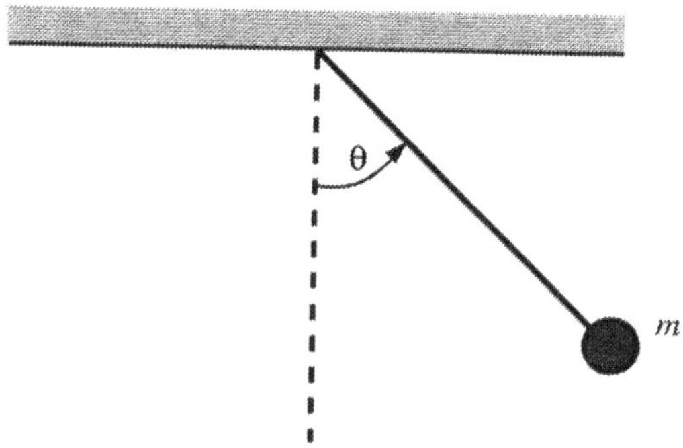

Figure 12.2. A simple pendulum.

where g is the gravitational acceleration.

To derive this, let us denote the amplitude at time t by θ and the mass of the pendulum by m. The kinetic energy of the pendulum at time t equals $\frac{1}{2}mL^2(d\theta/dt)^2$. The potential energy equals $-mgL\cos\theta$ (when we take this equal to zero when $\theta = \frac{1}{2}\pi$). Since at $\theta = \alpha$ (the highest level of the pendulum) the velocity vanishes, we have the following balance of energy:

$$\frac{1}{2}mL^2\left(\frac{d\theta}{dt}\right)^2 - mgL\cos\theta = -mgL\cos\alpha.$$

Solving for $d\theta/dt$ gives

$$\frac{d\theta}{dt} = \pm\sqrt{2g/L}\,\sqrt{\cos\theta - \cos\alpha}\,.$$

Observe that the mass m does not appear any more. We assume that $t = 0$ when $\theta = 0$, and that $d\theta/dt > 0$ at $t = 0$. Integrating from 0 to α yields

$$\int_0^\alpha \frac{d\theta}{\sqrt{\cos\theta - \cos\alpha}} = \sqrt{2g/L}\int_0^{t_\alpha} dt = \sqrt{2g/L}\,t_\alpha,$$

where t_α is the time corresponding with the maximal amplitude α. In other words, t_α is a quarter of the complete period. Hence $p = 4t_\alpha$. A new variable of integration ϕ, defined by

$$\sin\phi = \frac{\sin\frac{1}{2}\theta}{\sin\frac{1}{2}\alpha},$$

finally gives

$$p = 4\sqrt{L/g}\int_0^{\pi/2} \frac{d\phi}{\sqrt{1 - \sin^2\frac{1}{2}\alpha\sin^2\phi}} = 4\sqrt{L/g}\,K\!\left(\sin\frac{1}{2}\alpha\right).$$

Observe that when the amplitudes are small, that is, when $\alpha \to 0$, we have the limiting case $K(\sin \frac{1}{2}\alpha) \to \frac{1}{2}\pi$. We obtain in this case the harmonic oscillator, with period $p \sim 2\pi\sqrt{L/g}$.

12.1.2. Arithmetic Geometric Mean

An important feature of the theory of elliptic integrals is the connection with iterated number sequences based on the *arithmetic geometric mean* (AGM). In 1799 Gauss discovered by sheer luck that iteration by AGM can be linked with elliptic integrals. Consider the following recursions:

$$a_{n+1} = \frac{1}{2}(a_n + b_n), \quad b_{n+1} = \sqrt{a_n b_n}. \tag{12.5}$$

Assuming that $0 < b_0 \leq a_0$, we obtain from elementary properties of the geometric and arithmetic mean that

$$b_n \leq b_{n+1} \leq a_{n+1} \leq a_n,$$

and that

$$0 \leq a_{n+1} - b_{n+1} = \frac{(a_n - b_n)^2}{2\left(\sqrt{a_n} + \sqrt{b_n}\right)^2}.$$

Hence, a_n and b_n both converge (quadratically) to a common limit, which is uniquely determined by a_0 and b_0. Let us write $a = a_0$, $b = b_0$. Then we denote the common limit by $M(a, b)$. That is,

$$M(a, b) = \lim_{n\to\infty} a_n = \lim_{n\to\infty} b_n. \tag{12.6}$$

It is easily verified that this limit is homogeneous. This means that, taking $\lambda > 0$, we have

$$\lambda M(a, b) = M(\lambda a, \lambda b).$$

So M can be regarded as a function of one variable, and without loss of generality we can take $a = 1$. Furthermore,

$$M(a, b) = M\left(\frac{1}{2}(a + b), \sqrt{ab}\right). \tag{12.7}$$

In other words

$$M(1, b) = \frac{1}{2}(1 + b)M\left(1, \frac{2\sqrt{b}}{1 + b}\right). \tag{12.8}$$

It is quite remarkable that Gauss, after patient and careful computations, discovered that

$$\frac{1}{M\left(1, \sqrt{2}\right)} \quad \text{and} \quad \frac{2}{\pi}\int_0^1 \frac{dt}{\sqrt{1 - t^4}}$$

agree up to (at least) eleven digits. We will show that $M(1,x)$ can be expressed in terms of a complete elliptic integral of the first kind.

Theorem 12.1.

$$\frac{1}{M(1,x)} = \frac{2}{\pi}\int_0^{\pi/2} \frac{d\theta}{\sqrt{1-(1-x^2)\sin^2\theta}}, \qquad x>0. \tag{12.9}$$

Proof. Let

$$T(a,b) = \frac{2}{\pi}\int_0^{\pi/2} \frac{d\theta}{\sqrt{a^2\cos^2\theta + b^2\sin^2\theta}}.$$

When $a = b$ we have $T(a,a) = 1/a$. Next, the transformation $t = b\tan\theta$ yields

$$T(a,b) = \frac{1}{\pi}\int_{-\infty}^{\infty} \frac{dt}{\sqrt{(t^2+a^2)(t^2+b^2)}}.$$

Observe that the integrand is an even function of t and that we can take twice the integral over the interval $[0,\infty)$. With a further substitution $u = \frac{1}{2}(t - ab/t)$, and the intermediate results

$$u^2+ab = \frac{(t^2+ab)^2}{4t^2}, \quad u^2 + \left(\frac{a+b}{2}\right)^2 = \frac{(t^2+a^2)(t^2+b^2)}{4t^2}, \quad du = \frac{t^2+ab}{2t^2}\,dt$$

we obtain

$$T(a,b) = \frac{1}{\pi}\int_{-\infty}^{\infty} \frac{du}{\sqrt{(u^2+c^2)(u^2+d^2)}},$$

where $c = (a+b)/2, d = \sqrt{ab}$. Hence,

$$T(a,b) = T\left(\frac{1}{2}(a+b), \sqrt{ab}\right),$$

the same equation satisfied by $M(a,b)$. Now, generating the sequences $\{a_n\}$ and $\{b_n\}$ with $a_0 = a = 1$, $b_0 = b = x$ through the AGM-iteration, we observe that $T(a_n, b_n)$ does not depend on n. Interchanging the limit and integration, we obtain the result

$$T(1,x) = T(M(1,x), M(1,x)).$$

But $T(M,M) = 1/M$, and (12.9) is verified. ∎

Observe that from (12.9) the relation between $M(a,b)$ and the elliptic integrals follows. That is,

$$\frac{1}{M(1,x)} = \frac{2}{\pi} K\left(\sqrt{1-x^2}\right). \tag{12.10}$$

The functional equation (12.8) leads to the result

$$K(k) = \frac{2}{1+k'} K\left(\frac{1-k'}{1+k'}\right). \tag{12.11}$$

We rewrite this in the form

$$K(g) = \frac{2}{1+g'} K\left(\frac{1-g'}{1+g'}\right)$$

with $g' = \sqrt{1-g^2}$. Substituting $g(k) = 2\sqrt{k}\,/(1+k)$ (with the inverse relation $\sqrt{k} = (1-\sqrt{1-g^2}\,)/g)$ and using the intermediate results

$$\frac{1-g'}{1+g'} = \frac{1-\sqrt{1-g^2}}{1+\sqrt{1-g^2}} = \frac{\left(1-\sqrt{1-g^2}\right)^2}{g^2} = k, \quad \frac{2}{1+g'} = 1+k,$$

we obtain

$$K(k) = \frac{1}{1+k} K\left(\frac{2\sqrt{k}}{1+k}\right). \tag{12.12}$$

Since

$$\frac{1-k'}{1+k'} < k, \quad \frac{2\sqrt{k}}{1+k} > k$$

when $0 < k < 1$, (12.11) is called a downward transformation and (12.12) an upward transformation. Such transformations are quite interesting for numerical applications. The downward transformation produces a sequence of $K-$functions with decreasing argument. A combination of a few AGM-iterations and a power series expansion (based on the expansion of the hypergeometric function), yields a very efficient algorithm.

For the integrals of the second kind transformation formulas are available too. It takes more time to derive them. A few hints are given in Exercise 12.2. The results are (an upward and downward transformation, respectively)

$$E(k) = \frac{1+k}{2} E\left(\frac{2\sqrt{k}}{1+k}\right) + \tfrac{1}{2}k'^2 K(k), \tag{12.13}$$

$$E(k) = (1+k')E\left(\frac{1-k'}{1+k'}\right) - k'K(k). \tag{12.14}$$

12.2. Incomplete Elliptic Integrals

The incomplete integrals are the 'indefinite' elliptic integrals:

$$F(\phi, k) = \int_0^\phi \frac{d\theta}{\sqrt{1 - k^2 \sin^2 \theta}}, \qquad k \in [0, 1], \quad \phi \geq 0, \tag{12.15}$$

$$E(\phi, k) = \int_0^\phi \sqrt{1 - k^2 \sin^2 \theta}\, d\theta, \quad k \in [0, 1], \quad \phi \geq 0, \tag{12.16}$$

of the first and second kind, respectively; ϕ is called the *amplitude of the elliptic integrals*. When $\phi = \frac{1}{2}\pi$ the integrals are complete.

A final standard form is the *elliptic integral of the third kind*

$$\Pi(n; \phi, k) = \int_0^\phi \frac{1}{1 - n \sin^2 \theta} \frac{d\theta}{\sqrt{1 - k^2 \sin^2 \theta}} \qquad k \in [0, 1], \quad \phi \geq 0. \tag{12.17}$$

If $n > 1$ this integral should be interpreted as a Cauchy principal value integral. When we take $x = \sin \phi$ we obtain the following representations:

$$F(\phi, k) = \int_0^x \frac{dt}{\sqrt{(1 - t^2)(1 - k^2 t^2)}}, \tag{12.18}$$

$$E(\phi, k) = \int_0^x \frac{\sqrt{1 - k^2 t^2}}{\sqrt{1 - t^2}}\, dt, \tag{12.19}$$

$$\Pi(n; \phi, k) = \int_0^x \frac{1}{1 - nt^2} \frac{dt}{\sqrt{(1 - t^2)(1 - k^2 t^2)}}. \tag{12.20}$$

The functions in (12.18), (12.19) and (12.20) are considered as the standard forms of the elliptic integrals. As mentioned in the introduction to this chapter, the general form of an elliptic integral is $\int R(x, y)\, dx$, where $R(x, y)$ is a rational function of x and y, and

$$y^2 = a_0 + a_1 x + a_2 x^2 + a_3 x^3 + a_4 x^4.$$

By substituting for y^2 we can write $R(x, y)$ in the form

$$R(x, y) = R_1(x) + y^{-1} R_2(x),$$

where R_1 and R_2 are rational functions of x. Often it is a tedious job to express a given elliptic integral into one of the three standard forms, or into a combination of them. For instance, one has to know the zeros of y^2.

The locations of the limits of integration with respect to the zeros also play an important part in classifying elliptic integrals. Some examples are given in (12.32) and (12.33) below. A hint for the proper transformation is given in Exercise 12.7. Usually one resorts to published tables. The computer algebra systems *Maple* and *Mathematica* can also be used to solve this problem.

12.3. Elliptic Functions and Theta Functions

The idea of taking inverses of incomplete elliptic integrals is due to Abel, Jacobi and Gauss. A simple example of the inverse of an incomplete elliptic integral is the inverse of $F(\phi, k)$ when $k = 0$, that is, of

$$\phi = \arcsin x = \int_0^x \frac{dt}{\sqrt{1-t^2}}.$$

The inverse is $x = \sin\phi$. Observe that this relation is already is used as a substitution in passing from (12.15)–(12.17) to (12.18)–(12.20). When $k \neq 0$ the inverse function is less trivial, but we can proceed similarly. Let us consider the equation:

$$u = \int_0^x \frac{dt}{\sqrt{(1-t^2)(1-k^2t^2)}} \tag{12.21}$$

and let us concentrate on the relation between u and x, with k as a secondary parameter. The inverse relation of (12.21) is written as

$$x = \sin\phi = \operatorname{sn}(u, k). \tag{12.22}$$

When $k = 0$, as above, $u = \arcsin x$ and the sn–function reduces to the well-known sine function: $\operatorname{sn}(u, 0) = \sin(u) = \sin(\phi)$, a periodic function with respect to u, with real period 2π. A different limiting case follows when $k = 1$. Then we have $\operatorname{sn}(u, 1) = \tanh(u)$, a periodic function, with period $i\pi$. Has the period changed continuously when k changes from zero to unity? No! For general complex values of k the function $\operatorname{sn}(u, k)$ has two complex periods with respect to the variable u. This property is certainly not obvious. The theoretical background of this will not be considered here. We will discuss a few elementary aspects of the theory.

Two other functions are defined by

$$\operatorname{cn}(u, k) = \cos\phi, \quad \operatorname{dn}(u, k) = \sqrt{1 - k^2 \sin^2\phi} = \sqrt{1 - k^2 \operatorname{sn}^2(u, k)}. \tag{12.23}$$

Analogous definitions of these functions are:

$$u = \int_1^{\operatorname{cn}(u,k)} \frac{dt}{\sqrt{(1-t^2)(k'^2 + k^2t^2)}}, \tag{12.24}$$

$$u = \int_1^{\mathrm{dn}(u,k)} \frac{dt}{\sqrt{(1-t^2)(t^2-k'^2)}}. \tag{12.25}$$

The two limiting cases $k = 0$ and $k = 1$ again yield trivial periodic functions with respect to u.

12.3.1. Elliptic Functions

Let ω_1 and ω_2 be two real or complex numbers for which the ratio ω_1/ω_2 is not a real number. A function satisfying the relation

$$f(z + 2\omega_1) = f(z), \quad f(z + 2\omega_2) = f(z), \tag{12.26}$$

for all complex values of z at which $f(z)$ exists, is called a *doubly periodic function* of z with periods $2\omega_1$, $2\omega_2$. A doubly periodic function that is meromorphic in the finite part of the complex plane is called an *elliptic function.* A doubly periodic function f is completely defined by its restriction to a so called *fundamental parallelogram*, that is, a parallelogram with corners 0, $2\omega_1$, $2\omega_2$, $2\omega_1 + 2\omega_2$, or a translation thereof; see Figure 12.3.

The circular functions $\sin z, \cos z, \tan z, \sinh z$, and so on, can be interpreted as doubly periodic functions of which one period is infinitely large. The functions sn, cn, dn introduced in (12.24), (12.25) and (12.26) are the basic elements in the theory of the elliptic Jacobi functions.

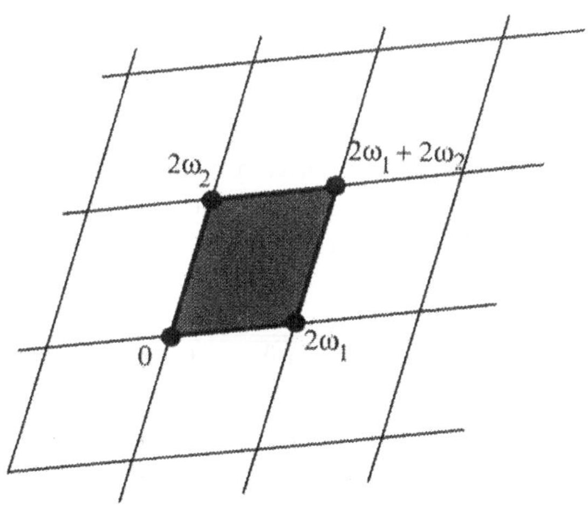

Figure 12.3. A fundamental parallelogram.

Another example of an elliptic function is *Weierstrass' function*:

$$\boxed{\wp(z) = \frac{1}{z^2} + \sum_{n,m}{}' \left[\frac{1}{(z - 2n\omega_1 - 2m\omega_2)^2} - \frac{1}{(2n\omega_1 + 2m\omega_2)^2} \right]} \tag{12.27}$$

where the double series is summed with respect to all integers n and m, except for $n = m = 0$. Is this an elliptic function? You are invited to prove this in Exercise 12.6.

Although Weierstrass' function is defined as a special case, it plays a crucial part in the theory. The fact is that any elliptic function can be written as a rational function of $\wp(z)$ and its derivative.

The function $\wp(z)$ is an even meromorphic function with Laurent series

$$\wp(z) = \frac{1}{z^2} + \frac{1}{20} g_2 z^2 + \frac{1}{28} g_3 z^4 + \cdots,$$

where the constants g_2 and g_3 are denoted by convention. They are defined by

$$g_2 = 60 \sum_{n,m}{}' \frac{1}{(2n\omega_1 + 2m\omega_2)^4}, \quad g_3 = 140 \sum_{n,m}{}' \frac{1}{(2n\omega_1 + 2m\omega_2)^6}.$$

It is quite simple to prove that $y = \wp(z)$ satisfies the differential equation

$$\left(\frac{dy}{dz} \right)^2 = 4y^3 - g_2 y - g_3 \tag{12.28}$$

(see Exercise 12.6). From this we derive a connection between Weierstrass' elliptic function and the elliptic integrals. Namely, a solution of the differential equation can be written in the form

$$z(y) = \int_y^\infty \frac{dt}{\sqrt{4t^3 - g_2 t - g_3}}.$$

The function $z(y)$ can be written in terms of an incomplete elliptic integral. The inverse of this integral is Weierstrass' function $y = \wp(z)$.

12.3.2. Theta Functions

A fundamental part of the theory of elliptic functions is constituted by a set of four functions. They were first investigated by Jacobi and are called *theta functions.* The definitions are:

$$
\begin{aligned}
\theta_1(z,q) &= 2\sum_{n=0}^{\infty}(-1)^n q^{(n+\frac{1}{2})^2}\sin(2n+1)z,\\
\theta_2(z,q) &= 2\sum_{n=0}^{\infty} q^{(n+\frac{1}{2})^2}\cos(2n+1)z,\\
\theta_3(z,q) &= 1+2\sum_{n=1}^{\infty} q^{n^2}\cos 2nz,\\
\theta_4(z,q) &= 1+2\sum_{n=1}^{\infty}(-1)^n q^{n^2}\cos 2nz,
\end{aligned}
\tag{12.29}
$$

where $|q| < 1$ and $z \in \mathbb{C}$.

From elementary properties of the circular functions it follows that (in the notation we drop the parameter q):

$$
\begin{aligned}
\theta_1\left(z+\frac{1}{2}\pi\right) &= \theta_2(z), \quad \theta_2\left(z+\frac{1}{2}\pi\right) = -\theta_1(z),\\
\theta_3\left(z+\frac{1}{2}\pi\right) &= \theta_4(z), \quad \theta_4\left(z+\frac{1}{2}\pi\right) = \theta_3(z)
\end{aligned}
$$

and also that

$$
\begin{aligned}
\theta_1(z+\pi) &= -\theta_1(z), \quad \theta_2(z+\pi) = -\theta_2(z),\\
\theta_3(z+\pi) &= \theta_3(z), \quad \theta_4(z+\pi) = \theta_4(z).
\end{aligned}
$$

We observe that θ_1 and θ_2 are periodic functions with respect to z with period 2π and that θ_3 and θ_4 have period π.

Finding other periods for the theta functions opens the channel to elliptic functions. We first write

$$
\begin{aligned}
\theta_1(z) &= -i\sum_{n=-\infty}^{\infty}(-1)^n q^{(n+\frac{1}{2})^2} e^{(2n+1)iz}, \quad \theta_3(z) = \sum_{n=-\infty}^{\infty} q^{n^2} e^{2niz},\\
\theta_2(z) &= \sum_{n=-\infty}^{\infty} q^{(n+\frac{1}{2})^2} e^{(2n+1)iz}, \quad \theta_4(z) = \sum_{n=-\infty}^{\infty}(-1)^n q^{n^2} e^{2iz}.
\end{aligned}
$$

Writing $q = e^{\pi i\tau}$, with $\Im\tau > 0$, we obtain after some manipulations

$$
\theta_1\left(z+\frac{1}{2}\pi\tau\right) = iq^{-\frac{1}{4}}e^{-iz}\theta_4(z), \quad \theta_4\left(z+\frac{1}{2}\pi\tau\right) = iq^{-\frac{1}{4}}e^{-iz}\theta_1(z).
$$

Hence,

$$
\theta_1(z+\pi\tau) = -q^{-1}e^{-2iz}\theta_1(z), \quad \theta_4(z+\pi\tau) = -q^{-1}e^{-2iz}\theta_4(z)
$$

and similarly

$$\theta_2(z+\pi\tau) = q^{-1}e^{-2iz}\theta_2(z), \quad \theta_3(z+\pi\tau) = q^{-1}e^{-2iz}\theta_3(z).$$

This does not yield periodic functions, but the so-called *quasi periodic functions.* However, it is evident that $\theta_1(z,q)/\theta_4(z,q)$ is periodic with respect to z, with period $\pi\tau$. Since τ is not real we have found for this ratio two periods: 2π and $\pi\tau$. The function $\theta_1(z,q)/\theta_4(z,q)$ is a meromorphic function of z. To verify this we need the zeros of the theta functions, which can be found quite easily. First from the definitions and next from the relations for $z+\frac{1}{2}\pi\tau$ and $z+\pi\tau$, we find that

$$\begin{aligned}
\theta_1(z,q) &= 0 \quad \text{if} \quad z = m\pi + n\pi\tau,\\
\theta_2(z,q) &= 0 \quad \text{if} \quad z = \frac{1}{2}\pi + m\pi + n\pi\tau,\\
\theta_3(z,q) &= 0 \quad \text{if} \quad z = \frac{1}{2}\pi + \frac{1}{2}\pi\tau + m\pi + n\pi\tau,\\
\theta_4(z,q) &= 0 \quad \text{if} \quad z = \frac{1}{2}\pi\tau + m\pi + n\pi\tau.
\end{aligned}$$

It requires some extra work to prove that these are the only zeros. This proof will be omitted. Also, all zeros are simple.

Some ratios of the theta functions yield the elliptic functions that we have introduced earlier. Without proof we give the relations:

$$\begin{aligned}
\operatorname{sn}(z,k) &= \frac{1}{\sqrt{k}}\,\frac{\theta_1(\zeta,q)}{\theta_4(\zeta,q)},\\
\operatorname{cn}(z,k) &= \sqrt{k'/k}\,\frac{\theta_2(\zeta,q)}{\theta_4(\zeta,q)},\\
\operatorname{dn}(z,k) &= \sqrt{k'}\,\frac{\theta_3(\zeta,q)}{\theta_4(\zeta,q)},
\end{aligned} \tag{12.30}$$

where

$$\zeta = \frac{z}{2K\pi}, \qquad q = e^{-\pi K'/K}$$

and K and K' are the complete elliptic integrals defined in (12.1) and (12.3).

Before discussing an interesting functional relation for theta functions we first pay attention to *Poisson's summation formula*, which usually turns up as an application of Fourier theory. We use the following version of this useful result.

Theorem 12.2. *Let f be of bounded variation and absolutely integrable on $\mathbb{R}$; let F be the Fourier transform of f, that is,*

$$F(t) = \int_{-\infty}^{\infty} f(y)e^{ity}\,dy.$$

Then we have

$$\sum_{n=-\infty}^{\infty} e^{ina} f(x+nb) = \frac{1}{b} \sum_{m=-\infty}^{\infty} F\left(\frac{2\pi m + a}{b}\right) e^{-ix(2\pi m+a)/b},$$

where a, b, x are real numbers, $b \neq 0$.

Proof. See ZYGMUND (1959, p. 68) or Exercise 12.8. ∎

It is sufficient to prove the theorem with $b = 1, a = 0, x = 0$ since the general case easily follows from this special case by redefining f. Hence, the basic form of the theorem reads

$$\sum_{n=-\infty}^{\infty} f(n) = \sum_{m=-\infty}^{\infty} F(2\pi m),$$

but the general form is quite convenient in applications. Forms of the theorem for the cosine or sine transform follow by taking for f an even or odd function, respectively.

In mathematical analysis, in particular when one is interested in transformations of series, Poisson's summation formula turns out to be an effective tool for improving convergence of series. A nice example, which directly leads to the theta functions, is the transformation

$$\sum_{n=-\infty}^{\infty} e^{-\pi n^2 s} = \frac{1}{\sqrt{s}} \sum_{m=-\infty}^{\infty} e^{-\pi m^2 s^{-1}}.$$

When s is small, the series on the left-hand side converges poorly, whereas, on the other hand, the series on the right converges strongly. A proof easily follows from well-known Fourier integrals. Take

$$f(y) = \exp(-\pi s y^2).$$

Then we have

$$\begin{aligned} F(t) &= \int_{-\infty}^{\infty} e^{-\pi s y^2 + ity}\, dy \\ &= e^{-t^2/(4\pi s)} \int_{-\infty}^{\infty} e^{-\pi s[y+it/(2\pi s)]^2}\, dy = \frac{1}{\sqrt{s}} e^{-t^2/(4\pi s)}. \end{aligned}$$

For the theta functions similar transformation formulas hold (the above example is a special case). Let us write $\theta_j(z|\tau) = \theta_j(z, q)$ with $q = e^{\pi i \tau}$. Then,

$$\begin{aligned} \theta_1(z|\tau) &= A\,\theta_1(z\tau'|\tau')/i, \\ \theta_2(z|\tau) &= A\,\theta_4(z\tau'|\tau'), \\ \theta_3(z|\tau) &= A\,\theta_3(z\tau'|\tau'), \\ \theta_4(z|\tau) &= A\,\theta_2(z\tau'|\tau'), \end{aligned} \tag{12.31}$$

where

$$A = \frac{1}{\sqrt{-i\tau}} e^{i\tau' z^2/\pi}, \quad \tau' = -\frac{1}{\tau},$$

and where the square root in $\sqrt{-i\tau}$ is positive when τ lies on the positive imaginary axis.

We have discussed a few interesting properties of theta functions and of the more general elliptic functions, and we have only lifted a corner of the veil. This fascinating theory was founded by Jacobi in 1829. He obtained his results by using algebraic methods. Later, the theory was based on more powerful methods of complex function theory (Cauchy's integral). Also, there are interfaces with the $q-$hypergeometric functions introduced in §5.8. The theta functions attract a great deal of attention in modern physics, in particular in the theory of solitons. Several *soliton equations*, for example the Korteweg-de Vries equation, can be solved in terms of (ratios of) theta functions. In classical physics theta functions solve the one dimensional heat equation (the diffusion equation)

$$\frac{\partial \theta}{\partial \tau} = h^2 \frac{\partial^2 \theta}{\partial z^2},$$

for particular values of h. This easily follows from the definitions of the theta functions.

12.4. Numerical Aspects

Published numerical algorithms for the standard elliptic functions can be found in BAKER (1992) and MOSHIER (1989). In general they are computed by using AGM-iterations, which converge very rapidly. When $|q| < 1$ the theta functions can be computed directly from the definitions (12.29); when $|q|$ is bounded away from unity the series convergence very rapidly. The functional relations in (12.31) may be used when $|q| > 1$. Due to the interesting convergence properties of the series that define the theta functions, it is also possible to base algorithms for the $\mathrm{sn}-, \mathrm{cn}-, \mathrm{dn}-$functions on the relations in (12.30).

For the complete and incomplete elliptic integrals the AGM-iterations may produce very fast algorithms. Also, the so-called *Landen transformation*, which we have not discussed in this chapter, is very useful. An interesting approach can be found in CARLSON (1987, 1988) where by means of a few concise, but very efficient, algorithms, the three standard elliptic integrals can be computed for real values of the arguments. Both papers supply Fortran programs for computing the following three basic forms:

$$
\begin{aligned}
R_F(x,y,z) &= \frac{1}{2}\int_0^\infty \frac{dt}{\sqrt{(t+x)(t+y)(t+z)}}, \\
R_J(x,y,z,\rho) &= \frac{3}{2}\int_0^\infty \frac{dt}{(t+\rho)\sqrt{(t+x)(t+y)(t+z)}}, \\
R_D(x,y,z) &= \frac{3}{2}\int_0^\infty \frac{dt}{(t+z)\sqrt{(t+x)(t+y)(t+z)}}.
\end{aligned} \tag{12.32}
$$

The three elliptic integrals can be written in terms of the above integrals (see Exercise 12.7):

$$
\begin{aligned}
F(\phi,k) &= \sin\phi R_F\left(\cos^2\phi, 1-k^2\sin^2\phi, 1\right), \\
E(\phi,k) &= F(\phi,k) - \frac{1}{3}k^2\sin^3\phi R_D\left(\cos^2\phi, 1-k^2\sin^2\phi, 1\right), \\
\Pi(\phi,k,n) &= F(\phi,k) - \frac{1}{3}n\sin^3\phi R_J\left(\cos^2\phi, 1-k^2\sin^2\phi, 1, 1+n\sin^2\phi\right).
\end{aligned} \tag{12.33}
$$

An important step in Carlson's algorithms is the transformation

$$
R_F(x,y,z) = 2R_F(x+\lambda, y+\lambda, z+\lambda) = R_F\left(\frac{x+\lambda}{4}, \frac{y+\lambda}{4}, \frac{z+\lambda}{4}\right), \tag{12.34}
$$

where $\lambda = \sqrt{xy} + \sqrt{yz} + \sqrt{zx}$. This transformation easily follows from changing the variable of integration in the first integral of (12.32). Transformation (12.34) will be repeated until the arguments of R_F are nearly equal. In that case a Taylor expansion of at most five terms is used to finish off the calculations. Observe that, when indeed the arguments are equal, we have $R_F(x,x,x) = 1/\sqrt{x}$.

12.5. Remarks and Comments for Further Reading

12.1. A recent and very attractive book on AGM-iterations is BORWEIN & BORWEIN (1987) (various aspects of the present chapter are derived from this book). It also gives various fast algorithms for the calculation of π and of elementary functions, theta functions, modular forms, etc.

12.2. The proof of Theorem 12.1 is due to NEWMAN (1985).

12.3. Finding the relation between a given elliptic integral and the three standard integrals may be drudgery. However, symbolic manipulation packages as *Maple* and *Mathematica* can handle several cases, and the developments in this area happen quickly. Excellent references for table look-up of elliptic integrals are available. See BYRD & FRIEDMAN (1954), GRADSHTEYN & RYZHIK (1980), and PRUDNIKOV et al. (1986).

12.4. More information on elliptic functions can be found in BOWMAN (1953) (for an introduction), LANG (1973) (for a modern mathematical treatment), LAWDEN (1989) with chapters on geometrical and physical applications. The classic on special functions WHITTAKER & WATSON (1927) gives also the basic theory of elliptic functions and theta functions, together with a note on their history (page 512. In FRICKE (1913) much more about the history of elliptic functions is given.

12.6. Exercises and Further Examples

12.1. Show that the elliptic integrals of the first and second kind satisfy the following relations:

$$\frac{dE}{dk} = \frac{E(k) - K(k)}{k}, \quad \frac{dK}{dk} = \frac{E(k) - k'^2 K(k)}{kk'^2}.$$

12.2. Differentiate the result in (12.12) with respect to k. This gives

$$(1+k)\dot{K}(k) + K(k) = \dot{g}(k)\dot{K}(g(k)) \tag{1}$$

(we write $\dot{K} = dK/dk$) and use the second result of the previous exercise to show that

$$E(g) = g\dot{g}^2\dot{K}(g) + \dot{g}^2 K(g).$$

Use again (1) for eliminating $\dot{K}(g)$. Next write $E(g(k))$ in terms of $K(k)$ and $E(k)$; this gives eventually (12.13). Derive (12.14) in a similar way, or write $g^{-1}(k)$ for k in (12.13).

12.3. The arc length of the lemniscate is given by

$$K(1/\sqrt{2}) = \frac{\Gamma^2\left(\frac{1}{4}\right)}{4\sqrt{\pi}}.$$

Compute $K(1/\sqrt{2})$ from

$$K(1/\sqrt{2}) = \sqrt{2}\int_0^1 \frac{dt}{\sqrt{(1-t^2)(2-t^2)}}$$

by using the change of variable $x^2 = t^2/(2-t^2)$. Show that

$$E(1/\sqrt{2}) = \frac{4\Gamma^2\left(\frac{3}{4}\right) + \Gamma^2\left(\frac{1}{4}\right)}{8\sqrt{\pi}}$$

by using in (12.2) the change of variable $x = \sqrt{1-t^2}$.

12.4. Show that an elliptic function that is analytic for each value of its argument equals a constant.

12.5. Let f be an elliptic function with fundamental parallelogram P; assume that f has no zeros or poles on the boundary B of P.

a. Show that the sum of the residues of f inside P equals zero. Use the periodicity of f to prove that $\int_B f(z)\,dz = 0$. Hence, each non-trivial elliptic function has at least two poles inside P. The number of poles (taking into account the multiplicity of the poles) is called the *order of the elliptic function*.

b. Show that the number of poles of f inside P equals the number of zeros of f inside P (taking into account the multiplicities of the poles and the zeros). *Hint*: f'/f is an elliptic function; there is a theorem from complex function theory that gives the relation between the number of zeros and poles of a function f and the integral of f'/f.

c. Let c be an arbitrary complex number, and let n be the order of f ($n \ge 2$). Then inside P there are exactly n values of z that satisfy the equation $f(z) = c$.

12.6. Prove that the Weierstrass function $\wp(z)$ defined in (12.27) is an even meromorphic function with double poles. Show also that

$$\frac{d\wp(z)}{dz} = -2\sum_{n,m} \frac{1}{(z - 2n\omega_1 - 2m\omega_2)^3}$$

is an odd elliptic function of order 3. Show that $\wp(z)$ is an even elliptic function of order 2. Finally, verify that $\wp(z)$ satisfies equation (12.28).

12.7. Prove the relations in (12.33). *Hint*: introduce in the first relation of (12.32) the variable of integration θ defined by $\cos^2\theta = (t+x)/(t+1)$, with $x = \cos^2\phi$.

12.8. Prove the *Poisson summation formula* in the form

$$\sum_{n=-\infty}^{\infty} f(x+n) = \sum_{m=-\infty}^{\infty} \left(\int_{-\infty}^{\infty} e^{2\pi imt} f(t)\,dt\right) e^{-2\pi ixm},$$

where f is of bounded variation; assume that the series at the left-hand side converges uniformly with respect to $x \in [0,1]$. *Hint*: observe that the left-hand side is a periodic function, with period unity. Expand this function in terms of a Fourier series.

13

Numerical Aspects of Special Functions

In this chapter we discuss a few basic tools for evaluating special functions. In BAKER (1992) and MOSHIER (1989) many details can be found on algorithms, with software programs written in C. Software is also available in packages for symbolic computations, as *Macsyma*, *Maple*, *Mathematica* and *Matlab*, and libraries as *IMSL* and *NAG*. Also well known is *Numerical Recipes*, see PRESS et al. (1992). Furthermore, several software packages are available through the electronic network. For instance, the collection *Netlib* is available at `netlib@ornl.gov` (send the message `help`). An excellent overview of the literature on numerical evaluation of special functions, with an overview of the available software, is given in LOZIER & OLVER (1994).

The special functions usually have several representations that may be used as starting points for numerical algorithms: series expansions, asymptotic expansions, integrals, differential equations, and recurrence relations. The series expansions for the functions of hypergeometric type (as are nearly all functions discussed in this book) may give efficient algorithms. The range of applicability is usually restricted by convergence or stability, however. Simple recurrence relations for the terms in the expansion improve the efficiency. For instance, the Gauss hypergeometric function $F(a, b; c; z)$ can be written as

$$F(a, b; c; z) = \sum_{n=0}^{\infty} t_n, \quad t_0 = 1, \quad t_{n+1} = \frac{z(a+n)(b+n)}{(c+n)(n+1)} t_n, \quad (n \geq 0),$$

which can be used when $|z| < 1$, although the speed of convergence and the stability may become problematic when a and/or b are large complex numbers.

In nearly all cases several methods of computation have to be combined in order to produce a safe and efficient algorithm for the function. An important tool will be discussed in this chapter: the method based on using the recurrence relations for the special functions. As will become clear, recurrence relations are a very powerful tool for computing a single special function or a sequence of functions. It is always necessary to know whether the recurrence relation is stable. That is, when we start the computation with initial values, which usually are not exact, we need to know the propagation of errors during the computations when repeatedly using the recurrence relation. Serious errors may be involved when unstable relations are used.

In the following sections we discuss several aspects of this topic. We start with a simple example, and later we give more information on how to deal with unstable recurrence relations. We also show how the situation is for the well-known recurrence relations of the special functions discussed in this book. In a final section we give some details on the evaluation of continued fractions.

13.1. A Simple Recurrence Relation

The recurrence relations

$$f_n = f_{n-1} - \frac{x^n}{n!}, \quad g_n = g_{n-1} + \frac{x^n}{n!}, \quad n = 1, 2, \ldots$$

with initial values $f_0 = e^x - 1$, $g_0 = 1$ have, according to (6.37) and (6.39) solutions in terms of incomplete gamma functions:

$$f_n = e^x \frac{\gamma(n+1, x)}{n!}, \quad g_n = e^x \frac{\Gamma(n+1, x)}{n!}, \quad n = 0, 1, 2, \ldots,$$

that is,

$$f_n = \sum_{m=n+1}^{\infty} \frac{x^m}{m!}, \quad g_n = e^x - f_n = \sum_{m=0}^{n} \frac{x^m}{m!}.$$

Assume that $x > 0$. Then, following our intuition, the recursion for g_n will not cause any problem, since two positive numbers are always added during the recursion. For the recurrence relation of f_n it is not clear, but there is a potential danger owing to the subtraction of two positive quantities. Note that the computation of the initial value f_0, for small values of x, may produce a large relative error, when the quantities e^x and 1 are simply subtracted. This problem repeats itself for each subsequent f_n that is computed by using the recurrence relation: in each step the next term of the Taylor series is subtracted from the exponential function. Apparently, this is a hopeless procedure for computing successive f_n (even when x is not small). On the other hand, the computation of successive g_n does not show any problem.

In the study of recurrence relations it may make sense to change the direction of the recursion. Writing the recursion for f_n and g_n in backward direction:

$$f_{n-1} = f_n + \frac{x^n}{n!}, \quad g_{n-1} = g_n - \frac{x^n}{n!}$$

then we note that for both solutions the roles are reversed: g_n is obtained by subtraction, whereas f_n is obtained by addition of positive numbers. In addition, $\lim_{n\to\infty} f_n = 0$. When we want to compute the sequence $f_0, f_1, \ldots, f_N$, then we might intuitively proceed as follows (assuming we do not have the faintest notion of stability of recurrence relations): choose a number M, $M \gg N$, and put $f_M = 0$ (that is, neglect the infinite part $\sum_{m=M+1}^{\infty} x^m/m!$ of the Taylor series of each required f_n. Next compute the values f_n for $n = M-1, M-2, \ldots 0$ by using the above backward recursion of f_n. If we choose M large enough, it will be clear that, in this way, $f_0, f_1, \ldots, f_N$ can be computed to any desired accuracy.

Just the same reasoning can be used for computing the sequence of incomplete gamma functions $\gamma(a,x), \gamma(a+1,x), \ldots, \gamma(a+N,x)$. Use the backward form of (6.39), that is, $a\gamma(a,z) = \gamma(a+1,z) + z^a e^{-z}$, with starting value $\gamma(a+M,x) = 0$. If we choose M large enough, we can compute $\gamma(a+n,x)$, $0 \le n \le N \ll M$ within any required relative accuracy. This statement can be supported by a simple error analysis; see, for instance, Van der Laan & Temme (1984, Ch. 3).

It can be easily verified that both f_n and g_n satisfy the recurrence relation

$$(n+1)y_{n+1} - (x+n+1)y_n + xy_{n-1} = 0.$$

Again, this relation is stable for the computation of g_n in the forward direction; it is stable for f_n in the backward direction. Note that the solutions of this recursion satisfy $f_n \to 0$, $g_n \to e^x$ as $n \to \infty$. Apparently, the solution which becomes ultimately small in the forward direction (small compared to the other solution), is the victim. A similar phenomenon occurs in the backward direction. This phenomenon will be explained and put in a general framework in the following sections.

13.2. Introduction to the General Theory

Consider the recurrence relation

$$y_{n+1} + a_n y_n + b_n y_{n-1} = 0, \quad n = 1,2,3,\ldots, \tag{13.1}$$

where a_n and b_n are given, with $b_n \neq 0$. Many special functions of mathematical physics satisfy such a relation. Equation (13.1) is also called a *linear homogeneous difference equation of the second order*. In analogy with the theory of differential equations, two linearly independent solutions f_n, g_n exist in general, with the property that any solution y_n of (13.1) can be written in the form

$$y_n = Af_n + Bg_n, \tag{13.2}$$

where A and B do not depend on n. We are interested in the special case that the pair $\{f_n, g_n\}$ satisfies

$$\lim_{n\to\infty} \frac{f_n}{g_n} = 0. \tag{13.3}$$

Then, for any solution (13.2) with $B \neq 0$, we have $f_n/y_n \to 0$ as $n \to \infty$. When $B = 0$ in (13.2), we call y_n a *minimal solution*; when $B \neq 0$, we call y_n a *dominant solution*. When we have two initial values y_0, y_1, assuming that f_0, f_1, g_0, g_1 are known as well, then we can compute A and B. That is,

$$A = \frac{g_1 y_0 - g_0 y_1}{f_0 g_1 - f_1 g_0}, \quad B = \frac{y_0 f_1 - y_1 f_0}{g_0 f_1 - g_1 f_0}.$$

The denominators are different from 0 when the solutions f_n, g_n are linearly independent.

When we assume that the initial values y_0, y_1 are to be used for generating a dominant solution, then A may, or may not, vanish; B should not vanish: $y_0 f_1 \neq y_1 f_0$. When however the initial values are to be used for the computation of a minimal solution, then the much stronger condition $y_0 f_1 = y_1 f_0$ should hold. It follows that, in this case, one and only one initial value can be prescribed; the other one follows from the relation $y_0 f_1 = y_1 f_0$. In the numerical approach this leads to the well-known instability phenomena for the computation of minimal solutions. The fact is that, when our initial values y_0, y_1 are not specified to an infinite precision, – and consequently B does not vanish exactly – the computed solution (13.2) always contains a fraction of g_n, the dominant solution. Hence, in the long run, our solution y_n does not behave as a minimal solution, although we assumed that we were computing a minimal solution. This happens even if all further computations are done exactly.

In applications it is important to know whether a given equation (13.1) has dominant and minimal solutions. Often this can be easily concluded from the asymptotic behavior of the coefficients a_n and b_n. The following useful theorem is due to Perron and taken from GAUTSCHI (1967). For a proof the reader is referred to the cited literature in that reference. Gautschi's paper contains a wealth of information and is considered as pioneering and is still authoritative.

Theorem 13.1. *Assume that for large values of n the coefficients a_n, b_n behave as follows:*

$$a_n \sim a n^{\alpha}, \quad b_n \sim b n^{\beta}, \quad ab \neq 0$$

with α and β real; assume that t_1, t_2 are the zeros of the characteristic polynomial $\Phi(t) = t^2 + at + b$ with $|t_1| \geq |t_2|$.

[1]. *If* $\alpha > \frac{1}{2}\beta$ *then the difference equation (13.1) has two linearly independent solutions* $y_{n,1}$ *and* $y_{n,2}$*, with the property*

$$\frac{y_{n+1,1}}{y_{n,1}} \sim -an^{\alpha}, \quad \frac{y_{n+1,2}}{y_{n,2}} \sim -\frac{b}{a}n^{\beta-\alpha}, \quad n \to \infty.$$

[2]. *If* $\alpha = \frac{1}{2}\beta$ *then the difference equation (13.1) has two linear independent solutions* $y_{n,1}$ *and* $y_{n,2}$*, with the property*

$$\frac{y_{n+1,1}}{y_{n,1}} \sim t_1 n^{\alpha}, \quad \frac{y_{n+1,2}}{y_{n,2}} \sim t_2 n^{\alpha}, \quad n \to \infty,$$

assuming that $|t_1| > |t_2|$. *If* $|t_1| = |t_2|$ *then we have*

$$\limsup_{n\to\infty} \left[|y_n|(n!)^{-\alpha}\right]^{\frac{1}{n}} = |t_1|$$

for each non-trivial solution of (13.1).

[3]. *If* $\alpha < \frac{1}{2}\beta$ *then*

$$\limsup_{n\to\infty} \left[|y_n|(n!)^{-\beta/2}\right]^{\frac{1}{n}} = \sqrt{|b|}$$

for each non-trivial solution of (13.1).

In case [1] and the first part of case [2] $f_n = y_{n,2}$ is a minimal solution of (13.1). In addition, in the first part of [2],

$$\lim_{n\to\infty} \frac{y_{n+1}}{n^{\alpha} y_n} = t_r, \quad r = 1 \quad \text{or} \quad r = 2,$$

where $r = 2$ holds for the minimal solution and $r = 1$ for any other solution. To verify this, we derive from [1]:

$$\frac{y_{n+1,2}}{y_{n+1,1}} \bigg/ \frac{y_{n,2}}{y_{n,1}} \sim \frac{b}{a^2} n^{\beta-2\alpha}, \quad n \to \infty.$$

The right-hand side converges to 0, since $\beta - 2\alpha < 0$. It follows that $y_{n,2}/y_{n,1}$ converges to 0. In the first part of [2] we have

$$\frac{y_{n+1,2}}{y_{n+1,1}} \bigg/ \frac{y_{n,2}}{y_{n,1}} \sim \frac{t_2}{t_1}, \quad n \to \infty.$$

Since $|t_1| > |t_2|$ we again conclude that $y_{n,2}/y_{n,1}$ converges to 0.

The second part of case [2] and case [3] of the theorem does not give information on the minimal and dominant solutions. As can be seen from the examples below we then need extra asymptotic information about the solutions of the difference equation (13.1).

13.3. Examples

We give an overview of the most important recurrence relations for special functions and the stability aspects including the maximal and minimal solutions of the particular relation. The quantities f_n, g_n denote the minimal and maximal solutions, respectively. In Example 9, the Jacobi polynomials, the situation is different when $x \in [-1, 1]$, because in that case f_n and g_n have similar asymptotic behavior.

1. *Bessel functions*
Recurrence relation:

$$y_{n+1} - \frac{2n}{z} y_n + y_{n-1} = 0, \quad z \neq 0.$$

Solutions:

$$f_n = J_n(z), \quad g_n = Y_n(z).$$

This is covered by case [1] of the theorem, with

$$a = -\frac{2}{z}, \quad \alpha = 1, \quad b = 1, \quad \beta = 0.$$

Claim of the theorem:

$$\frac{f_{n+1}}{f_n} \sim \frac{z}{2n}, \quad \frac{g_{n+1}}{g_n} \sim \frac{2n}{z}.$$

Known asymptotic behavior:

$$f_n \sim \frac{1}{\sqrt{2\pi n}} \left(\frac{ez}{2n}\right)^n, \quad g_n \sim -\sqrt{\frac{2}{\pi n}} \left(\frac{ez}{2n}\right)^{-n}, \quad n \to \infty.$$

Similar results hold for the *modified Bessel functions* $I_n(z)$ and $K_n(z)$.

2. *Legendre functions*, recursion with respect to the order.
Recurrence relation:

$$y_{m+1} + \frac{2mz}{\sqrt{z^2-1}} y_m + (m+\nu)(m-\nu-1) y_{m-1} = 0.$$

Solutions:

$$f_m = P_\nu^m(z), \quad g_m = Q_\nu^m(z),$$
$$\Re z > 0, \quad \nu \in \mathbb{C} \quad \nu \neq -1, -2, \ldots, \quad z \notin (0, 1].$$

This is covered by case [2] of the theorem, with

$$a = \frac{2z}{\sqrt{z^2-1}}, \quad \alpha = 1, \quad b = 1, \quad \beta = 2.$$

$$t_1 = -\sqrt{\frac{z+1}{z-1}}, \quad t_2 = \frac{1}{t_1}, \quad |t_1| > 1 > |t_2|.$$

Claim of the theorem:

$$\lim_{m\to\infty} \frac{f_{m+1}}{m f_m} = t_2, \quad \lim_{m\to\infty} \frac{g_{m+1}}{m g_m} = t_1.$$

3. *Legendre functions*, recursion with respect to the degree.
Recurrence relation:

$$y_{n+1} - z\frac{2n+2\nu+1}{n+\nu-\mu+1} y_n + \frac{n+\nu+\mu}{n+\nu-\mu+1} y_{n-1} = 0.$$

Solutions:

$$f_n = Q^{\mu}_{\nu+n}(z), \quad g_n = P^{\mu}_{\nu+n}(z), \quad \Re z > 0.$$

This is covered by case [2] of the theorem, with

$$a = -2z, \quad \alpha = 0, \quad b = 1, \quad \beta = 0.$$

$$t_1 = z + \sqrt{z^2-1}, \quad t_2 = \frac{1}{t_1}, \quad |t_1| > 1 > |t_2|.$$

Claim of the theorem:

$$\lim_{n\to\infty} \frac{f_{n+1}}{f_n} = t_2, \quad \lim_{n\to\infty} \frac{g_{n+1}}{g_n} = t_1.$$

4. *Coulomb wave functions*
Recurrence relation:

$$L\sqrt{(L+1)^2+\eta^2}\, y_{L+1} - (2L+1)\left[\eta + \frac{L(L+1)}{\rho}\right] y_L + (L+1)\sqrt{L^2+\eta^2}\, y_{L-1} = 0.$$

Solutions:

$$f_L = F_L(\eta,\rho), \quad g_L = G_L(\eta,\rho), \quad \eta \in \mathbb{R}, \quad \rho > 0.$$

This is covered by case [1] of the theorem, with

$$a = -\frac{2}{\rho}, \quad \alpha = 1, \quad b = 1, \quad \beta = 0.$$

$$t_1 = z + \sqrt{z^2 - 1}, \quad t_2 = \frac{1}{t_1}, \quad |t_1| > 1 > |t_2|.$$

Claim of the theorem:

$$\frac{g_{L+1}}{g_L} \sim \frac{2L}{\rho}, \quad \frac{f_{L+1}}{f_L} \sim \frac{\rho}{2L}, \quad L \to \infty.$$

Known asymptotic behavior:

$$f_L \sim C_L(\eta)\rho^{L+1}, \quad g_L \sim \frac{1}{2LC_L(\eta)\rho^L}, \quad C_L(\eta) \sim \frac{e^{-\frac{1}{2}\pi\eta}}{\sqrt{2}} \left[\frac{e}{2(L+1)}\right]^{L+1},$$

as $L \to \infty$.

5. *Incomplete beta functions*

Recurrence relation:

$$y_{n+1} - \left(1 + \frac{n+p+q-1}{n+p}x\right) y_n + \frac{n+p+q-1}{n+p} x y_{n-1} = 0.$$

Solutions:

$$f_n = I_x(p+n, q) = \frac{B_x(p+n, q)}{B(p+n, q)}, \quad g_n = 1, \quad 0 \leq x < 1.$$

This is covered by case [2] of the theorem, with

$$a = -(1+x), \quad \alpha = 0, \quad b = x, \quad \beta = 0, \quad t_1 = 1, \quad t_2 = x.$$

Claim of the theorem:

$$\lim_{n\to\infty} \frac{f_{n+1}}{f_n} = x.$$

Known asymptotic behavior:

$$f_n \sim \frac{(1-x)^{q-1} n^{q-1} x^{p+n}}{\Gamma(q)}, \quad n \to \infty.$$

6. *Repeated integrals of the error function*

Recurrence relation:

$$y_{n+1} + \frac{z}{n+1} y_n - \frac{1}{2(n+1)} y_{n-1} = 0.$$

Solutions:

$$f_n = e^{z^2} i^n \mathrm{erfc}\, z, \quad g_n = (-1)^n e^{z^2} i^n \mathrm{erfc}(-z),$$

where, for all $z \in \mathbb{C}$,

$$i^n\operatorname{erfc} z = \int_z^\infty i^{n-1}\operatorname{erfc} t\,dt, \quad i^0\operatorname{erfc} z = \operatorname{erfc} z, \quad i^{-1}\operatorname{erfc} z = \frac{2}{\sqrt{\pi}}e^{-z^2}.$$

This is covered by case [3] of the theorem, with

$$a = z, \quad \alpha = -1, \quad b = -\frac{1}{2}, \quad \beta = -1, \quad t_1 = 1, \quad t_2 = x.$$

Claim of the theorem:

$$\limsup_{n\to\infty} \left[|y_n|\sqrt{n!}\right]^{\frac{1}{n}} = \frac{1}{\sqrt{2}}, \quad \text{for both } y_n = f_n \text{ and } y_n = g_n.$$

Known asymptotic behavior:

$$i^n\operatorname{erfc} z \sim \frac{2^{-n}}{\Gamma\left(\frac{n}{2}+1\right)}e^{-\frac{1}{2}z^2 - z\sqrt{2n}}.$$

Hence

$$(-1)^n\frac{f_n}{g_n} \sim e^{-2z\sqrt{2n}}, \quad n \to \infty.$$

Similar results hold for *parabolic cylinder functions*.

7. *Confluent hypergeometric functions*, recursion with respect to a
Recurrence relation:

$$(n + a + 1 - c)y_{n+1} + (c - z - 2a - 2n)y_n + (a + n - 1)y_{n-1} = 0.$$

Solutions:

$$f_n = \frac{\Gamma(a+n)}{\Gamma(a)}U(a+n, c, z), \quad g_n = \frac{\Gamma(a+n)}{\Gamma(a+n+1-c)}M(a+n, c, z).$$

This is covered by case [2] of the theorem, with

$$a = -2, \quad \alpha = 0, \quad b = 1, \quad \beta = 0, \quad t_1 = t_2 = 1.$$

Claim of the theorem:

$$\limsup_{n\to\infty} |y_n|^{\frac{1}{n}} = 1.$$

Known asymptotic behavior (c_i do not depend on n):

$$f_n \sim c_1 n^{\frac{1}{2}c - \frac{3}{4}} e^{-2\sqrt{nz}}, \quad g_n \sim c_2 n^{\frac{1}{2}c - \frac{3}{4}} e^{+2\sqrt{nz}}, \quad n \to \infty.$$

8. *Confluent hypergeometric functions*, recursion with respect to c
Recurrence relation:

$$zy_{n+1} + (1 - c - n - z)y_n + (c + n - a - 1)y_{n-1} = 0.$$

Solutions:

$$f_n = \frac{\Gamma(c+n-a)}{\Gamma(c+n)} M(a, c+n, z), \quad g_n = U(a, c+n, z).$$

This is covered by case [1] of the theorem, with

$$a = -\frac{1}{z}, \quad \alpha = 1, \quad b = \frac{1}{z}, \quad \beta = 1.$$

Claim of the theorem:

$$\frac{f_{n+1}}{f_n} \sim 1, \quad \frac{g_{n+1}}{g_n} \sim \frac{n}{z} \quad n \to \infty.$$

Known asymptotic behavior:

$$f_n \sim n^{-a}, \quad g_n \sim z^{1-c-n} \frac{\Gamma(c+n-1)}{\Gamma(a)}, \quad n \to \infty.$$

9. *Jacobi polynomials*
Recurrence relation:

$$\begin{aligned}
&(2n+2)(n+\alpha+\beta+1)(2n+\alpha+\beta)y_{n+1} \\
&- \ (2n+\alpha+\beta+1)\left[(2n+\alpha+\beta+2)(2n+\alpha+\beta)x + \alpha^2 - \beta^2\right] y_n \\
&+ \ 2(n+\alpha)(n+\beta)(2n+\alpha+\beta+2)y_{n-1} = 0.
\end{aligned}$$

Solutions:

$$f_n = Q_n^{(\alpha,\beta)}(x), \quad g_n = P_n^{(\alpha,\beta)}(x), \quad x \in \mathbb{C}.$$

This is covered by case [2] of the theorem, with

$$a = -2x, \quad \alpha = 0, \quad b = 1, \quad \beta = 0.$$

$$t_1 = x + \sqrt{x^2-1}, \quad t_2 = x - \sqrt{x^2-1},$$

$$|t_1| = |t_2| = 1 \text{ if } x \in [-1,1], \quad |t_1| > 1, |t_2| < 1 \text{ if } x \notin [-1,1].$$

Claim of the theorem:

$$x \in [-1,1]: \ \limsup_{n\to\infty} |y_n|^{\frac{1}{n}} = 1, \quad \text{for both } y_n = f_n \text{ and } y_n = g_n.$$

$$x \notin [-1,1]: \ \frac{g_{n+1}}{g_n} \sim t_1, \quad \frac{f_{n+1}}{f_n} \sim t_2.$$

Known asymptotic behavior:

$$x \in (-1,1): \quad P_n^{(\alpha,\beta)}(\cos\theta) \sim \sqrt{\frac{2}{\pi n \sin\theta}} \cos\left[\left(n+\frac{1}{2}\right)\theta - \frac{1}{4}\pi\right],$$

$$x \notin [-1,1]: \quad P_n^{(\alpha,\beta)}(x) \sim \frac{1}{\sqrt{n}}\phi(x)t_1^n, \quad Q_n^{(\alpha,\beta)}(x) \sim \frac{1}{\sqrt{n}}\psi(x)t_2^n,$$

where $\phi(x)$ and $\psi(x)$ do not depend on n.

The first six examples are discussed in detail in GAUTSCHI (1967). Examples 7 and 8 are considered in TEMME (1983). From the results for Jacobi polynomials (which also hold for Gegenbauer, Legendre and Chebyshev polynomials) we conclude that, when x is outside the interval of orthogonality $[-1,1]$, the polynomials can be computed by recurrence, without the risk of instabilities, in the forward direction with the initial values

$$P_0^{(\alpha,\beta)}(x) = 1 \quad P_1^{(\alpha,\beta)}(x) = \frac{1}{2}(\alpha-\beta) + \frac{1}{2}(\alpha+\beta+2)x.$$

When $x \in [-1,1]$ the theorem does not give a clear statement. From a further study of the asymptotic behavior of the Jacobi polynomials and that of the second solution $Q_n^{(\alpha,\beta)}(x)$, the *Jacobi function of the second kind* (see SZEGÖ (1974,p. 224)), it can be concluded that $P_n^{(\alpha,\beta)}(x)$ is not a minimal solution of the recurrence relation. Only the usual rounding errors have to be taken into account.

13.4. Miller's Algorithm

From the previous discussion, it appears that the numerical computation of the minimal solution of a recurrence relation (13.1) with initial values f_0 and f_1 is quite problematic. One has to accept that the results are completely wrong after a few recursion steps. Of course, it depends on the required absolute or relative accuracy as to how much risk can be incurred, but in general one should be very careful.

From the asymptotic behavior of the minimal and a dominant solution, one can usually conclude whether recursion for the minimal solution is dangerous. If, for instance in Example 6, the real part of z is very small (and positive), then the ratio $|f_n/g_n|$ is small only for quite large values of n. Hence, the dominance of the dominant solution becomes significant only for large values of n. One can compute the first values f_n with only a slight loss of accuracy.

In this section we discuss an algorithm for computing a sequence of values

$$f_0, f_1, \ldots, f_N \tag{13.4}$$

of a minimal solution; N is a non-negative integer. Obviously, we can apply (13.1) in the backward direction; in that case f_n becomes a dominant solution and g_n the minimal solution. Then we need two initial values f_N and f_{N-1}. *Miller's algorithm* does not need these values, and uses a smart idea for the computation of the required sequence (13.4). The algorithm works for many interesting cases and gives an efficient method for computing the sequence (13.4) numerically.

Assume we have a relation of the form

$$\sum_{n=0}^{\infty} \lambda_n f_n = s, \quad s \neq 0. \tag{13.5}$$

The series should be convergent and λ_n and s should be known. As will become clear, the series in (13.5) plays a role in normalizing the required minimal solution. The series may be finite; we only require that at least one coefficient λ_n is different from zero. When just one coefficient, say λ_j is different from zero, we assume that the value f_j is available.

In Miller's algorithm a starting value ν is chosen, $\nu > N$, and a solution $\{y_n^{(\nu)}\}$ of (13.1) is computed with the false initial values

$$y_{\nu+1}^{(\nu)} = 0, \quad y_\nu^{(\nu)} = 1. \tag{13.6}$$

The right-hand sides may be replaced by other values; at least one value should be different from zero. In some cases a judicious choice of these values may improve the convergence of the algorithm. The computed solution, with (13.6) as initial values, is a linear combination of the solutions f_n and g_n introduced earlier. A simple computation gives

$$y_n^{(\nu)} = \frac{g_{\nu+1} f_n - f_{\nu+1} g_n}{g_{\nu+1} f_\nu - f_{\nu+1} g_\nu}, \quad n = 0, 1, \ldots, \nu + 1.$$

This can be verified by checking the relations in (13.6). We write this in the form

$$y_n^{(\nu)} = p_\nu f_n + q_\nu g_n. \tag{13.7}$$

We observe that $y_n^{(\nu)}/p_\nu = f_n - [f_{\nu+1}/g_{\nu+1}]g_n$ and from (13.3) it follows that

$$\lim_{\nu\to\infty} \frac{y_n^{(\nu)}}{p_\nu} = f_n, \quad 0 \le n \le N. \tag{13.8}$$

Apparently, when ν is large enough, an approximation of f_n can be obtained from the quantities $y_n^{(\nu)}$ and p_ν. However, in general, p_ν is not known. At this moment the normalizing relation (13.5) becomes of interest. We compute

$$s^{(\nu)} = \sum_{n=0}^{\nu} \lambda_n y_n^{(\nu)}, \quad f_n^{(\nu)} = \frac{s}{s^{(\nu)}} y_n^{(\nu)}. \tag{13.9}$$

Replacing in the series, on account of (13.8), $y_n^{(\nu)}$ with $p_\nu f_n$, we then obtain $p_\nu \sim s^{(\nu)}/s$. It follows that we can consider $f_n^{(\nu)}$ as an approximation to f_n, if ν is large enough. That is, we assume that the circumstances are favorable, and that we can conclude that

$$f_n = \lim_{\nu\to\infty} f_n^{(\nu)}, \quad n = 0, 1, \ldots, N. \tag{13.10}$$

This claim will be founded by introducing extra conditions.

From (13.9) we obtain for the relative error (when $f_n \neq 0$)

$$\varepsilon_n^{(\nu)} = \frac{f_n^{(\nu)} - f_n}{f_n} = \frac{s/s^{(\nu)}\, y_n^{(\nu)} - f_n}{f_n} = \frac{s(p_\nu + q_\nu g_n/f_n) - s^{(\nu)}}{s^{(\nu)}}.$$

We rewrite this:

$$\varepsilon_n^{(\nu)} = \frac{\sigma_\nu - \rho_{\nu+1}/\rho_n + \tau_\nu}{1 - \sigma_\nu - \tau_\nu} \tag{13.11}$$

with

$$\sigma_\nu = \frac{1}{s}\sum_{m=\nu+1}^{\infty} \lambda_m f_m, \quad \rho_n = \frac{f_n}{g_n}, \quad \tau_\nu = \frac{\rho_{\nu+1}}{s}\sum_{m=0}^{\nu} \lambda_m g_m. \tag{13.12}$$

When introducing (13.5) we assumed that the series converges. Hence, $\sigma_\nu \to 0$ as $\nu \to \infty$. Also, (see (13.3)) we assumed that $\rho_\nu \to 0$. From this we infer that the relative error $\varepsilon_n^{(\nu)}$ of (13.11) converges to zero (as $\nu \to \infty$), if and only if τ_ν converges to zero. Under this final condition, the limit in (13.10) holds.

For the numerical part of the method it is important to obtain an estimate of $\varepsilon_n^{(\nu)}$ for large values of ν. In many cases it is not easy to obtain a strict estimate; usually some terms in (13.12) can be approximated by replacing the series with their dominant terms. Taking in the first series only the first term, and in the second series the final term, we obtain

$$\sigma_\nu \simeq \frac{1}{s}\lambda_{\nu+1} f_{\nu+1}, \quad \tau_\nu \simeq \frac{1}{s}\rho_{\nu+1}\lambda_\nu g_\nu.$$

With these approximations (13.11) reads

$$\begin{aligned}\varepsilon_n^{(\nu)} &\simeq \frac{1}{s}\lambda_{\nu+1} f_{\nu+1} + \frac{f_{\nu+1}}{g_{\nu+1}}\frac{\lambda_\nu g_\nu}{s} - \frac{f_{\nu+1}}{g_{\nu+1}}\frac{g_n}{f_n} \\ &\simeq \frac{1}{s}\lambda_{\nu+1} f_{\nu+1} - \frac{f_{\nu+1}}{g_{\nu+1}}\frac{g_n}{f_n},\end{aligned}$$

since usually the second term on the first right-hand side is less important than the first term. A further step is to replace in this estimate n by N, because, when the N–th element in the sequence in (13.4) is accurate, the situation will only improve for the remaining values. Reasoning in this way, we finally arrive at

$$\varepsilon_n^{(\nu)} \simeq \frac{1}{2}\lambda_{\nu+1} f_{\nu+1} - \frac{f_{\nu+1}\, g_N}{g_{\nu+1}\, f_N}. \tag{13.13}$$

By using asymptotic estimates of the dominant and minimal solutions, the estimation of ν can be executed, perhaps numerically. The estimate of the error in (13.13) reflects two aspects of the algorithm for favorable convergence. The first term on the right-hand side of (13.13) indicates that the series in (13.5) should converge quickly. The second term indicates that the extent of dominance of g_n with respect to f_n is very significant.

In GAUTSCHI (1967) this algorithm is discussed in great detail (in a slightly different form). Gautschi estimates the starting point of the backward recursion by using asymptotic estimates of the special functions involved. In OLVER (1967) a direct numerical approach is used for obtaining a good starting point. Olver also considers inhomogeneous recurrence equations. Both methods are summarized in VAN DER LAAN & TEMME (1984). An excellent monograph for the numerical aspects of recurrence relations, including Miller's algorithm, is WIMP (1984).

Example 13.1. In MILLER (1952) the above method has been introduced for computing the modified Bessel functions $I_n(x)$. The recurrence relation for these functions reads (see (8.34))

$$I_{n+1}(x) + \frac{2n}{x} I_n(x) - I_{n-1}(x) = 0. \tag{13.14}$$

A normalizing condition (13.5) is (see (9.33))

$$e^x = I_0(x) + 2I_1(x) + 2I_2(x) + 2I_3(x) \ldots .$$

That is, $s = e^x$, $\lambda_0 = 1$, $\lambda_n = 2\,(n \geq 1)$. We take $x = 1$ and initial values (13.6) with $\nu = 9$ and obtain Table 13.1.
The column on the right is obtained by dividing the results of the middle column by (see (13.8) and (13.9))

$$p_9 \simeq \sum_{n=0}^{9} \lambda_n y_n^{(9)} / e^1 = 1.8071328986 \times 10{+}8$$

The underlined digits in the third column are correct. See also ABRAMOWITZ & STEGUN (1964, p. 428).

Table 13.1. Computing the Modified Bessel Functions $I_n(x)$ for $x = 1$ by Using (13.14) in Backward Direction

n	$y_n^{(13.9)}$	$I_n(1)$
0	$2.2879\ 49300 \times 10^{+8}$	$1.26606\ 587801 \times 10^{-0}$
1	$1.0213\ 17610 \times 10^{+8}$	$5.65159\ 104106 \times 10^{-1}$
2	$2.4531\ 40800 \times 10^{+7}$	$1.35747\ 669794 \times 10^{-1}$
3	$4.0061\ 29000 \times 10^{+6}$	$2.21684\ 249288 \times 10^{-2}$
4	$4.9434\ 00000 \times 10^{+5}$	$2.73712\ 022160 \times 10^{-3}$
5	$4.9057\ 00000 \times 10^{+4}$	$2.71463\ 156012 \times 10^{-4}$
6	$4.0640\ 00000 \times 10^{+3}$	$2.24886\ 614761 \times 10^{-5}$
7	$2.8900\ 00000 \times 10^{+2}$	$1.59921\ 829887 \times 10^{-6}$
8	$1.8000\ 00000 \times 10^{+1}$	$9.96052\ 919710 \times 10^{-8}$
9	$1.0000\ 00000 \times 10^{+0}$	$5.53362\ 733172 \times 10^{-9}$
10	$0.0000\ 00000 \times 10^{+0}$	$0.00000\ 000000 \times 10^{-0}$

13.5. How to Compute a Continued Fraction

We describe one method for computing continued fractions. See GAUTSCHI (1967) for different approaches and LORENTZEN & WAADELAND (1992) for more details on the theory of continued fractions.

A continued fraction of the form

$$K = b_0 + \frac{a_1}{b_1+}\,\frac{a_2}{b_2+}\,\frac{a_3}{b_3+}\cdots$$

can be computed as follows. One defines two sequences $\{A_n\}$, $\{B_n\}$ by writing

$$A_n = b_n A_{n-1} + a_n A_{n-2}, \quad B_n = b_n B_{n-1} + a_n B_{n-2}, \quad n = 1, 2, 3, \ldots,$$

with $A_{-1} = B_0 = 1$, $A_0 = b_0$, $B_{-1} = 0$. Then the so-called n–th convergent

$$K_n = b_0 + \frac{a_1}{b_1+}\,\frac{a_2}{b_2+}\,\frac{a_3}{b_3+}\cdots\frac{a_n}{b_n}$$

satisfies

$$K_n = \frac{A_n}{B_n}, \quad n = 1, 2, 3, \ldots .$$

When $\lim_{n\to\infty} K_n$ exists, the infinite continued fraction is said to be convergent. When a_i and b_i are positive then

$$K_{2n} < K_{2n+2}, \quad K_{2n+1} < K_{2n-1}.$$

So, in this case and if the limit K exists, one has an inclusion of K, which is very convenient in numerical evaluations.

We mention a few numerical aspects and pitfalls.

- The recursion relations for A_n and B_n may be unstable.
- Various transformations of the continued fraction are available to speed up the convergence.
- The quantities A_n and B_n may grow very fast and may not be representable on the computer when n becomes large, although the ratios K_n may be representable. Scaling A_n and B_n by a suitable factor is a simple remedy.
- The convergence of K_n as n becomes large may be very peculiar. In GAUTSCHI (1977) an example of anomalous convergence of a continued fraction for a ratio of Kummer functions is given,albeit that the method of evaluation in that paper is not based on the above recursion relations.

Bibliography

[1] M. ABRAMOWITZ AND I.A. STEGUN (1964), *Handbook of mathematical functions with formulas, graphs and mathematical tables*, Nat. Bur. Standards Appl. Series, **55**, U.S. Government Printing Office, Washington, D.C. (paperback edition published by Dover, New York).

[2] D.E. AMOS (1986), Algorithm 644: A portable package for Bessel functions of a complex argument and nonnegative order *ACM Trans. Math. Softw.*, **12**, 265–273.

[3] E. ARTIN (1964), *The gamma function*, Holt, Rinehart and Winston, New York.

[4] R. ASKEY (1980), Ramanujan's extension of the gamma and beta functions, *Amer. Math. Monthly*, **87**, 346–359.

[5] R. ASKEY AND J.A. WILSON (1985), *Some basic hypergeometric polynomials that generalize Jacobi polynomials*, *Memoirs Amer. Math. Soc.*, **319**.

[6] W. VAN ASSCHE (1987), *Asymptotics for orthogonal polynomials*, LNM **1265**, Springer-Verlag, Berlin.

[7] L. BAKER (1992), *C Mathematical function handbook*, McGraw-Hill, New York.

[8] E.W. BARNES (1908), A new development of the theory of hypergeometric functions, *Proc. London Math. Soc.* (2), **6**, 141–177.

[9] BATEMAN MANUSCRIPT PROJECT (1953), *Higher transcendental functions* 3 vols., *Tables of integral transforms* 2 vols., (A. Erdélyi, et al., eds.), McGraw-Hill, New York.

[10] G. BAUER (1859), Von den Coefficienten der Reihen von Kugelfunctionen einer Variablen, *Journal für Math.*, **56**, 101–121.

[11] J. BERNOULLI (1713), *Ars Conjectandi*, Basel. Reprinted on pp. 106–286 in Vol. 3 of *Die Werke von Jakob Bernoulli*, Birkhaüser Verlag, Basel, 1975. Academic Press, New York.

[12] M.V. BERRY (1989), Uniform asymptotic smoothing of Stokes' discontinuities, *Proc. R. Soc. Lond. A*, **422**, 7–21.

[13] M.V. BERRY (1991), Infinitely many Stokes smoothings in the gamma function, *Proc. R. Soc. Lond. A* **434**, 465–472.

[14] F.W. BESSEL (1824), Untersuchung des Theils der planetarischen Störungen, welcher aus der Bewegung der Sonne entsteht, *Berliner Abh.*, 1–52.

[15] J. BINET (1839), Mémoire sur les intégrales Eulériennes et sur leur application à la théorie des suites, ainsi qu'à l'évaluation de fonctions des grands nombres, *Journal de l'École Polytechn.*, cahier 27, 123–343.

[16] N. BLEISTEIN AND R.A. HANDELSMAN (1975), *Asymptotic expansions of integrals*, Holt, Rinehart and Winston, New York.

[17] G. BOOLE (1860), *A treatise on the calculus of finite differences*, Cambridge (later editions 1872, 1880, London).

[18] J.M. BORWEIN AND P.B. BORWEIN (1987), *Pi and the AGM*, John Wiley, New York.

[19] J.M. BORWEIN, P.B. BORWEIN AND K. DILCHER (1989), Pi, Euler numbers, and asymptotic expansions, *Amer. Math. Monthly*, **96**, 681–687.

[20] F. BOWMAN (1953), *Introduction to elliptic functions*, English Universities Press, London.

[21] W.G.C. BOYD AND T.M. DUNSTER (1986), Uniform asymptotic solutions of a class of second-order linear differential equations having a turning point and a regular singularity, with an application to Legendre functions, *SIAM J. Math. Anal.*, **17**, 422–450.

[22] W.G.C. BOYD (1994), Gamma function asymptotics by an extension of the method of steepest descents, *Proc. R. Soc. Lond.*, A**447**, 609–630.

[23] H. BOHR AND J. MOLLERUP (1922), *Laereborg i matematisk analyse*, Vol. III, Copenhagen.

[24] C. BREZINSKI et al. (1985), *Polynômes orthogonaux et applications*, Proceedings, Bar le Duc 1984, Springer-Verlag, LNM 1171, Berlin.

[25] T.J.I'A BROMWICH (1926), *An introduction to the theory of infinite series*, Macmillan, London.

[26] N.G. DE BRUIJN (1961), *Asymptotic methods in analysis*, John Wiley, New York.

[27] H. BUCHHOLZ (1969), *The confluent hypergeometric function*, Springer-Verlag, Berlin.

[28] B.M. BUDAK, A.A. SAMARSKII AND A.N. TIKHONOV (1964), *A collection of problems on mathematical physics*, Pergamon Press, Oxford.

[29] J.C. BURKILL (1956), *The theory of ordinary differential equations*, Oliver and Boyd, Edinburgh.

[30] P.F. BYRD AND M.D. FRIEDMAN (1954), *Handbook of elliptic integrals for engineers and physicists*, Springer-Verlag, Berlin.

[31] B.C. CARLSON (1987), A table of elliptic integrals of the second kind, *Math. Comp.*, **49**, 595–606 (with a supplement on S13–S17).

[32] B.C. CARLSON (1988), A table of elliptic integrals of the third kind, *Math. Comp.*, **51**, 267–280 (with a supplement on S1–S5).

[33] H.S. CARSLAW AND J.C. JAEGER (1959), *Conduction of heat in solids*, Oxford University Press, London and New York.

[34] A.L. CAUCHY (1827), *Exercises de Mathémathiques, seconde année*, 91–92.

[35] T.S. CHIHARA (1978), *An introduction to orthogonal polynomials*, Gordon and Breach, New York.

[36] C.W. CLENSHAW (1962), *Chebyshev series for mathematical functions*, *Nat. Phys. Lab. Math. Tables*, **5**. H.M. Stationery Office, London.

[37] C.W. CLENSHAW AND S.M. PICKEN (1966), *Chebyshev series for Bessel functions of fractional order*, *Nat. Phys. Lab. Math. Tables*, **8**. H.M. Stationery Office, London.

[38] E.A. CODDINGTON AND N. LEVINSON (1955), *Theory of ordinary differential equations*, McGraw-Hill, New York.

[39] L. COMTET (1974), *Advanced combinatorics*, D. Reidel, Dordrecht, Holland.

[40] E.T. COPSON (1935), *Theory of functions of a complex variable*, Oxford University Press, London and New York.

[41] P.J. DAVIS (1959), Leonhardt Euler's integral: A historical profile of the Gamma function, *Amer. Math. Monthly*, **66**, 849–869.

[42] L. DEBNATH (1995), *Integral transforms and their applications*, CRC Press, Boca Raton.

[43] A.R. DIDONATO AND A.H. MORRIS, JR. (1986), Computation of the incomplete gamma function ratios and their inverses, *ACM Trans. Math. Softw.*, **12**, 377–393.

[44] A.R. DIDONATO AND A.H. MORRIS,JR. (1987), Algorithm 654. Fortran subroutines for computing the incomplete gamma function ratios and their inverses, *ACM Trans. Math. Softw.*, **13**, 318–319.

[45] A.R. DIDONATO AND A.H. MORRIS,JR. (1992), Algorithm 708. Significant digit computation of the incomplete beta function ratios, *ACM Trans. Math. Softw.*, **18**, 360–373.

[46] K. DILCHER, L. SKULA AND I.SH. SLAVUTSKII (1991), *Bernoulli numbers*, Bibliography (1713–1990), Queen's papers in pure and applied mathematics, **87**, Queen's University, Kingston, Ontario.

[47] R.B. DINGLE (1973), *Asymptotic expansions: Their derivation and interpretation*, Academic Press, New York.

[48] T.M. DUNSTER (1989), Uniform asymptotic expansions for Whittaker's confluent hypergeometric functions, *SIAM J. Math. Anal.*, **20**, 744–760.

[49] T.M. DUNSTER (1990), Uniform asymptotic solutions of second-order linear differential equations having a double pole with complex exponent and a coalescing turning point, *SIAM J. Math. Anal.*, **21**, 1594–1618.

[50] T.M. DUNSTER (1991), Conical functions with one or both parameters large, *Proc. Royal Soc. Edinburgh*, **119A**, 311–327.

[51] J. DUTKA (1995), On the early history of Bessel functions, *Arch. Hist. Exact Sciences*, **49**, 105–134.

[52] H.M. EDWARDS (1974), *Riemann's zeta function*, Academic Press, New York and London.

[53] A. ERDÉLYI (1956), *Asymptotic expansions*, Dover, New York.

[54] A. ERDÉLYI (1961), General asymptotic expansions of Laplace integrals, *Arch. Rational Mech. Anal.* **7**, 1–20.

[55] A. ERDÉLYI AND M. WYMAN (1963), The asymptotic evaluation of certain integrals, *Arch. Rational Mech. Anal.* **14**, 217–260.

[56] L. EULER (1732) (published in 1738), Methodus generalis summandi progressiones, *Comm. Acad. Imp. Petrop*, **6**, 68–97.

[57] L. EULER (1748), *Introductio in Analysis Infinitorum*, Vol. I, Lausanne.

[58] L. EULER (1771) (published in 1772), Evolutio formulae integralis

$$\int x^{f-1}(\log x)^{m/n}\,dx$$

integratione a valore $x = 0$ ad $x = 1$, *Nov. Comm. Acad. Petrop.*, **16**, 91–139.

[59] J. FAVARD (1935), Sur les polynômes de Tchebicheff, *Comptes Rendus de l'Académie des Sciences*, **200**, 2052–2053.

[60] J. FIELDS (1966), The uniform asymptotic expansion of the ratio of two gamma functions, *Proc. Edinburgh Math. Soc.*, **15**, 43–45.

[61] L. FOX AND I.B. PARKER (1968), *Chebyshev polynomials in numerical analysis*, Oxford University Press.

[62] C.L. FRENZEN (1987), Error bounds for the asymptotic expansions of the ratio of two gamma functions, *SIAM J. Math. Anal.*, **18**, 890–896.

[63] C.L. FRENZEN (1990), Error bounds for a uniform asymptotic expansion of the Legendre function $Q_n^{-m}(\cosh z)$, *SIAM J. Math. Anal.*, **21**, 523–535.

[64] C.L. FRENZEN (1992), Error bounds for the asymptotic expansions of the ratio of two gamma functions with complex argument, *SIAM J. Math. Anal.*, **23**, 505–511.

[65] C.L. FRENZEN AND R. WONG (1985), A uniform asymptotic expansion of the Jacobi polynomials with error bounds, *Can. J. Math.*, **37**, 979–1007

[66] C.L. FRENZEN AND R. WONG (1988), Uniform asymptotic expansions of Laguerre polynomials, *SIAM J. Math. Anal.*, **19**, 1232–1248.

[67] G. FREUD (1971), *Orthogonal polynomials*, Pergamon, Oxford.

[68] R. FRICKE (1913), Elliptische Funktionen, *Encyklopädie der mathematischen Wissenschaften*, **2**, pt. 2, 181–348, B.G. Teubner, Leipzig.

[69] G. FROBENIUS (1873), Über die Integration der linearen Differentialgleichungen durch Reihen, *J. reine angew. Math.*, **76**, 214–235.

[70] C.F. GAUSS (1812), *Comment. Gotting.*, **2**, p. 30.

[71] C.F. GAUSS (1876), *Ges. Werke*, **III**, 123–163, 207–229.

[72] W. GAUTSCHI (1967), Computational aspects of three-term recurrence relations, *SIAM Review*, **9**, 24–82.

[73] W. GAUTSCHI (1977), Anomalous convergence of a continued fraction for ratios of Kummer functions, *Mat. Comp.*, **31**, 994–999.

[74] W. GAUTSCHI (1979), A computational procedure for incomplete gamma functions, *ACM Trans. Math. Softw.*, **5**, 466–481.

[75] W. GAUTSCHI (1990), Computational aspects of orthogonal polynomials, 181–216 in *Orthogonal polynomials*, P. Nevai (ed.), NATO ASI Series, **C-294**, Kluwer Academic Publishers.

[76] M. GODEFROY (1901), *La fonction gamma; théorie, histoire, bibliographie*, Thèse, Paris.

[77] E. GOURSAT (1881), Sur l'équation différentielle linéaire qui admet pour intégrale la série hypergéométrique, *Ann. Sci. École Norm. Sup.* (2) **10**, S3–S142.

[78] I.S. GRADSHTEYN AND I.W. RYZHIK (1980), *Table of integrals, series, and products*, Academic Press, New York.

[79] H. HANKEL (1863), *Dissertation*, p. 23; Leipzig; see also the paper Die Euler'schen Integrale bei unbeschränkter Variabilität des Argumentes, *Zeitschrift für Math. und Phys.* (1864), **9**, 1–21.

[80] E. HEINE (1846), Über die Reihe

$$1+\frac{(q^\alpha-1)(q^\beta-1)}{(q-1)(q^\gamma-1)}x+\frac{(q^\alpha-1)(q^{\alpha+1}-1)(q^\beta-1)(q^{\beta+1}-1)}{(q-1)(q^2-1)(q^\gamma-1)(q^{\gamma+1}-1)}x^2+\dots,$$

J. reine angew. Math., **32**, 210–212

[81] E. HEINE (1847), Untersuchungen über die Reihe

$$1+\frac{(q^\alpha-1)(q^\beta-1)}{(q-1)(q^\gamma-1)}x+\frac{(q^\alpha-1)(q^{\alpha+1}-1)(q^\beta-1)(q^{\beta+1}-1)}{(q-1)(q^2-1)(q^\gamma-1)(q^{\gamma+1}-1)}x^2+\dots,$$

J. reine angew. Math., **34**, 285–328.

[82] P. HENRICI (1977), *Applied and computational analysis*, Vol. II: Special functions, integral transformations, asymptotics, continued fractions, John Wiley, New York.

[83] G.W. HILL (1977), Algorithm 518. Incomplete Bessel function I_0: The von Mises distribution, *ACM Trans. Math. Softw.*, **3**, 279–284.

[84] H. HOBSON (1931), *The theory of spherical and ellipsoidal harmonics*, Cambridge University Press, London and New York.

[85] H. HOCHSTADT (1971), *The functions of mathematical physics*, John Wiley, New York.

[86] O. HÖLDER (1886), Über die Eigenschaft der Gammafunktion keiner algebraischen Differentialgleichung zu genügen, *Math. Annalen*, **28**, 1–13.

[87] A. HURWITZ (1882), *Zeitschrift für Math. und Phys.*, **27**, 86–101.

[88] E.L. INCE (1956), *Ordinary differential equations*, Dover, New York.

[89] C. JORDAN (1947), *The calculus of finite differences*, Chelsea, New York.

[90] C.H. KNESSL AND J.B. KELLER (1991), Stirling number asymptotics from recursion equations using the ray method, *Studies Appl. Math.*, **84**, 43–56.

[91] K. KNOPP (1946), *Theory and application of infinite series*, Blackie, London.

[92] D.E. KNUTH (1992), Two notes on notation, *Amer. Math. Monthly*, **99**, 403–422.

[93] T.H. KOORNWINDER (1990a), Jacobi functions as limit cases of $q-$ultraspherical polynomials, *J. Math. Anal. Appl.*, **148**, 44–54.

[94] T.H. KOORNWINDER (1990b), *Handling hypergeometric series in Maple*, CWI Report AM-R9014, Amsterdam.

[95] T.H. KOORNWINDER (1994), *$q-$Special functions, a tutorial*, Mathematical Reprint Series, Report 94-08, Department of Mathematics and Computer Science, University of Amsterdam. See also §§3 and 4 in T.H. Koornwinder, Compact quantum groups and $q-$special functions, in *Representations of Lie groups and quantum groups*, V. Baldoni and M.A. Picardello (eds.), Pitman Research Notes in Mathematics Series 311, Longman Scientific & Technical, 46-128, 1994.

[96] E.E. KUMMER (1836), Über die hypergeometrischen Reihe

$$1 + \frac{\alpha.\beta}{1.\gamma}x + \frac{\alpha(\alpha+1)\beta(\beta+1)}{1.2.\gamma(\gamma+1)}x^2 + \frac{\alpha(\alpha+1)(\alpha+2)\beta(\beta+1)(\beta+2)}{1.2.3.\gamma(\gamma+1)(\gamma+2)}x^3 + \ldots,$$

Journal für Math., **15**, 39–83 and 127–172.

[97] E.E. KUMMER (1837), De integralibus quibusdam definitis et seriebus infinitis *J. reine angew. Math.*, **17**, 228–242.

[98] C.G. VAN DER LAAN AND N.M. TEMME (1984), *Calculation of special functions: the gamma function, the exponential integrals, and error-like functions*, CWI Tract 10, Mathematisch Centrum, Amsterdam.

[99] S. LANG (1973), *Elliptic functions*, Addison-Wesley, Reading, MA.

[100] H.A. LAUWERIER (1974), *Asymptotic analysis*, MC Tract 54, Mathematisch Centrum, Amsterdam.

[101] D.F. LAWDEN (1989), *Elliptic functions and applications*, Springer-Verlag, Berlin.

[102] N. LEBEDEV (1965), *Special functions and their applications*, Prentice-Hall, Englewood Cliffs, New Jersey.

[103] N. LEBEDEV, I.P. SKALSKAYA AND Y.S. UFLYAND (1965), *Problems of mathematical physics*, Prentice-Hall, Englewood Cliffs, New Jersey.

[104] A.M. LEGENDRE (1785), *Mémoires par divers savans*, **10**.

[105] A.M. LEGENDRE (1809a), *Mémoires de l'Institut de France*, **10**, p. 476.

[106] A.M. LEGENDRE (1809b), *Mémoires de l'Institut de France*, **10**, p. 485.

[107] N. LEVINSON AND R.M. REDHEFFER (1970), *Complex variables*, Holden-Day, San Fransisco.

[108] L. LORENTZEN AND H. WAADELAND (1992), *Continued fractions with applications*, North-Holland, Elsevier, Amsterdam.

[109] E.R. LOVE (1970), Changing the order of integration, *J. Austral. Math. Soc.*, **9**, 421–432.

[110] D.W. LOZIER AND F.W.J. OLVER (1994), Numerical evaluation of special functions, in W. Gautschi (ed.), *Mathematics of computation 1943–1993: A half-century of computational mathematics*, PSAM, AMS, Providence.

[111] Y.L. LUKE (1962), *Integrals of Bessel functions*, McGraw-Hill, New York.

[112] Y.L. LUKE (1969), *The special functions and their approximations*, Vol. I–II, Academic Press, New York.

[113] Y.L. LUKE (1975), *Mathematical functions and their approximations*, Academic Press, New York.

[114] W. MAGNUS, F. OBERHETTINGER AND R.P. SONI (1966), *Formulas and theorems for the special functions of mathematical physics*, Springer-Verlag, Berlin.

[115] J.I. MARCUM (1960), A statistical theory of target detection by pulse radar, *IRE Trans. Inform. Theory*, **6**, 59–268.

[116] G. MATVIYENKO (1993), On the evaluation of Bessel functions, *Appl. Comp. Harm. Anal.*, **1**, 116–135.

[117] J.C.P. MILLER (1950), On the choice of standard solutions for a homogeneous linear differential equation of the second order, *Quart. J. Mech. Appl. Math.*, **3**, 225–235.

[118] J.C.P. MILLER (1952), *Bessel functions*, vol. II in Mathematical Tables, Cambridge University Press, London and New York.

[119] J.C.P. MILLER (1952), On the choice of standard solutions to Weber's equation, *Proc. Cambridge Philos. Soc.*, **48**, 428–435.

[120] L.M. MILNE-THOMSON (1933), *The calculus of finite differences*, Macmillan, London.

[121] R. VON MISES (1918), Über die 'Ganzzahligkeit' der Atomgewicht und verwandte Fragen, *Physikalische Z.*, **19**, 490–500.

[122] P.M. MORSE AND H. FESCHBACH (1953), *Methods of theoretical physics*, McGraw-Hill, New York.

[123] L. MOSER AND M. WYMAN (1958), Stirling numbers of the second kind, *Duke Math. J.*, **25**, 29–43.

[124] L. MOSER AND M. WYMAN (1958), Asymptotic development of the Stirling numbers of the first kind, *J. London Math. Soc.*, **33**, 133–146.

[125] S.L. MOSHIER (1989), *Methods and programs for mathematical functions*, Ellis Horwood Limited, New York.

[126] G. NÉMETH (1992), *Mathematical approximation of special functions: Ten papers on Chebyshev expansions*, Nova Science Publishers, New York.

[127] P. NEVAI (1990), *Orthogonal polynomials*, P. Nevai (ed.), NATO ASI Series, **C-294**, Kluwer Academic Publishers.

[128] D.J. NEWMAN (1985), A simplified version of the fast algorithms of Brent and Salamin, *Math. Comp.*, **44**, 207–210.

[129] A.F. NIKIFOROV AND V.B. UVAROV (1988), *Special functions of mathematical physics*, Birkhäuser, Basel.

[130] N.E. NÖRLUND (1924), *Vorlesungen über Differenzenrechnung*, Springer-Verlag, Berlin.

[131] F. OBERHETTINGER (1957), *Tabellen zur Fourier Transformation*, Springer-Verlag, Berlin.

[132] F. OBERHETTINGER (1972), *Tables of Bessel transforms*, Springer-Verlag, Berlin.

[133] F. OBERHETTINGER (1973), *Tables of Laplace transforms*, Springer-Verlag, Berlin.

[134] F. OBERHETTINGER (1973), *Fourier expansions. A collection of formulas*, Springer-Verlag, Berlin.

[135] F. OBERHETTINGER (1974), *Tables of Mellin transforms*, Springer-Verlag, Berlin.

[136] F. OBERHETTINGER AND W. MAGNUS (1949), *Anwendung der elliptischen Funktionen in Physik und Technik*, Springer-Verlag, Berlin.

[137] A.B. OLDE DAALHUIS AND F.W.J. OLVER (1994), Exponentially-improved asymptotic solutions of ordinary differential equations. II, *Proc. Roy. Soc. Lond. A*, **445**, 39–56.

[138] F.W.J. OLVER (1959), Uniform asymptotic expansions for Weber parabolic cylinder functions of large order, *J. Research NBS*, **63B**, 131–169.

[139] F.W.J. OLVER (1967), Numerical solutions of second order linear difference equations, *J. Res. Nat. Bur. Standards*, **71B**, 11–29.

[140] F.W.J. OLVER (1974), *Asymptotics and special functions*, Academic Press, New York.

[141] F.W.J. OLVER (1975), Legendre functions with both parameters large, *Philos. Trans. Roy. Soc. London Ser A*, **278**, 175–185.

[142] F.W.J. OLVER (1980), Whittaker functions with both parameters large: Uniform approximations in terms of parabolic cylinder functions, *Proc. Roy. Soc. Edinburgh*, **86A**, 213–234.

[143] F.W.J. OLVER (1991a), Uniform, exponentially improved, asymptotic expansions for the generalized exponential integral. *SIAM J. Math. Anal.* **22,** 1460–1474.

[144] F.W.J. OLVER (1991b), Uniform, exponentially improved, asymptotic expansions for the confluent hypergeometric function and other integral transforms, *SIAM J. Math. Anal.*, **22**, 1475–1489.

[145] F.W.J. OLVER (1993), Exponentially-improved asymptotic solutions of ordinary differential equations. I: The confluent hypergeometric function. *SIAM J. Math. Anal.* **24,** 756–767.

[146] F.W.J. OLVER AND J.M. SMITH (1983), Associated Legendre functions on the cut, *J. Computational Physics*, **51**, 502–518.

[147] A.V. OPPENHEIM AND R.W. SCHAFER (1975), *Digital signal processing*, Prentice-Hall, Englewood Cliffs, New Jersey.

[148] E. PAPPERITZ (1885), *Math. Annalen*, **25**, 212–221.

[149] R.B. PARIS AND A.D. WOOD (1992), Exponentially–improved asymptotics for the gamma function, *J. Comp. Appl. Math.*, **41**, 135–143.

[150] R.B. PARIS AND A.D. WOOD (1995), Stokes phenomenon demystified, *IMA Bulletin*, **31**, no.1-2, 21–28.

[151] F.E. PRYM (1877), Zur Theorie der Gammafunktion, *J. reine angew. Math.*, **82**, 165–172.

[152] W. H. PRESS, S. A. TEUKOLSKY, W. T. VETTERLING AND B. P. FLANNERY (1992), *Numerical recipes. The art of scientific computing*, Cambridge University Press, second edition; diskettes and example books available. Editions exist in Basic (1991), C (1992), Fortran (1992), Macintosh Fortran (1988) and Pascal (1989).

[153] A.P. PRUDNIKOV, YU.A. BRYCHKOV AND O.I. MARICHEV (1986), *Integrals and series*, 4 Volumes, Gordon and Breach, New York.

[154] E.D. RAINVILLE (1960), *Special Functions*, Macmillan, New York.

[155] B. RIEMANN (1857), Beiträge zur Theorie der durch die Gauss'sche Reihe $F(\alpha,\beta,\gamma,x)$ darstellbare Funktionen, *Abh. d. k. Ges. d. Wiss. zu Göttingen*, **7**; see also *Complete Works*, 67–83, Teubner, Leipzig (1892) or Dover, New York (1953).

[156] B. RIEMANN (1859), Über die Anzahl der Priemzahlen unter einer gegebene Größe, *Berliner Monatsberichte*, 671–680; see also *Complete Works*, 136–144, Teubner, Leipzig (1892) or Dover, New York (1953).

[157] T.J. RIVLIN (1990), *Chebyshev polynomials: From approximation theory to algebra and number theory*, 2nd Edition, John Wiley, New York.

[158] G.H. ROBERTSON (1969), Computation of the noncentral chi-square distribution, *Bell System Tech. J.*, **48**, 201–207.

[159] L. ROBIN (1957–1959), *Fonctions sphériques de Legendre et fonctions sphéroïdales*, Vol. I–III, Gauthiers-Villars, Paris.

[160] W. RUDIN (1976), *Principles of mathematical analysis*, McGraw-Hill, New York.

[161] L. SAALSCHÜTZ (1887-1888), Bemerkungen über die Gammafunktionen mit negativem Argument, *Zeitschrift für Math. und Phys.*, **32**, 246–250, **33**, 362–371.

[162] J.B. SEABORN (1991), *Hypergeometric functions and their applications*, Springer-Verlag, Berlin.

[163] P.N. SHIVAKUMAR AND R. WONG (1988), Error bounds for a uniform asymptotic expansion of the Legendre function $P_n^{-m}(\cosh z)$, *Q. Appl. Math.*, **46**, 473–488.

[164] L.J. SLATER (1960), *Confluent hypergeometric functions*, Cambridge University Press, London and New York.

[165] I.N. SNEDDON (1956), *Special functions of mathematical physics and chemistry*, Oliver and Boyd, Edinburgh.

[166] I.N. SNEDDON (1972), *The use of integral transforms*, McGraw-Hill, New York.

[167] J. STIRLING (1730), *Methodus differentialis*, London. English translation, *The differential method*, 1749.

[168] A. STROUD AND D. SECREST (1966), *Gaussian quadrature formulas*, Prentice-Hall, Englewood Cliffs, New Jersey.

[169] R. SPIRA (1971), Calculation of the gamma function by Stirling's formula, *Math. Comp.*, **25**, 317–322.

[170] J.C.F. STURM (1836), Mémoire sur les équations différentielles linéaires du second ordre, *Journal de Math.*, **1**,106–186.

[171] G. SZEGÖ (1975), *Orthogonal polynomials*, 4th edition, Amer. Math. Soc. Colloq. Publ. **23**, Providence, R.I.

[172] N.M. TEMME (1975), Uniform asymptotic expansions of the incomplete gamma functions and the incomplete beta function *Math. Comput.*, **29**, 1109–1114.

[173] N.M. TEMME (1976), Remarks on a paper of A. Erdélyi: "Asymptotic evaluations of integrals involving a fractional derivative" (*SIAM J. Math. Anal.*, **5** (1974), 159–171), *SIAM J. Math. Anal.*, **7**, 767–770.

[174] N.M. TEMME (1978), Uniform asymptotic expansions of confluent hypergeometric functions, *J. Inst. Math. Appl.*, **22**, 215–223.

[175] N.M. TEMME (1979), The asymptotic expansion of the incomplete gamma functions, *SIAM J. Math. Anal.* **10**, 757–766.

[176] N.M. TEMME (1982), The uniform asymptotic expansion of a class of integrals related to cumulative distribution functions, *SIAM J. Math. Anal.*, **13**, 239–253.

[177] N.M. TEMME (1983a), The numerical computation of the confluent hypergeometric function $U(a,b,z)$, *Numer. Math.*, **41**, 63–82.

[178] N.M. TEMME (1983b), Uniform asymptotic expansions of Laplace integrals, *Analysis*, **3**, 221–249.

[179] N.M. TEMME (1985), Laplace type integrals: Transformation to standard form and uniform asymptotic expansions, *Q. Appl. Math.*, **43**, 103–123.

[180] N.M. TEMME (1986), A double integral containing the modified Bessel function: Asymptotics and computation, *Math Comp.*, **47**, 683–691.

[181] N.M. TEMME (1987), Incomplete Laplace integrals: Uniform asymptotic expansion with application to the incomplete beta function, *SIAM J. Math. Anal.*, **18**, 1638–1663.

[182] N.M. TEMME (1987), On the computation of the incomplete gamma functions for large values of the parameters, 479–489 in *Algorithms for approximation*, Proceedings of the IMA-Conference, Shrivenham, July 15–19, 1985, E.J.C. Mason and M.G. Cox (eds.), Oxford, Clarendon.

[183] N.M. TEMME (1990), Asymptotic estimates for Laguerre polynomials, *ZAMP*, **41**, 114–126.

[184] N.M. TEMME (1990), Polynomial asymptotic estimates of Gegenbauer, Laguerre, and Jacobi polynomials, 455–476 in *Asymptotic and computational analysis*, Proceedings, R. Wong (ed.), *Lect. Notes Pure Appl. Math.*, **124**, Marcel Dekker, New York.

[185] N.M. TEMME (1993), Asymptotic estimates of Stirling numbers, *Stud. Appl. Math.*, **89**, 233–243.

[186] N.M. TEMME (1993), Asymptotic and numerical aspects of the noncentral chi-square distribution, *Computers Math. Applic.*, **25**, 55–63.

[187] N.M. TEMME (1994), A set of algorithms for the incomplete gamma functions, *Probability in the Engineering and Informational Sciences*, **8**, 291–307.

[188] S. THANGAVELU (1993), *Lectures on Hermite and Laguerre expansions*, Mathematical Notes, Princeton University Press, Princeton, NJ.

[189] E.C. TITCHMARSH (1939), *The theory of functions*, Oxford University Press, London and New York.

[190] E.C. TITCHMARSH (1951), *The theory of the Riemann Zeta-function*, Oxford University Press, London and New York.

[191] F.G. TRICOMI AND A. ERDÉLYI (1951), The asymptotic expansion of a ratio of gamma functions, *Pacific J. Math.*, **1**, 133–142.

[192] F. URSELL (1984), Integrals with a large parameter: Legendre functions of large degree and fixed order, *Math. Proc. Camb. Philos. Soc.*, **95**, 367–380.

[193] E. WAGNER (1990), Asymptotische Entwicklungen der Gausschen hypergeometrischen Funktion für unbeschränkte Parameter, *Zeitschrift Anal. Anw.*, **9**, 351–360.

[194] G.N. WATSON (1944), *A treatise on the theory of Bessel functions*, Cambridge University Press, London and New York.

[195] K. WEIERSTRASS (1856), *Journal für Math.*, **51**,1–60.

[196] E.T. WHITTAKER AND G.N. WATSON (1927), *A course of modern analysis*, Cambridge University Press, London and New York.

[197] J. WIMP (1984), *Computation with recurrence relations*, Pitman, Boston.

[198] R. WONG (1989), *Asymptotic approximations of integrals*, Academic Press, New York.

[199] A. ZYGMUND (1959), *Trigonometric series* Cambridge University Press, London and New York (in 1988 a paperback edition appeared in the series *Cambridge mathematical library*).

Notations and Symbols

The Cauchy Principal Value Integral

Let f be continuous on the interval $[a, b]$ and $a < \xi < b$. Then we define the *Cauchy principal value integral*

$$\fint_a^b \frac{f(x)}{x - \xi}\, dx = \lim_{\varepsilon \to 0} \left[\int_a^{\xi - \varepsilon} \frac{f(x)}{x - \xi}\, dx + \int_{\xi + \varepsilon}^b \frac{f(x)}{x - \xi}\, dx \right].$$

Polar Coordinates

Let

$$x = r\cos\theta, \quad y = r\sin\theta.$$

Then

$$\frac{\partial u}{\partial x} = \frac{\partial u}{\partial r}\frac{\partial r}{\partial x} + \frac{\partial u}{\partial \theta}\frac{\partial \theta}{\partial x}, \quad \frac{\partial u}{\partial y} = \frac{\partial u}{\partial r}\frac{\partial r}{\partial y} + \frac{\partial u}{\partial \theta}\frac{\partial \theta}{\partial y}$$

$$\begin{aligned}
\frac{\partial r}{\partial x} &= \cos\theta, & \frac{\partial^2 r}{\partial x^2} &= \frac{\sin^2\theta}{r} \\
\frac{\partial r}{\partial y} &= \sin\theta, & \frac{\partial^2 r}{\partial y^2} &= \frac{\cos^2\theta}{r} \\
\frac{\partial \theta}{\partial x} &= -\frac{\sin\theta}{r}, & \frac{\partial^2 \theta}{\partial x^2} &= \frac{\sin 2\theta}{r^2} \\
\frac{\partial \theta}{\partial y} &= \frac{\cos\theta}{r}, & \frac{\partial^2 \theta}{\partial y^2} &= -\frac{\sin 2\theta}{r^2}
\end{aligned}$$

With these formulas it is not difficult to transform the Laplace equation

$$\Delta u = \nabla^2 u = \frac{\partial^2 u}{\partial x^2} + \frac{\partial^2 u}{\partial y^2} = 0$$

in terms of *polar coordinates*:

$$\frac{\partial^2 u}{\partial r^2} + \frac{1}{r}\frac{\partial u}{\partial r} + \frac{1}{r^2}\frac{\partial^2 u}{\partial \theta^2} = 0.$$

A more extensive treatment of coordinate transformations, in connection with separation of variables and partial differential equations, is given in Chapter 10.

General Notations

$\mathbb{N}$	the set of natural numbers	$\{1, 2, 3, \ldots\}$
$\mathbb{Z}$	the set of integers	$\{\ldots, -3, -2, -1, 0, 1, 2, 3, \ldots\}$
$\mathbb{Q}$	the set of rational numbers	$\{r \mid r = p/q,\ p,\ q \in \mathbb{Z},\ q \neq 0\}$
$\mathbb{R}$	the set of real numbers	$\{x \mid -\infty < x < \infty\}$
$\mathbb{C}$	the set of complex numbers	$\{z \mid z = x + iy,\ x,\ y \in \mathbb{R}\}$

$\Re z = x$, $\Im z = y$ are the real and imaginary parts of $z = x + iy$.

$$\delta_{m,n} = \begin{cases} 1, & \text{if } m = n; \\ 0, & \text{if } m \neq n, \end{cases} \qquad m,\ n \in \mathbb{Z}, \text{ Kronecker delta function.}$$

Index

A

B

C

Printed and bound by CPI Group (UK) Ltd, Croydon, CR0 4YY
16/09/2026
14963414-0003